International Workshop on Multi-Rate Processes and Hysteresis 2006

AA002640

Journal of Physics: Conference Series Volume 55

Cork, Ireland
3-8 April 2006

Editors:

Michael P. Mortell
Robert E. O'Malley
Alexei Pokrovskii
Vladimir Sobolev

ISBN: 978-1-61738-259-8

Printed from e-media with permission by:

Curran Associates, Inc.
57 Morehouse Lane
Red Hook, NY 12571
www.proceedings.com

Some format issues inherent in the e-media version may also appear in this print version.

Copyright© (2006) by the Institute of Physics
All rights reserved.

Printed by Curran Associates, Inc. (2010)

For permission requests, please contact the Institute of Physics
at the address below.

Institute of Physics
Dirac House, Temple Back
Bristol BS1 6BE UK

Phone: 44 1 17 929 7481
Fax: 44 1 17 920 0979

techtracking@iop.org

TABLE OF CONTENTS

International Workshop on Multi-Rate processes and Hysterisis

Michael P Mortell *et al* 2006 *J. Phys.: Conf. Ser.* **55** doi: 10.1088/1742-6596/55/1/E01 (Help)

Michael P Mortell, Robert E O'Malley, Alexei V Pokrovskii and Vladimir A Sobolev

We are interested in singular perturbation problems and hysteresis as common strongly nonlinear phenomena that occur in many industrial, physical and economic systems. The wording `strongly nonlinear' means that linearization will not encapsulate the observed phenomena. Often these two types of phenomena are manifested for different stages of the same or similar processes. A number of fundamental hysteresis models can be considered as limit cases of time relaxation processes, or admit an approximation by a differential equation which is singular with respect to a particular parameter. However, the amount of interaction between practitioners of theories of systems with time relaxation and systems with hysteresis (and between the `relaxation' and `hysteresis' research communities) is still low.

In recent years Ireland has become a home for a series of prestigious International Workshops in Singular Perturbations and Hysteresis:

- *International Workshop on Hysteresis and Multi-scale Asymptotics* (University College Cork, Ireland, 17–21 March 2004). Proceedings are published in *Journal of Physics: Conference Series 22.*
- *International Workshop on Relaxation Oscillations and Hysteresis* (University College Cork, Ireland, 1–6 April 2002). The related collection of invited lectures, was published as a volume *Singular Perturbations and Hysteresis*, SIAM, Philadelphia, 2005.
- *International Workshop on Geometrical Methods of Nonlinear Analysis and Semiconductor Laser Dynamics* (University College Cork, Ireland, 5–5 April 2001). A collection of invited papers has been published as a special issue of Proceedings of the Russian Academy of Natural Sciences: Nonlinear dynamics of laser and reacting systems.

Among the aims of these workshops were

- to bring together leading experts in time relaxation and hysteresis phenomena in applied problems;
- to discuss important problems in areas such as reacting systems, semiconductor lasers, shock phenomena in economic modelling, fluid mechanics, etc with an emphasis on hysteresis and singular perturbations;
- to learn and to share modern techniques in areas of common interest.

The `International Workshop on Multi-Rate Processes and Hysteresis' (University College Cork, Ireland, April 3–8, 2006) brought together more than 50 scientists, actively researching in the areas of dynamical systems with hysteresis and singular perturbations, to analyze these phenomena that occur in many industrial, physical and economic systems.

The Workshop has been sponsored by the University College Cork (UCC), the Boole Centre for Research in Informatics, UCC, Cork, the School of Mathematical Sciences UCC, Cork, Science Foundation Ireland and the Irish Mathematical Society. The supportive affiliation of the UK and Republic of Ireland SIAM Section is gratefully acknowledged.

The Editors and the Organizers of the Workshop wish to place on record their sincere gratitude to Mr Andrew Zhezherun of University College Cork for both the assistance which he provided to all the presenters at the Workshop, and for the careful formatting of all the manuscripts prior to their being forwarded to the Publisher.

More information about the Workshop can be found at
http://euclid.ucc.ie/murphys2006.htm

Michael P Mortell, Robert E O'Malley, Alexei Pokrovskii and Vladimir Sobolev
Editors

Institute of Physics Publishing — Journal of Physics: Conference Series **55** (2006) 1–11
doi:10.1088/1742-6596/55/1/001 — International Workshop on Multi-Rate Processes and Hysteresis

Existence of multiple cycles in a van der Pol system with hysteresis in the inductance

B Appelbe
Department of Applied Mathematics, University College Cork, Ireland
E-mail: b.appelbe@student.ucc.ie

Abstract. The van der Pol oscillator has interesting properties central to the study of electrical circuits and nonlinear dynamics. This paper demonstrates the effect of modelling the van der Pol oscillator with hysteresis in the inductance element, a feature often omitted, using the derivative of a Preisach nonlinearity. The system is then solved using a recently developed algorithm and it is shown that multiple limit cycles are present.

1. Introduction

Many integrated circuits contain resistors that act as normal resistors for high currents but offer negative resistance at lower currents. This characteristic was first observed in the behaviour of the van der Pol oscillator where vacuum tubes caused non-linear resistance to occur. In such systems negative resistance causes small current oscillations to increase while large current oscillations decrease due to the normal resistance. This is a famous Hopf bifurcation phenomenon [1] in which the oscillations tend towards a stable limit cycle. Fig. 1 shows a triode oscillating circuit that demonstrates this effect. Given $J_a = S(x)$, where x represents the grid voltage, Kirchoff's laws may be applied to the circuit to give

$$x'' + \frac{R}{L}(1 - \frac{M}{RC}\frac{dS}{dx})x' + \frac{1}{LC}x = 0 \tag{1}$$

where x' is the first derivative of x with respect to time. It is commonly assumed that $S(x) = \sigma x - \sigma_3 \frac{x^3}{3}$, where σ and σ_3 are normalizing coefficients. Substituting into eqn. (1) above results in an equation of the form

$$x'' + \lambda(1 - \frac{d}{\lambda}x^2)x' + \omega_0^2 x = 0 \tag{2}$$

where

$$\lambda = \frac{R}{L} - \frac{M\sigma}{LC}$$

$$d = \frac{M\sigma_3}{LC}$$

and

$$\omega_0^2 = \frac{1}{LC}$$

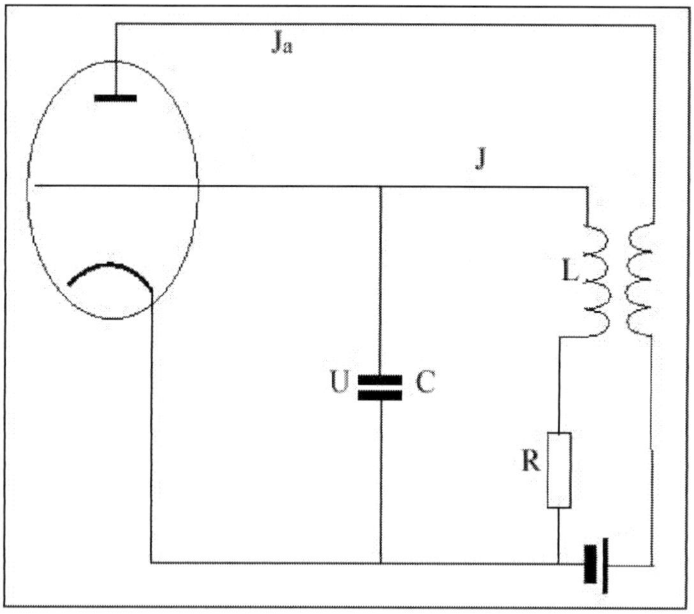

Figure 1. Van der Pol circuit with resistance, R, capacitance, C, self-inductance, L, anode-current, J_a and current, J.

This is the widely-studied classical van der Pol equation, first proposed by Balthaser van der Pol in the 1920's [3] with many later studies of its dynamical behaviour and the effect of an external forcing term including [4] and [5]. Taking λ as the control parameter (due to a variation in R, for example) results in a Hopf bifurcation at $\lambda = 0$. For $\lambda > 0$ the system has one stable limit cycle of non-zero amplitude and an unstable fixed point at the origin.

It is proposed that a more accurate model of the van der Pol circuit can be achieved by the inclusion of hysteresis effects in the inductance element, L. In particular, a hysteretic relationship between the magnetic induction and the magnetic field is known to exist. Thus, modifying eqn. (2) above we get the following system

$$y' = -x + \lambda y - dx^2 y \quad (3)$$

$$y = \xi x' + \upsilon (P(x))' \quad (4)$$

where $(P(x))'$ is the derivative of the Preisach non-linearity (described in section 2 and [7]), the hysteron to be used in this investigation. If we let $\upsilon = 0$ then we have the same system as eqn. (2) with $\xi = LC$ and $\lambda = -\lambda$.

The relationship between system (3), (4) and the classical van der Pol system (eqn. (2)) is similar to that between the oscillation of the ferromagnetic pendulum and the standard pendulum. The problem of the ferromagnetic pendulum is studied in detail in Bertotti and Mayergoyz's comprehensive 3-volumed study of hysteresis [2]. Here it is shown that the dynamical behaviour of the ferromagnetic pendulum is extremely rich. In particular, it is noted that the system contains multiple limit cycles. The similarities between system (3), (4) and the ferromagnetic pendulum suggest that such behaviour may also be found in the van der Pol oscillator with hysteresis. The paper is mainly devoted to experimental verification of this suggestion.

Section 2 of this paper will describe the mathematical and numerical model of the van der Pol system with hysteresis. Advancements in the study of hysteresis have resulted in an algorithm that can efficiently solve the system with hysteresis. This algorithm can be easily implemented in widely-used software such as MATLAB, thereby making the study of hysteretic behaviour accessible to many.

Section 3 will then offer results on the dynamical behaviour of the system. In particular, it will be shown that the system with hysteresis differs significantly from that without due to the presence of multiple limit cycles. It is concluded that the introduction of a hysteresis operator causes multiple limit cycles to appear in the van der Pol oscillator. Moreover, several Hopf bifurcations can exist in the system with hysteresis. For some point $\lambda_1 > 0$, an additional Hopf bifurcation point results in a second stable limit cycle appearing, giving a region with multiple limit cycles. The system then returns to one stable limit cycle at the point $\lambda_2 > \lambda_1$.

2. Mathematical and Numerical model

Eqn. (3) and eqn. (4) above summarise the mathematical model of the system. It is a second order differential equation with hysteresis.

2.1. Preisach model of hysteresis

The non-ideal relay forms the basis of the Preisach model. This is a simple hysteretic transducer, the definition of which can be found in [6] . The non-ideal relay, $R_{\alpha,\beta}$, expresses the idea that, as x increases, the system will not turn on until a threshold, β, is reached, while as x decreases, the system will turn off only when a different threshold, α, is reached where $\alpha < \beta$.

Say we now take a bundle of such relays where α is the minimum value of all thresholds and β the maximum. We can now have multiple states for the transducer. The transducer is fully on for $x > \beta$ and fully off for $x < \alpha$. Each change of state corresponds to some relay,$R^{w}_{\alpha,\beta}$, being turned on or off. Placing an infinite number of relays in the range $[\alpha, \beta]$ results in a transducer with continually changing state. This gives us the Preisach nonlinearity (as defined in [7])

$$P(t) = \int\int_{\alpha\leq\beta} \mu(\alpha,\beta)\gamma_{\alpha,\beta}x(t)\, d\alpha d\beta \tag{5}$$

where $x(t)$ is the input to the system; $\mu(\alpha, \beta)$ is a weighting function which allows different relays or bundles of relays to have a greater or lesser effect and $\gamma_{\alpha,\beta}$ is a term containing the "local memory" of the system. $\gamma_{\alpha,\beta}$ is the set of local extrema of $x(t)$. Geometrically, $P(t)$ is represented by the area enclosed by the lines $x = \alpha, y = \beta, x = y$ and the piecewise linear curve, L. L has a total of ω links parallel to the axes with the coordinates of the end points of links contained in the set η. The alternating horizontal and vertical links correspond to maximum and minimum local extrema of x, respectively. An algorithmic expression of the Preisach nonlinearity may be found in [6].

When including the Preisach nonlinearity in the van der Pol oscillator we are interested in the derivative, $(P(x))'$. Furthermore, a uniform weight function, $\mu(\alpha, \beta) = 1$, is applied, giving the following expression for the Preisach nonlinearity

$$(P(x))' = (\eta_\omega - \eta_{\omega-1})x' \tag{6}$$

where $\eta_\omega - \eta_{\omega-1}$ indicates the length of the last link in $\eta(\omega)$.

2.2. Numerical Model

The algorithm used for integration of eqn. (3) and eqn. (4) is based on that devised by Flynn and Rasskazov, details of which can be found on pages 53-55 of [6]. However, due to the presence of the ξ term in eqn. (4), the algorithm can be modified. Rearranging eqn. (4) we get

$$x' = \frac{y}{\xi + \upsilon(\eta_\omega - \eta_{\omega-1})} \tag{7}$$

The Flynn and Rasskazov algorithm uses a linear integration step for eqn. (7) at any time t where $\eta_\omega - \eta_{\omega-1} = 0$. Such a step is only necessary for situations in which $\xi = 0$. For all $\xi \neq 0$, the usual higher order integration step can be carried out. Therefore, when applying Flynn and Rasskazov's algorithm to the van der Pol oscillator the "LinearStep" function may be omitted.

We can express the system of equations to be integrated in the vector form

$$\frac{dx}{dt} = F(x) \tag{8}$$

Let $\nu_n(x_0)$ denote the n-th point on the numerical trajectory obtained by numerical integration. Then, given the initial conditions x_0, t_0 and $\nu_0(x_0) = x_0$, the point $\nu_{n+1}(x_0)$ can be found using the classical Runge-Kutta method with a time step h

$$\nu_{n+1}(x_0) = \nu_n(x_0) + \frac{h}{6}(\mathbf{k}_1(\nu_n(x_0), h) + 2\mathbf{k}_2(\nu_n(x_0), h) + 2\mathbf{k}_3(\nu_n(x_0), h) + \mathbf{k}_4(\nu_n(x_0), h)) + \omega_n \tag{9}$$

where

$$\begin{aligned}
\mathbf{k}_1(x, \tau) &= F(x) \\
\mathbf{k}_2(x, \tau) &= F(x + \frac{\tau}{2}\mathbf{k}_1(x, \tau)) \\
\mathbf{k}_3(x, \tau) &= F(x + \frac{\tau}{2}\mathbf{k}_2(x, \tau)) \\
\mathbf{k}_4(x, \tau) &= F(x + \frac{\tau}{2}\mathbf{k}_3(x, \tau))
\end{aligned}$$

and ω_n is the numerical error of the method. This is a 4th order integration method. It is important to note that, although $L = \eta_\omega - \eta_{\omega-1}$ is a function of x, L remains constant throughout each time step. Using an adaptive L (i.e. $L_2 = f(x + \frac{k_1^x}{2})$, $L_3 = f(x + \frac{k_2^x}{2})$, $L_4 = f(x + \frac{k_3^x}{2})$) can cause non-convergence of the solution at $y \approx 0$.

2.3. Execution of the Numerical Model

The numerical model was executed using MATLAB version 7.0.1. The values assigned to the parameters are given in table 1.

The relatively large absolute values of α and β are chosen to ensure that the Preisach nonlinearity is always present in the system. We wish to avoid the situation of $x < \alpha$ or $x > \beta$ as both result in $(P(x))' = 0$. While such cases are physically feasible we are concerned here with the more interesting case of a continuously changing Preisach nonlinearity.

The values λ and ξ are treated as the control parameters of the system and so are varied across the range given above. Given $\lambda = \frac{R}{L} - \frac{M\sigma}{LC}$ and $\xi = LC$ it is reasonable that the physical circuit can be changed to reflect changes in the control parameters. In fact, the resistance, R, capacitance, C, self-inductance, L and mutual inductance, M, may all be altered as long as the relationship $\frac{M}{LC}$ remains constant. In this way, λ and ξ may be altered but not d.

A time step, $h = 0.001$, is used for the numerical integration of the system, while the initial conditions required for integrating the system are x_0, y_0 and η_0.

Table 1. Parameter values assigned to van der Pol system

α	β	λ	d	ξ	v
-100	100	$[0,2]$	1	$(0,0.1]$	1

3. Limit cycles

The behaviour of the system is largely characterized by the number, stability and amplitude of limit cycles. Thus, the numerical model discussed in Section 2 is used to search for such cycles in the van der Pol system. Because the trajectory of the system is non-circular and not centered at the origin (see fig. 2) but oscillates about the origin the following convention is adopted as a measure of the amplitude: distance from the origin to the point where the trajectory crosses the positive x-axis.

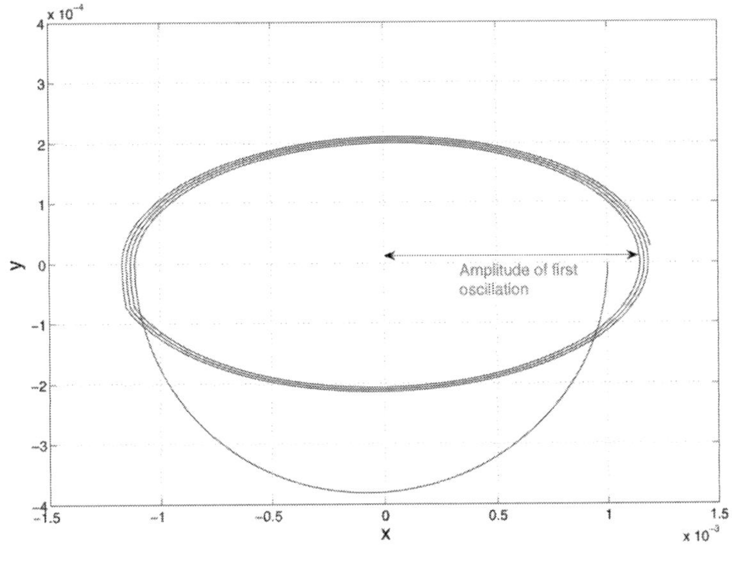

Figure 2. Trajectory of the system with $\lambda = 0.2$, $\xi = 0.03$ and the initial conditions $x_0 = 0.001$, $y_0 = 0$, $\eta_0 = [0.1, 0.001]$. The vertical scale is an order of magnitude less than the horizontal scale. Therefore, the trajectory is roughly elliptical in shape with a major axis much greater than the minor axis. Note also how the trajectory intersects itself at two points during the first oscillation.

3.1. Searching for Limit Cycles

Identifying and characterising limit cycles for a given λ and ξ is achieved by carrying out a two-step process:

Step 1. The range $[0, x_n]$ along the x-axis is discretised to a medium resolution, typically $\Delta x = 0.01$. Each node in this set is then selected in turn as the initial voltage of the system

(x_0) with other initial conditions set to

$$y_0 = 0$$
$$\eta_0 = x_0$$

The system is then integrated for five full oscillations with the amplitude of each oscillation, ψ, being measured. The fourth and fifth amplitudes are compared and a value assigned

$$\theta = \begin{cases} 1 & \text{if} \quad \psi_5 \geq \psi_4 \\ -1 & \text{if} \quad \psi_5 < \psi_4 \end{cases} \tag{10}$$

The case $\theta = 1$ indicates a trajectory with increasing amplitude while $\theta = -1$ indicates a trajectory with decreasing amplitude. In this way, each of the grid points, x_n, has a corresponding θ_n and a corresponding amplitude ψ_{5n}. We can let $x_n = \psi_{5n}$. If we have $\theta_{p-1} = 1$ and $\theta_p = -1$, we are in the region of a stable limit cycle since $\psi_{p-1} < \psi_p$. Conversely, say we have $\theta_{q-1} = -1$ and $\theta_q = 1$, we are in the region of an unstable limit cycle since $\psi_{q-1} < \psi_q$. Fig. 3 shows typical results of this step in which the two stable and one unstable limit cycles have been located to within an accuracy of 0.01.

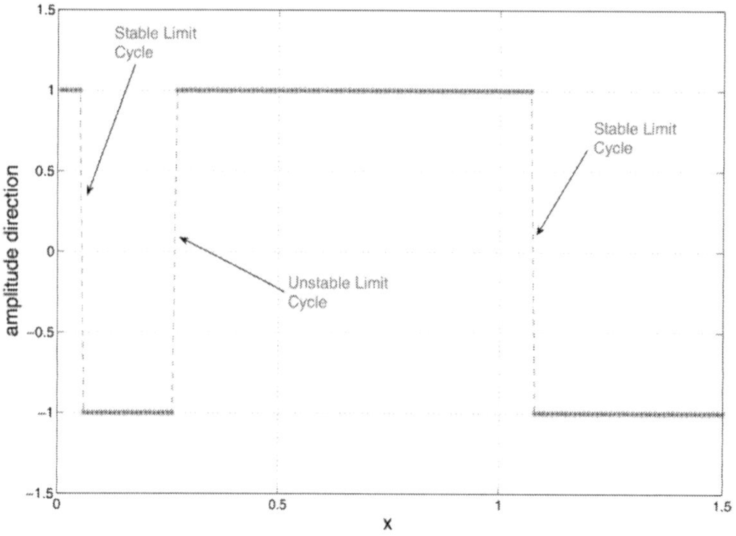

Figure 3. Plot of θ versus x for $\lambda = 0.7$ and $\xi = 0.05$. The grid resolution, $\Delta x = 0.01$. Where the curve jumps from +1 to -1 indicates the presence of a stable limit cycle while from -1 to +1 indicates an unstable limit cycle.

Step 2. Identifying the amplitude of a stable limit cycle given x_{p-1} and x_p: The system is solved with $x_0 = x_{p-1}$. The change in amplitude, $\Delta\psi$, is measured for each oscillation

$$\Delta\psi = \psi_n - \psi_{n-1} \tag{11}$$

Integration continues until $\Delta\psi$ becomes less than a specified tolerance, in our case $\Delta\psi < 10^{-5}$. Similarly, the system is solved with $x_0 = x_p$ until $\Delta\psi < 10^{-5}$. This gives two approximations for the stable limit cycle, a lower and an upper bound. Assuming the upper and lower bounds are sufficiently close, the final approximation for the location of the stable limit cycle can be found by averaging the lower and upper bounds.

Identifying the amplitude of an unstable limit cycle given x_{q-1} and x_q: It has been confirmed in step 1 that x_{q-1} corresponds to a region where oscillations have decreasing amplitude while x_q corresponds to a region of increasing amplitude. Therefore, a bisection method is used whereby the midpoint, $x_{q-\frac{1}{2}}$, of $[x_{q-1}, x_q]$ is found and the direction of its amplitude, θ, established as in step 1. If $\theta = 1$, indicating an increasing amplitude, then the interval of interest becomes $[x_{q-1}, x_{q-\frac{1}{2}}]$. Otherwise, the amplitude is decreasing and the interval of interest is $[x_{q-\frac{1}{2}}, x_q]$. In this way, the interval $[x_{q-1}, x_q]$ on which the unstable limit cycle lies can be reduced until its length is less than a specific tolerance.

This method of locating limit cycles is based on an important assumption that should be discussed. It assumes that, as the system is being solved, ψ is monotonically increasing or decreasing. This allows us to predict the locations to which the trajectory is being attracted or from which it is being repelled. However, as is clear from fig. 2, this assumption is incorrect. It is possible for the trajectory of the system to intersect itself.

Observation has shown that intersection is limited to two particular times in the evolution of the trajectory. The first is shown in fig. 2. The first oscillation of the system may intersect later oscillations. Such intersections do not occur after the first oscillation is complete. It is proposed that the initial conditions of the system may give rise to such a trajectory. Secondly, intersection of the trajectory is observed to occur as the system approaches the stable limit cycle, where $\Delta\psi$ becomes extremely small. The trajectory tends to correct such intersections so that the location of the limit cycle is not altered. It is believed that this second type of intersection appears as a result of errors in the numerical integration.

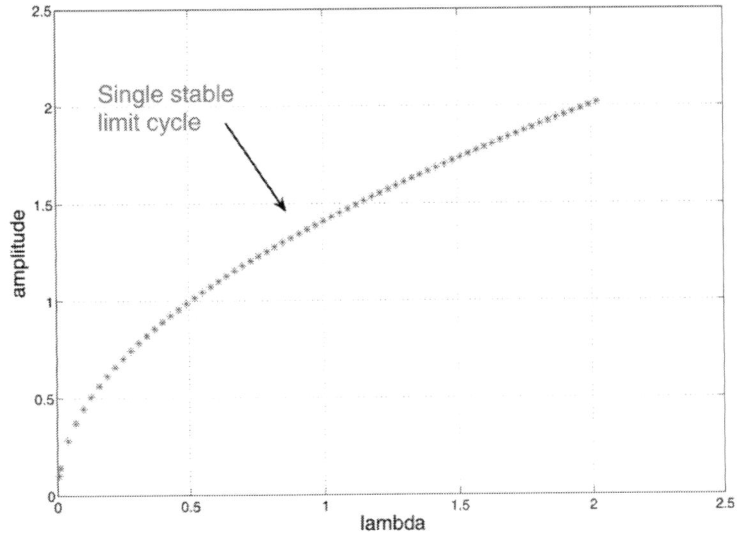

Figure 4. The classical van der Pol oscillator has 1 stable limit cycle. The plot shows the amplitude of this limit cycle for $\lambda \in (0, 2.02]$ and $\xi = 0.1$.

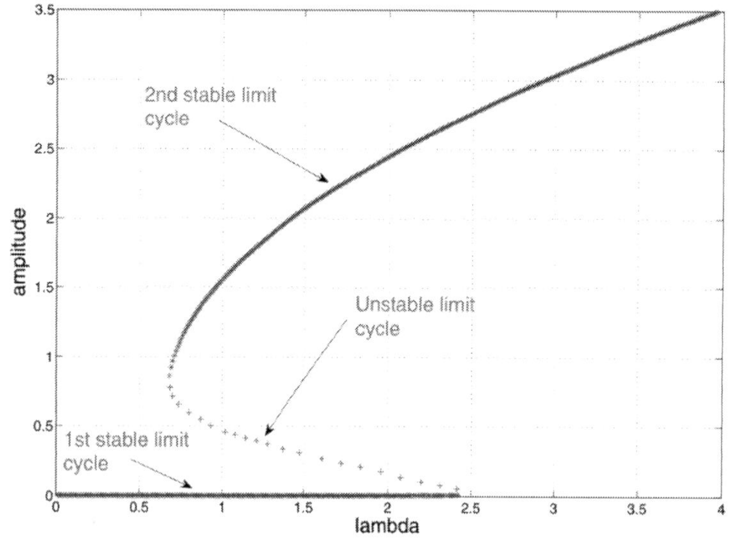

Figure 5. The amplitude of the limit cycles as a function of λ for $\xi = 0.005$.

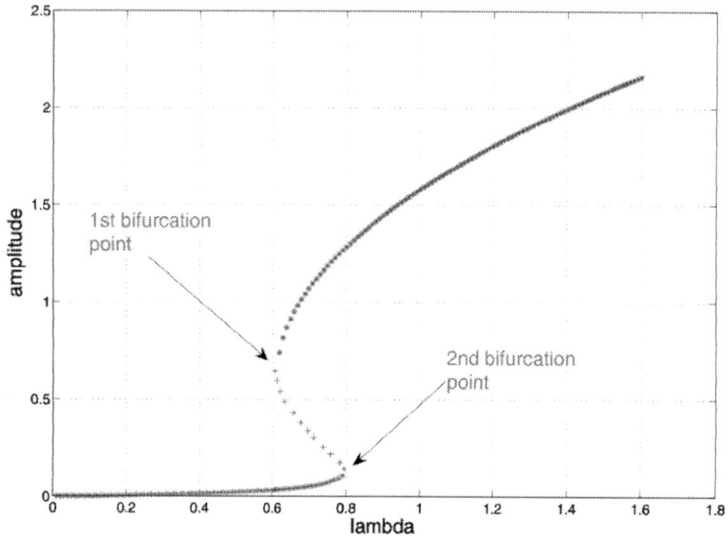

Figure 6. The amplitude of the limit cycles as a function of λ for $\xi = 0.05$.

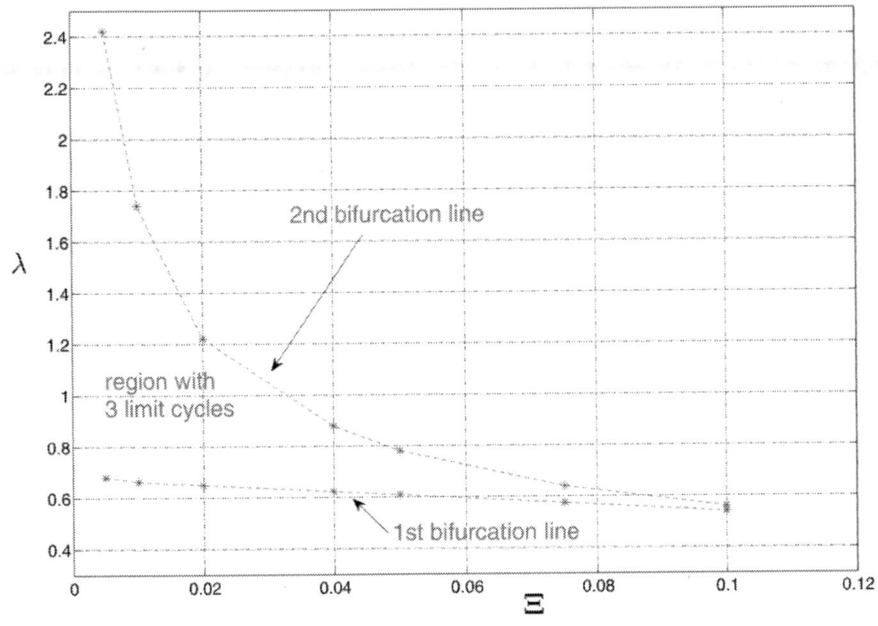

Figure 7. The range of λ with multiple limit cycles for a given ξ. Only 1 stable limit cycle exists for any point below the 1st bifurcation line or above the 2nd bifurcation line. ξ is on the x axis with λ on the y axis.

Empirically, it has been shown that ψ monotonically increases (or decreases) towards stable limit cycles and away from unstable limit cycles for all oscillations apart from the first oscillation and oscillations for which $\Lambda\psi$ becomes extremely small. Thus, using a minimum of five oscillations before measuring direction and ensuring that tolerance values for $\Delta\psi$ are greater than those values at which the second type of intersection occur appears accurate for measuring the direction of limit cycles.

3.2. Limit Cycles of the Classical van der Pol System

Say we let $v = 0$ in eqn. (4). This gives the classical van der Pol system with no hysteresis. Taking λ as the control parameter and $\xi = 0.1$ produces the results shown in fig. 4. There is one stable limit cycle in the classical van der Pol system. This remains true for variations in all other parameters of the system.

3.3. Limit Cycles of the van der Pol System with Hysteresis

Including hysteresis in our model, i.e. by letting $v = 1$, causes the behaviour of the system to differ from the classical van der Pol oscillator. Similar to before, we let ξ = a constant and identify the magnitude of the limit cycles over a range of λ. The resulting situation is shown in fig. 5 and fig. 6 for $\xi = 0.005$ and 0.05, respectively. In both cases two stable and one unstable limit cycles are identified. As λ is increased from zero, there is initially one stable limit cycle of relatively small amplitude. When the first bifurcation point is reached a second stable limit cycle of relatively large amplitude appears, as well as an unstable limit cycle separating the two. When the second bifurcation point is reached the first stable limit cycle and the unstable limit cycle disappear. For all values of λ greater than the second bifurcation point the system has

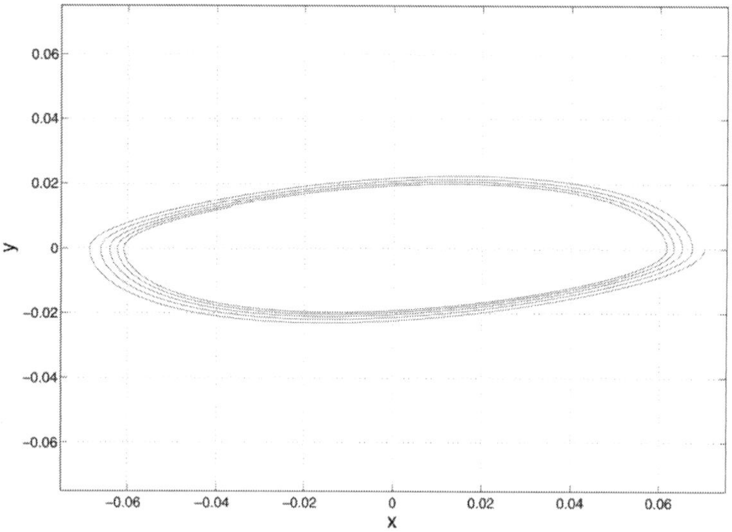

Figure 8. Trajectory approaching the first stable limit cycle for $\lambda = 0.7$ and $\xi = 0.05$.

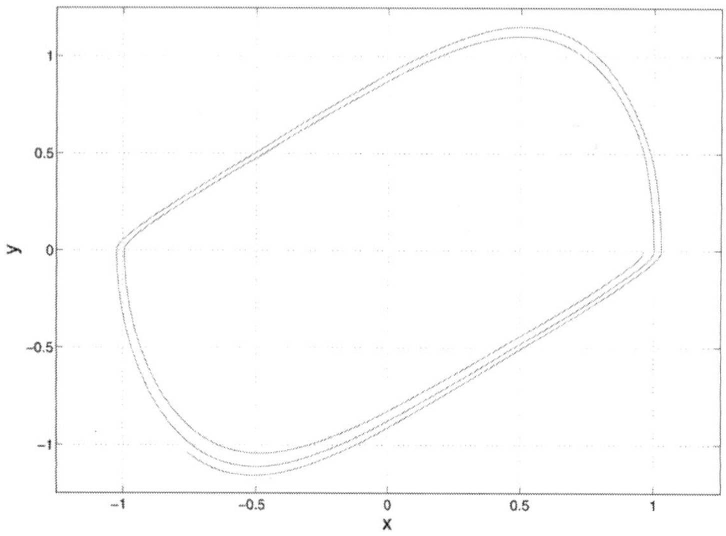

Figure 9. Trajectory approaching the second stable limit cycle for $\lambda = 0.7$ and $\xi = 0.05$.

one stable limit cycle of ever increasing amplitude.

Comparing fig. 5 and fig. 6 it will be noted that for greater values of ξ the value, λ, of the both the first and second bifurcation point has decreased. The magnitude of the decrease of the second bifurcation point is much larger than that of the first. In effect, as ξ increases the region in which we have multiple limit cycles is decreasing. This behaviour is verified in fig. 7 in which the range of λ containing multiple limit cycles for a given ξ is plotted. For $\xi = 0.005$, the range of λ with multiple limit cycles is 1.42, while for $\xi = 0.1$ the range has been reduced to 0.02. As ξ increases the relative effect of the Preisach nonlinearity in eqn. (4) decreases. The system is tending towards the classical van der Pol system with one stable limit cycle.

Clearly, there is a large difference in the amplitude of the first and second stable limit cycle. Further differences may also be observed, as can be seen from fig. 8 and fig. 9. The typical trajectory of the system near the first stable limit cycle is shown in fig. 8. It is elliptical in shape with a major axis in the x direction and minor axis in the y direction. However, the trajectory of the second stable limit cycle, as shown in fig. 9, is quite different. In this case the amplitude measured along the y axis is almost equivalent to that measured along the x axis, while the point of maximum amplitude of the trajectory lies at approximately 60° to the x axis. Thus, we can say that the rate of increase of the amplitude of the stable limit cycle is greatest in the direction at 60° to the x axis and lowest in the direction at -30° to the x axis.

4. Conclusion

This paper has demonstrated the feasibility of modelling the van der Pol oscillator system with hysteresis which can then be solved using an algorithm based on standard higher order numerical integration methods. Investigations using this algorithm have shown that the classical van der Pol system is characterised by one stable limit cycle while the system with hysteresis has multiple limit cycles. In particular, it has been shown that a system with two stable and one unstable limit cycles is common for a range of resistance values of the circuit.

References

[1] Kuznetsov Y 2004 *Elements of Applied Bifurcation Theory* (Springer-Verlag)

[2] Bertotti G and Mayergoyz I (ed) 2006 *The Science of Hysteresis* vol 1 (Elsevier)

[3] van der Pol B 1926 On relaxation oscillations *Phil. Mag. (7)* vol 2

[4] Kennedy M P and Chua L O 1986 Van der Pol and chaos *IEEE Trans. Circuits Syst. (10)* vol CAS-33

[5] Jian-Xue X and Jun J 1996 The global bifurcation characteristics of the forced van der Pol oscillator *Chaos, Solitons and Fractals (1)* vol 7

[6] Flynn D and Rasskazov O 2005 On the integration of an ODE involving the derivative of a Preisach nonlinearity *J. Phys.: Conf. Series* **22**, ed M P Mortell, R E O'Malley, A Pokrovskiĭ and V Sobolev 43–55

[7] Mayergoyz I 2003 *Mathematical Models of Hysteresis and Their Applications* (Elsevier)

[8] Lindfield G and Penny J 2000 *Numerical Methods using MATLAB* (Prentice Hall)

Institute of Physics Publishing
doi:10.1088/1742-6596/55/1/002

Journal of Physics: Conference Series **55** (2006) 12–27
International Workshop on Multi-Rate Processes and Hysteresis

Complex links between codimension-2 bifurcations in an electronic oscillator based on hysteresis

F Bizzarri[1], O De Feo[2,3], D Stellardo[1] and M Storace[1]

[1] Biophysical and Electronic Engineering Department, University of Genoa
Via Opera Pia 11a, I-16145 Genova, Italy
[2] Department of Microelectronic Engineering, University College Cork
Butler Building, North Mall, Cork, Ireland
[3] Laboratory of Nonlinear Systems, Swiss Federal Institute of Technology Lausanne
EPFL IC ISC LANOS, BC 216, Station 14, CH-1015, Lausanne, Switzerland

E-mail: `Marco.Storace@unige.it`, `Oscar.DeFeo@ucc.ie`

Abstract. The possible dynamics of a model of a hysteresis-based circuit oscillator are analyzed when varying its characteristic's parameters. By combining brute-force simulations with continuation methods it is shown that the complexity of the scenario obtained through brute-force simulation is largely justified by the presence of a few codimension-2 bifurcation points, which organize the whole bifurcation scenario.

1. Introduction

In most of the electronic hysteresis-based oscillators already proposed in the literature [1–5], the hysteresis is provided through Schmitt-trigger-like nonlinear circuit elements, which are not hysteretic in a strict sense [6–8].

Similarly to another oscillator already considered in the literature [1], the hysteresis oscillator analyzed in this paper is made up of a linear circuit fed back through a hysteretic nonlinearity. In contrast to [1], the hysteretic cell considered here provides a real hysteretic relationship between two of the circuit state variables [9]; indeed, owing to the *local memory* property [7] of the hysteretic cell, the trajectories are allowed to enter within the hysteresis cycle, thus engendering dynamics richer than those exhibited by most of the known circuit oscillators.

The circuit considered here has been extensively studied in the last decade. The studies have been organized into two threads, according to the model adopted for the nonlinear resistive elements contained in the hysteretic cell. In the first stage, such elements were considered as ideal switches and modeled by means of piecewise-linear characteristics. In this case, the bifurcation analysis of the circuit was carried out through one-dimensional return maps [10, 11]. At the next stage, the nonlinear resistors were modeled by more realistic smooth functions. In this case, the bifurcation analysis was carried out initially through brute-force simulations [12, 13]. However, this approach is strongly dependent on the initial conditions used for the simulations and reveals neither unstable invariant sets nor coexisting invariant sets, even if stable. Recently, to overcome the problems related to brute-force simulations, the dependence of the oscillator dynamics on bifurcation parameters has been addressed through harmonic balance methods [14] and continuation methods combined with normal form theory [15, 16] [17, 18].

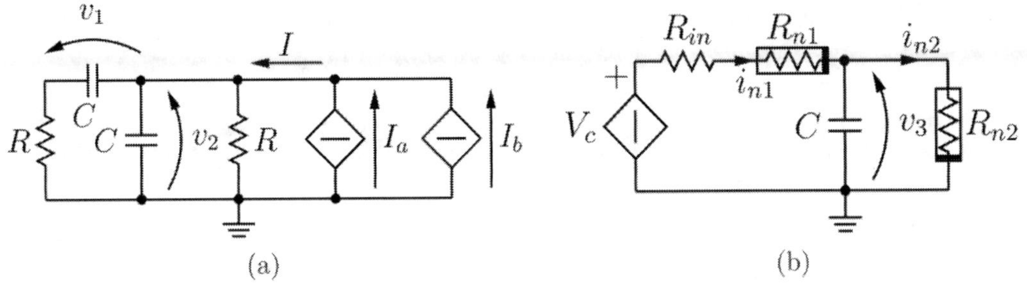

Figure 1. Electrical scheme of the hysteresis oscillator. (a) Linear part. (b) Hysteretic cell.

Up to now, the variety of dynamical behaviors observable for the smooth model has not yet been justified. In particular, it has not yet been discovered whether the richness of the observable dynamical behaviors is due to a few characteristic nonlinear phenomena organizing the overall scenario in parameter space. In this paper, we report the bifurcation analysis of some of the periodic solutions of the smooth oscillator model. By combining results from brute-force simulations and continuation methods, it is shown that the bifurcation curves are organized in parameter space by a few peculiar codimension-2 bifurcation points.

2. The system

The hysteresis oscillator considered here is composed of two sub-circuits, which are shown in Fig. 1. The linear part of the circuit, shown in Fig. 1(a), is the same as in [1], whereas the hysteretic cell, shown in Fig. 1(b), is the simplest version of the hysteresis circuit model proposed in [9].

The two linear voltage-controlled current sources I_a and I_b and the linear voltage-controlled voltage source V_c are driven as follows:

$$I_a = G_a(v_1 + v_2); \quad I_b = -G_b v_3; \quad V_c = v_1; \tag{1}$$

where the transconductances G_a and G_b are usually taken as bifurcation parameters. The nominal values of the other circuit parameters are: $R = 30$ kΩ, $R_{in} = 100$ Ω, and $C = 1$ nF.

The only nonlinear elements of the circuit are the nonlinear resistors R_{n1} and R_{n2}. Each of these is made up of two diodes, in series with a constant ideal voltage source, connected in

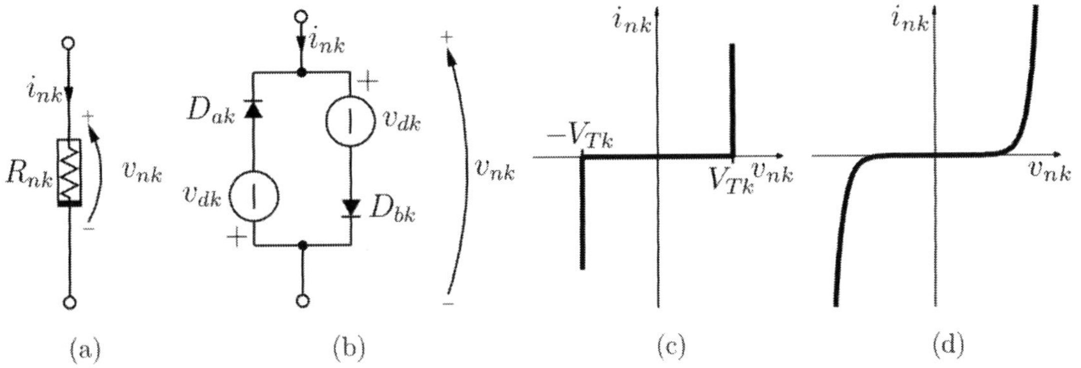

Figure 2. Circuit nonlinearity. (a) Nonlinear resistor R_{nk} ($k = 1, 2$); (b) its equivalent circuit; its (c) piecewise-linear and (d) smooth characteristics.

antiparallel, as shown in Figs. 2(a) and 2(b). These nonlinearities will be described in greater detail in the next subsection.

2.1. *The hysteretic cell*

The input to the hysteretic cell is the voltage v_1, copied by the voltage-controlled voltage source V_c, and the output is the voltage v_3.

To understand more easily the behavior of the hysteretic cell, it is convenient to start by modeling the nonlinear resistors through the piecewise-linear characteristic shown in Fig. 2(c), where the threshold voltage V_{Tk} can be thought of as the sum of the constant voltage v_{dk} and the ideal diode knee-voltage (see Fig. 2(b)). In this case, for any given working condition,

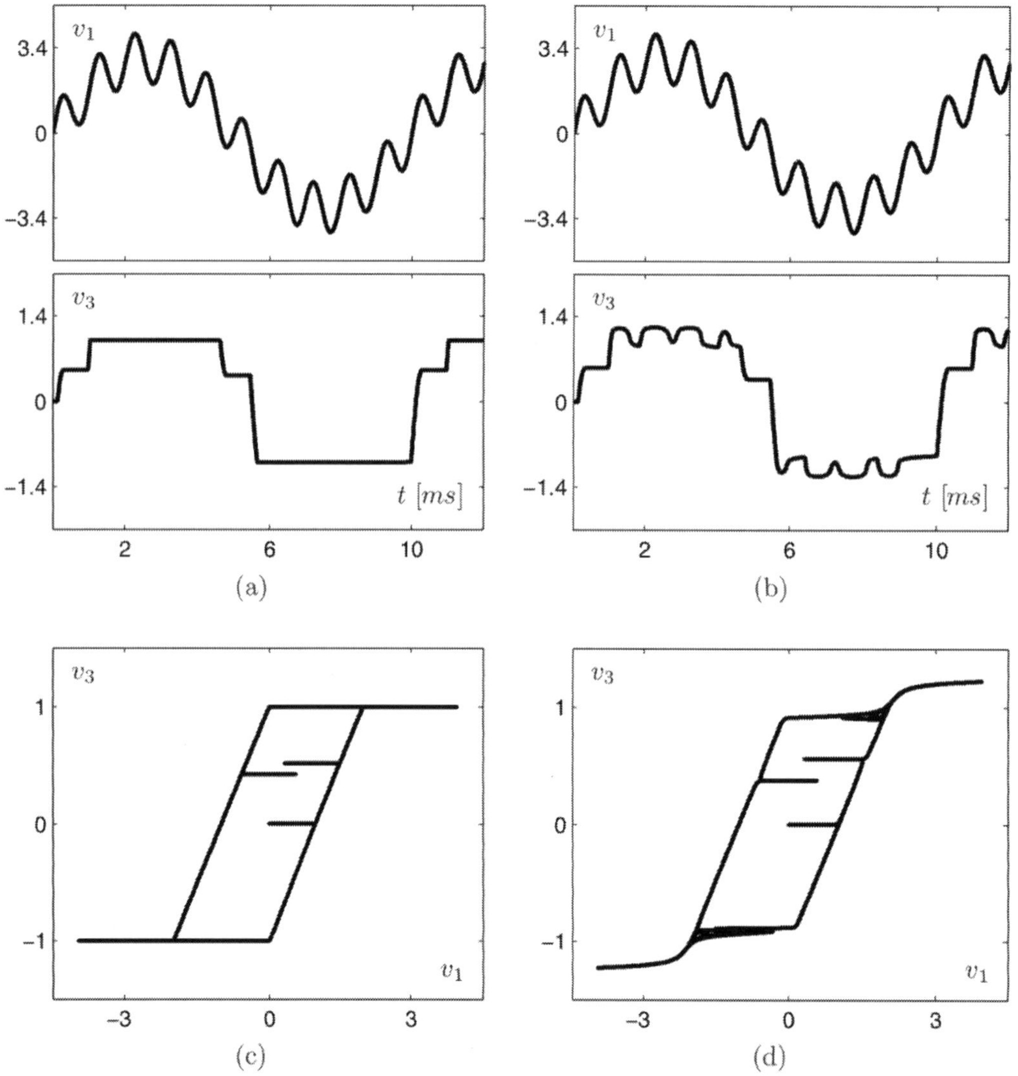

Figure 3. Hysteretic input–output relationships for the subcircuit of Fig. 1(b), by assuming $v_1 = V_1 sin(2\pi f_1 t) + V_2 sin(2\pi f_2 t)$ (with $V_1 = 3$, $V_2 = 3$ V, $f_1 = 100$ Hz, and $f_2 = 1$ kHz). (a,b): Driving and response signal's time series in the case of the PWL and the smooth model, respectively. (c,d): Input (v_1) output (v_3) relationships corresponding to (a,b), respectively.

each nonlinear resistor R_{nk} ($k = 1, 2$) is either an open circuit, $i_{nk} = 0$ when $|v_{nk}| < V_{Tk}$, or imposes a constant voltage $v_{nk} = sign(i_{nk})V_{Tk}$ whenever a nonzero current i_{nk} flows through it. Consequently, at any given time, the hysteretic cell can be modeled by an equivalent first order linear circuit with natural frequency (eigenvalue) $\lambda = -\frac{1}{R_{in}C}$.

If the maximum frequency λ_{MAX} in the spectrum of the input is much lower than $|\lambda|$, we can assume that the state of the cell *instantaneously* follows the input variations. This provides an approximately *rate independent* hysteretic input–output relationship. The *rate-independence* condition, *i.e.*, $\lambda_{MAX} \ll |\lambda|$, can be expressed in terms of the circuit parameters as $R_{in} \ll \frac{1}{C\lambda_{MAX}}$.

An example of the input-output relationship exhibited by the hysteretic cell (for $V_{T1} = V_{T2} = 1$ V) is shown in Fig. 3(c), where the plot of v_3 (output) versus v_1 (input) is shown. The corresponding time series of driving and response signals are shown in Fig. 3(a).

More correctly, the diodes can be modeled through their *smooth* driving-point constitutive equation $i_d = I_S\,(exp(v_D/V_0) - 1)$, where v_D is the voltage across the diode, $I_S = 1$ fA, and $V_0 = 25.9$ mV at room temperature[1]. In this case, the piecewise-linear characteristics of Fig. 2(c) are replaced by the more realistic smooth functions shown in Fig. 2(d) and, correspondingly, the shape of the hysteresis cycle changes as shown in Fig. 3(d), where the constant voltage sources of Fig. 2(b) are fixed to $v_{d1} = v_{d2} = 0.5$ V.

In the following, we shall assume $v_{d1} = 0$ V and $v_{d2} = 0.8$ V.

2.2. *The normalized system*

By modeling each diode by its smooth constitutive equation, and by normalizing the time variable and the voltages across the capacitors, the state equations of the circuit in Fig. 1 can be written as:

$$\begin{cases} \dot{x}_1 = -(x_1 + x_2) & \text{(2a)} \\ \dot{x}_2 = (2 + p_1)(x_1 + x_2) - x_2 - p_2\, x_3 & \text{(2b)} \\ \dot{x}_3 = p_3\,(\Psi - p_4 \sinh(x_3)) & \text{(2c)} \\ x_1 - x_3 = \operatorname{asinh}\left(\dfrac{\Psi}{p_5}\right) + \Psi & \text{(2d)} \end{cases}$$

where $\tau = \frac{t}{RC}$, $x_i = \frac{v_i}{V_0}$ ($i = 1, \ldots, 3$), and $\dot{x} = \frac{dx}{d\tau}$.

The bifurcation parameters become $p_1 = RG_a - 3$ and $p_2 = RG_b$, whereas $p_3 = R/R_{in} = 300$, $p_5 = 2R_{in}I_S/V_0 = 77.22E - 12$, and $p_4 = p_5 \exp(-\frac{v_{d2}}{V_0}) = 2.97E - 24$ are held fixed. The only nonlinear equation of the ODE system is Eq. (2c), where Ψ is implicitly defined by Eq. (2d).

2.3. *System properties*

It is evident from Eqs. (2) that: (i) the equilibria of the system lie on the plane $x_1 = -x_2$; (ii) their positions depend on p_2 only; and (iii) the origin $E_0 = (0, 0, 0)$ is a trivial equilibrium for all parameter values.

On the other hand, the stabilities of the equilibria depend on both parameters p_1 and p_2, as follows from the Jacobian matrix of the system, that can be written only partially in explicit

[1] Actually, a more accurate model of each diode would also include dynamical features induced by parasitic elements. However, their effects are negligible when reasonable oscillations frequencies are considered. Hence, for the sake of simplicity, we shall neglect all the parasitic capacitors in our model.

form:

$$A = \begin{bmatrix} 1 & -1 & 0 \\ 2+p_1 & 1+p_1 & -p_2 \\ \dfrac{p_3\sqrt{p_5^2+\Psi^2}}{1+\sqrt{p_5^2+\Psi^2}} & 0 & -p_3\left(\dfrac{p_3\sqrt{p_5^2+\Psi^2}}{1+\sqrt{p_5^2+\Psi^2}} + p_4\cosh(x_3)\right) \end{bmatrix} \tag{3}$$

System (2) is $\mathbb{Z}_2$-equivariant [18], namely it is invariant under the transformation $(x_1, x_2, x_3) \to (-x_1, -x_2, -x_3)$ or, in other words, there is a mirror symmetry with respect to the origin of the state space. As a consequence, pitchfork (or symmetry breaking) bifurcations break the $\mathbb{Z}_2$-symmetry of *fixed* (*i.e.*, invariant under the symmetry transformation) invariant sets and give rise to a *symmetric* pair of asymmetric invariant sets that are mirror image copies of each other. Henceforth, we shall denote as F-sets and S-sets the fixed and the symmetric invariant sets, respectively. Whilst E_0 is of F-type, in some regions of the parameter space there are two other non-trivial S-type equilibria, denoted as E_+ and E_-. In the following, we refer to $E_\pm$ to denote both of them. The same holds, *mutatis mutandis*, for other S-sets.

3. Observed families of periodic solutions

If we properly choose the values of the bifurcation parameters, we can observe different kinds of periodic behaviors. Such behaviors can be grouped naturally into families indexed according to three integer numbers (n^+, n^0, n^-), representing the numbers of revolutions (turns) of the

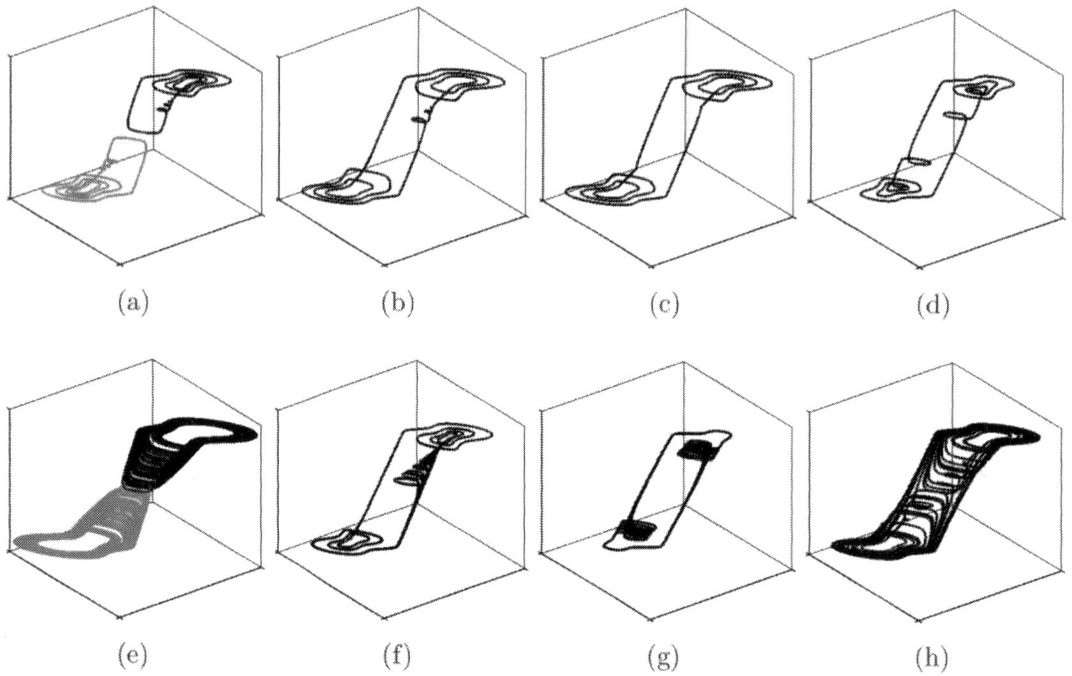

Figure 4. Families of periodic and chaotic solutions in the state space (x_1, x_2, x_3); the scale is the same for all plots. Limit cycles: (a) – small (visiting half state space) S-type $(4, 4, 0)$ $\pm$ solutions (shown in black and gray, respectively); (b) – large (visiting the whole state space) S-type $(3, 3, 3)$ (only the + solution is shown); (c) – F-type $(3, 0, 3)$; (d) – F-type $(4, 2, 4)$. (e-h) – chaotic solutions belonging to the families of periodic solutions shown in (a-d), respectively.

periodic solution (limit cycle) in the upper[2], central, and lower regions of the state-space, respectively. Examples of these periodic solutions, both F- and S-type, are shown in the first row of Fig. 4, where $x_1 \in [-180, 180]$, $x_2 \in [-210, 210]$, and $x_3 \in [-70, 70]$. Furthermore, as shown in Fig. 4(e-h), for each kind of periodic solution there are corresponding (shaped like) chaotic behaviors.

Experimentally, the families of periodic solutions are organized in the parameter space as shown in Fig. 5, where the colouring is organized controlling the Red Green and Blue (RGB) components proportionally to the indices n^-, n^0, and n^+, respectively. In the figure, the transition from bluish to white colours marks a symmetry breaking of F-type attractors; the system asymptotically converges, in the light-blue region, to the one or the other of a pair of S-type attractors, similar to those shown in Fig. 4(e), whereas in the white region it converges to an F-type attractor, similar to that shown in Fig. 4(h).

The diagram turns out to be split in two parts by the central "sail"-shaped region (white, grey, dark-blue and black), where up to four stable invariants coexist [15,16]. At the left of this "sail", there are many grey and blue "hooks", that mark the presence of periodic windows in a dominant chaotic regime, whereas at the right of the "sail", one can see many other "U"-shaped bi-colour periodic windows, that will be described in detail in Sec. 4.5.

It is neither clear nor evident how the families of solutions are linked to each other and how the chaos is organized by such periodic families.

In order to understand the links between families, we will carry out a bifurcation analysis based on the numerical continuation of limit cycles using the package AUTO2000 [17], starting from cycles with low values of n^+, n^0, and n^-. According to this strategy, we will also follow, moving in the parameter space, unstable cycles that may exist also in the pale pink region or

[2] Upper, lower, and central are meant with respect to the x_3 variable.

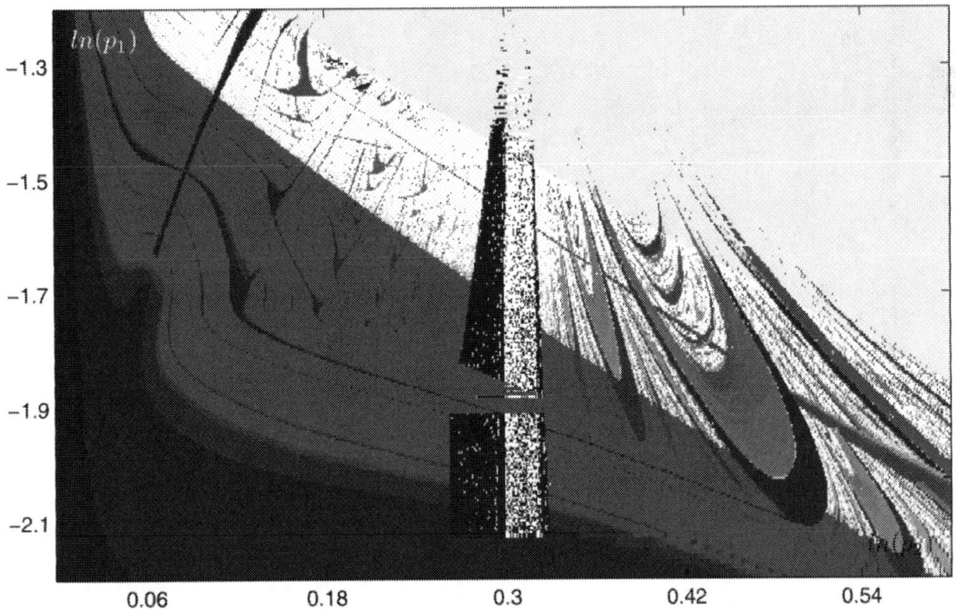

Figure 5. Brute-force simulations bifurcation scenario ($ln(p_1)$ *vs.* $ln(p_2)$). The colouring is obtained controlling the Red Green and Blue (RGB) components proportionally to the n^-, n^0, and n^+ indices, respectively. The pale pink colour corresponds to unbounded dynamics.

even outside the portion of parameter plane shown in Fig. 5.

4. Bifurcation analysis: Linking the families

The previously studied bifurcations [15] do not justify the presence of the families of limit cycles, how they are linked, and why they lead to chaotic attractors.

We start from the limit cycle (C_0) emerging from one of the previously studied bifurcations (the Hopf H_0 of the origin E_0) and we see how a few codimension-2 bifurcation points involving C_0 justify all the families and the transitions between them. In the sketches provided in the next subsections, the dots denote unstable equilibria, with E_0 blue and $E_{\pm}$ green, the solid (dashed) closed lines denote stable (unstable) limit cycles, with C_0 always blue.

4.1. Bogdanov-Takens: Existence of a global bifurcation

As sketched in Fig. 6, at very high values of p_1, at the intersection between H_0 and the pitchfork of the origin P_E [15], E_0 undergoes a Bogdanov-Takens (double zero) bifurcation (point BT in Fig. 6). Hence, because of the $\mathbb{Z}_2$ symmetry of the system, from the intersection point BT a global bifurcation curve Ht originates, on which the cycle C_0 degenerates to a double heteroclinic connection between E_+ and E_-.

Figure 7 shows the taming approach of the limit cycle C_0 to its heteroclinic bifurcation. The cycle is continued with respect to p_2, for $ln(p_1) = 5$. While approaching the heteroclinic bifurcation, C_0 tends to (i) grow, (ii) increase its period, and (iii) settle down on the plane $x_1 = -x_2$, containing the equilibria. These properties can be easily inferred from Fig. 7.

However, due to the saddle (real) nature of $E_{\pm}$, BT does not locally justify the existence of so many families of periodic solutions.

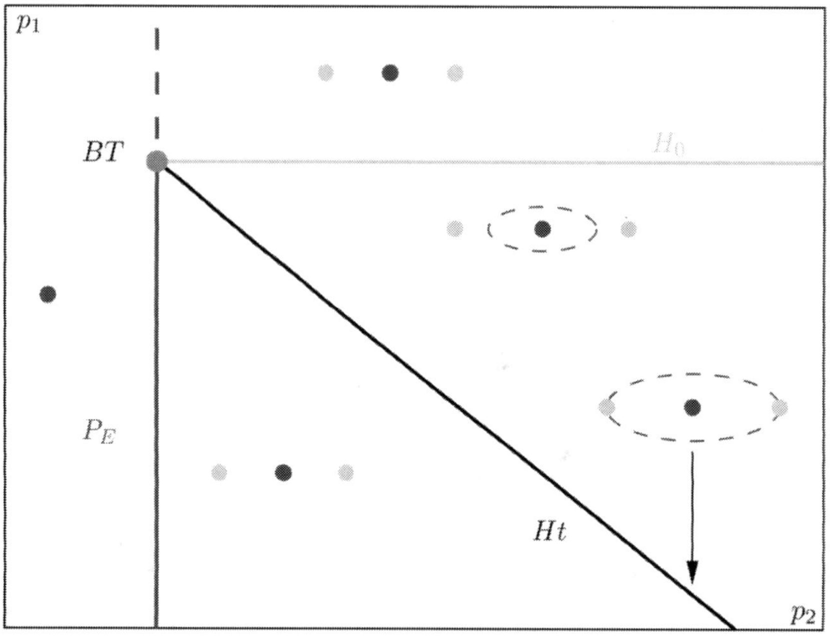

Figure 6. System unfolding around the codimension-2 bifurcation BT.

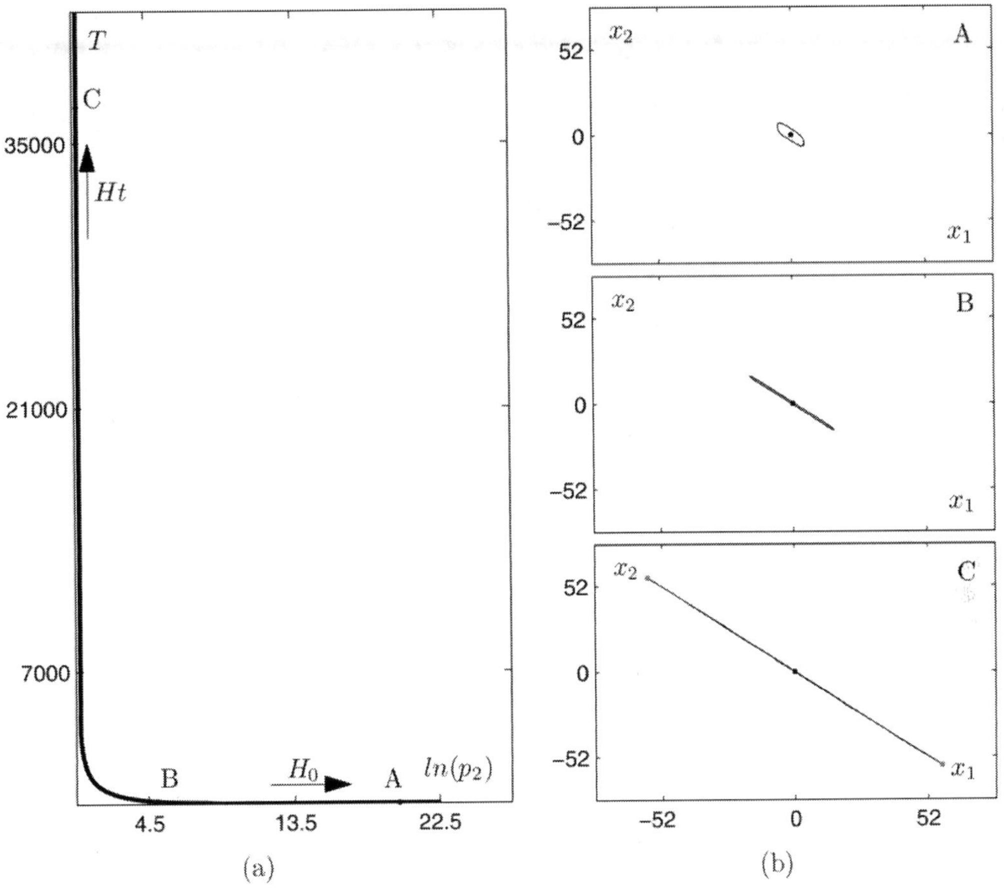

Figure 7. Numerical continuation of limit cycle C_0 with respect to p_2, for $ln(p_1) = 5$. (a) Period vs. p_2. (b) Limit cycle C_0, together with equilibria E_0 (in black) and $E_\pm$ (in grey), in the state plane (x_1, x_2) at parameter values $ln(p_2) = 20$ (A), $ln(p_2) = 5$ (B), and $ln(p_2) \simeq 0.262$ (C).

4.2. Fold-Pitchfork: Organization of the symmetry breaking

The heteroclinic bifurcation, being global, is a good candidate for organizing more complex behaviors. Indeed, by continuing Ht far-away from BT, we discovered a degenerate point FP, marking a codimension-2 Fold-Pitchfork bifurcation for the limit cycle C_0. At this point, the structurally unstable double heteroclinic connection undergoes a concurrent symmetry breaking. From this point, four further bifurcation curves originate:

1 A pitchfork of limit cycles (bifurcation curve P_C), along which the S-type limit cycles $C_\pm$ (in green in the sketches) emerge from the F-type limit cycle C_0;

2 A fold of limit cycles (bifurcation curve F_0), along which the two F-type limit cycles C_0 (in blue) and, say, C_h (in magenta) collide and disappear;

3–4 A pair of coinciding homoclinic bifurcations (bifurcation curve Hm), where the two S-type limit cycles appearing at P_C collide with the equilibria $E_\pm$, thus breaking and disappearing.

The aforesaid bifurcation lines are arranged around the point FP as shown in Fig. 8, where sketches of the bifurcating invariant sets are shown in the corresponding regions.

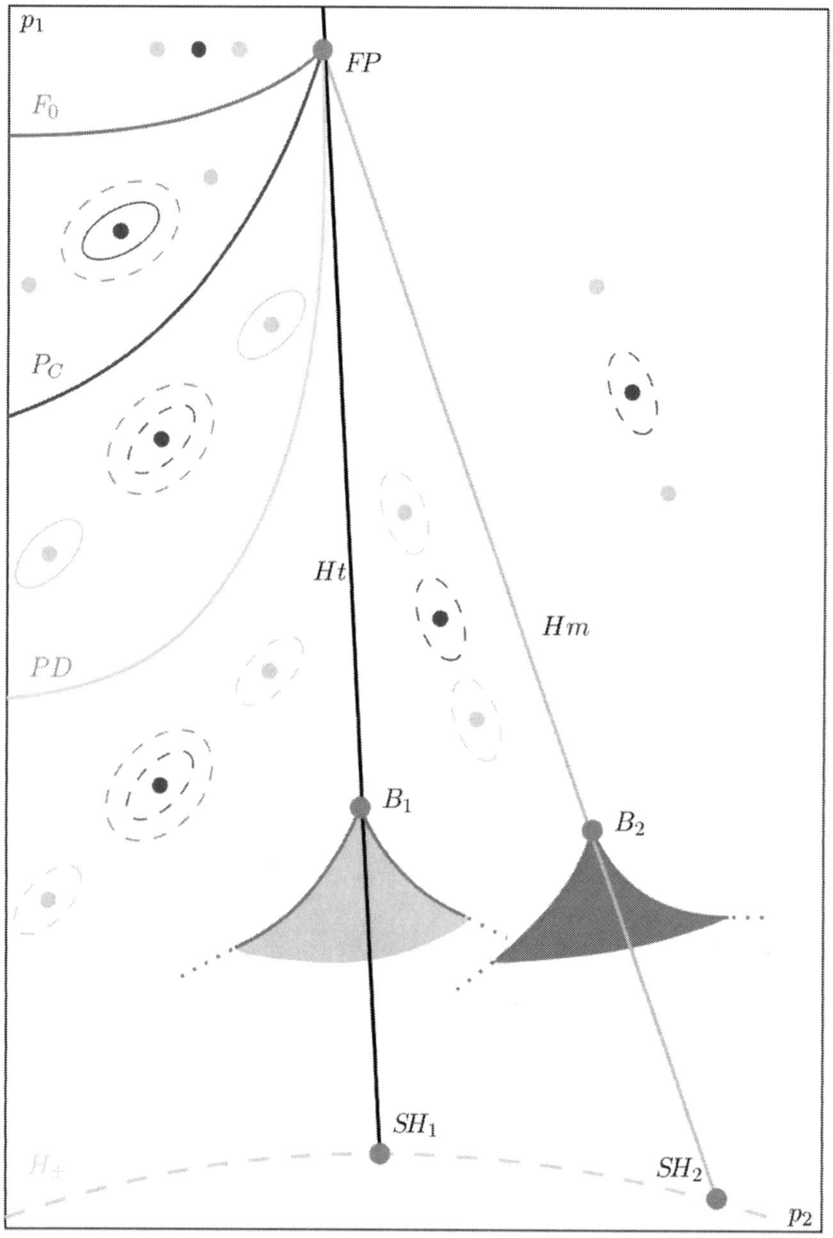

Figure 8. Arrangement of the bifurcation diagram in the neighborhoods of the Fold-Pitchfork, Belyakov, and Shil'nikov-Hopf codimension-2 bifurcations.

Less evident is the existence of a period doubling bifurcation curve (PD), originating in FP, involving the S-type limit cycles. This suggests the non-orientability of the flow around these cycles in the neighborhood of FP and, consequently, a more complex scenario [20], which, however, we did not investigate further. Indeed, in this complex scenario, the saddle (real) nature of the equilibria $E_\pm$ does not justify the existence of chaotic solutions.

To give an idea of the scenario linking Ht and F_0, Fig. 9 shows the continuation of the limit cycles C_0 and C_h with respect to p_2, for $ln(p_1) = 0.35$.

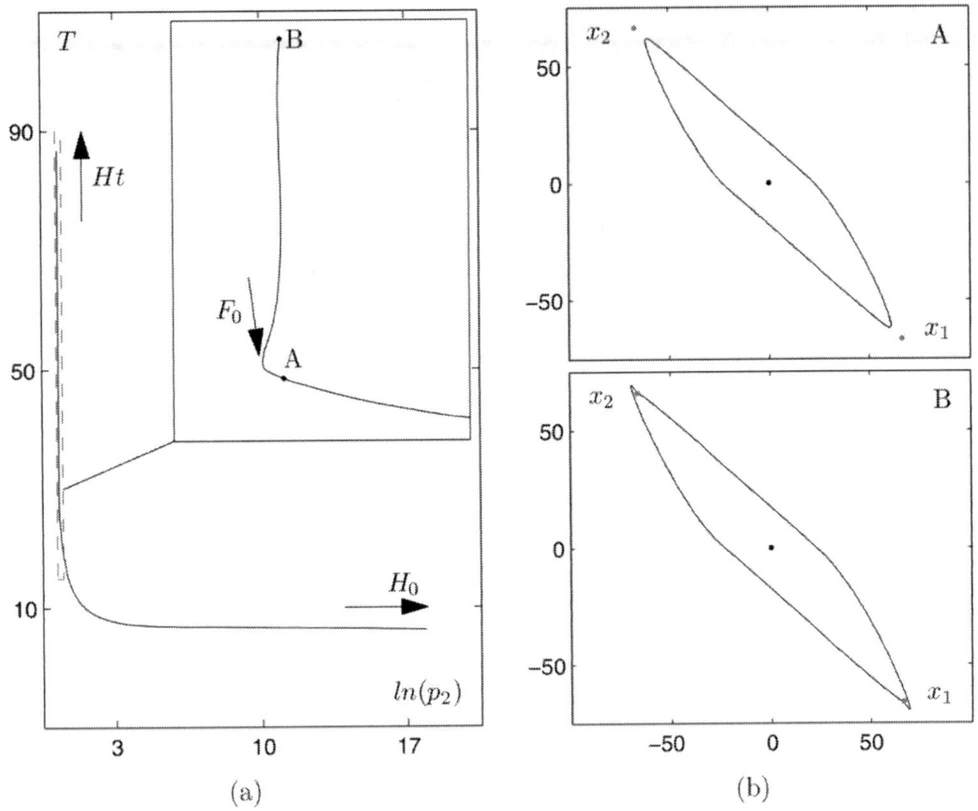

Figure 9. Numerical continuation of limit cycles C_0 and C_h with respect to p_2, for $ln(p_1) = 0.35$. (a) Period vs. p_2. (b) Two limit cycles (C_0 and C_h) coexisting at $ln(p_2) = 0.311$, together with equilibria E_0 (in black) and $E_\pm$ (in grey), in the state plane (x_1, x_2) .

4.3. Belyakov: Organization of chaotic solutions

Proceeding further, Ht undergoes a Belyakov degeneracy, where the equilibria $E_\pm$ bearing the double heteroclinic connection change from saddle (real) to saddle-focus. Theory predicts several families, of infinite cardinality, of bifurcation curves originating in this point [19, 20]. According to the well-known Shil'nikov theorem [18], these curves are predicted to accumulate exponentially on Ht, and the outer ones delimit a region (light gray region in Fig. 8) where wild (chaotic) trajectories can be observed [21]. The hypothetical chaotic trajectories organized by this Belyakov point would mimic the shape of the F-type limit cycles as shown in Figs. 4(b) and 4(f). Similarly, the homoclinic curve Hm originating in FP undergoes the same degeneracy and, consequently, S-type wild trajectories can be observed in the parameter space (dark gray region in Fig. 8). As an example, Fig. 10 shows the continuation of one of the family of infinite limit cycles, linked to C_0 and C_h, with respect to p_2, for $ln(p_1) = -0.8$.

Whilst the Belyakov points justify the existence of chaotic solutions associated with both F- and S-type limit cycles, they do not explain the existence of different families of limit cycles characterized by different numbers $n^{+,0,-}$ of turns (revolutions).

4.4. Shil'nikov-Hopf: Organization of the $n^{+,-}$-turns

Both Ht and Hm bifurcation curves end on the Hopf bifurcation curve ($H_\pm$) of the equilibria $E_\pm$ (Fig. 8). The end points $SH_{1,2}$ are codimension-2 bifurcations known as Shil'nikov-Hopf

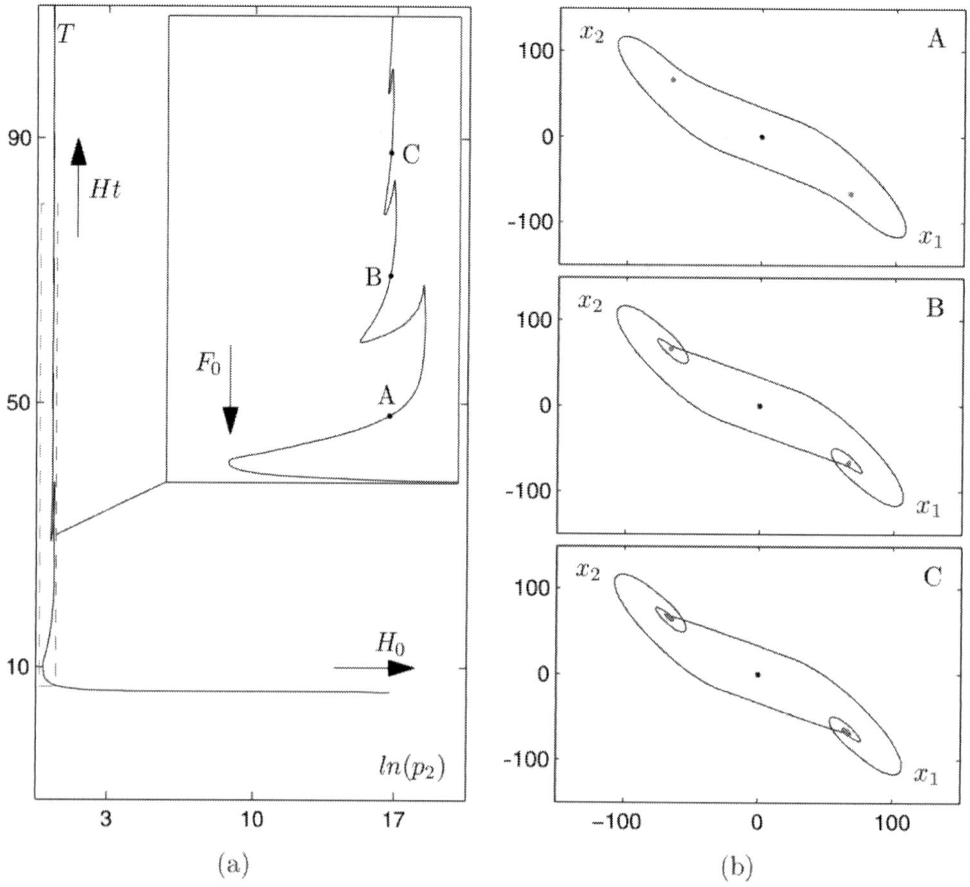

Figure 10. Numerical continuation of the family of infinite limit cycles originating from C_0 and C_h with respect to p_2, for $ln(p_1) = -0.8$. (a) Period vs. p_2. (b) Example of three of the infinite limit cycles coexisting at $ln(p_2) = 0.312$, together with equilibria E_0 (in black) and $E_\pm$ (in grey), in the state plane (x_1, x_2).

bifurcations. At this point, the homoclinic loop breaks into a double heteroclinic connection between a limit cycle and an equilibrium point [20] and, similarly, the double heteroclinic connection between equilibria breaks into a more complex multiple heteroclinic connection between equilibria and cycles. The theory predicts a very complex scenario near such a degeneracy [20], where infinite, non necessarily accumulating, bifurcations organize the existence of limit cycles characterized by different numbers of turns around the pair of limit cycles that have emerged on $H_\pm$. Consequently, since the limit cycles appear in the upper [lower] region of the state-space, these degeneracies justify the existence of F-type (from SH_1 and Ht) as well as S-type (from SH_2 and Hm) limit cycles with different numbers $n^{+,-}$ of turns. However, none of these bifurcation structures justifies the formation of the central (n^0) turns.

4.5. A blue-sky catastrophe: Formation of the n^0-turns

The continuation of F-type limit cycles with respect to p_2, at values of p_1 below the level of the two SH points, usually indicates a scenario such as the one shown in Figs. 11 and 12. This characteristic scenario, with both period and length increasing towards infinity, suggests the existence of the so-called *blue-sky* catastrophe, where two limit cycles collide transversally, with the trajectory of one cycle winding around the other one. In our case, the trajectories tend

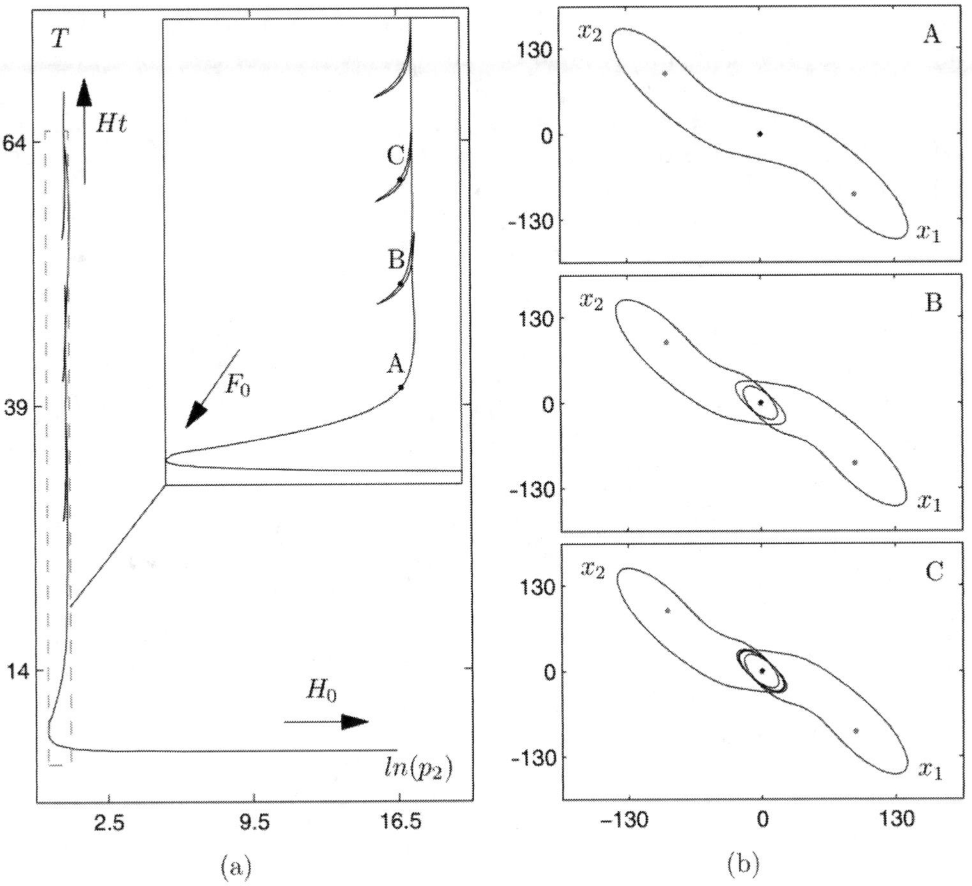

Figure 11. Numerical continuation of the family of infinite limit cycles originating from C_0 and C_h with respect to p_2, for $ln(p_1) = -1.2$. (a) Period vs. p_2. (b) Three of the limit cycles coexisting at $ln(p_2) = 0.48$, together with E_0 (in black) and $E_\pm$ (in grey), in the plane (x_1, x_2).

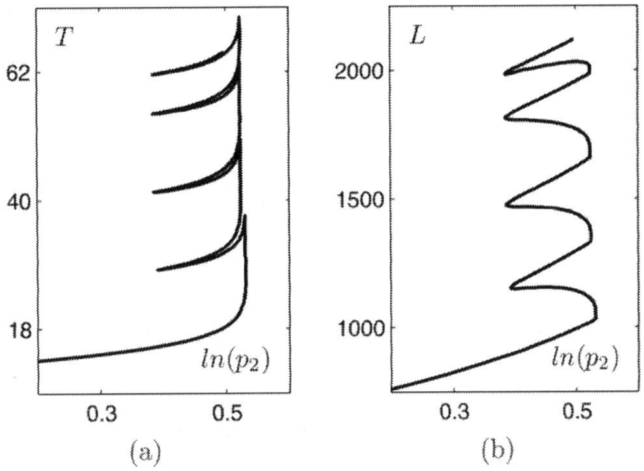

Figure 12. Blue-sky scenario for a F-type limit cycle ($ln(p_1) = -1.2$): (a) – Period *vs.* $ln(p_2)$; (b) – Length *vs.* $ln(p_2)$.

to roll up around the limit cycle C_0, thus justifying the formation of the central (n^0) turns. Unfortunately, blue-sky bifurcations cannot be numerically continued, at least by resorting to standard numerical packages. Extensive simulations suggest the existence of an infinity of them, one for each family of cycles with different $n^{+,-}$ turns. These bifurcation loci appear to be U-shaped with the upper part converging towards the degeneracy FP, and they coincide with the transitions from the bluish to the greyish regions in the "boomerang"-like shaped regions on the right side of the diagram of Fig. 5. Figure 13 shows a detail of one of the "boomerang"-like regions, where the blue-sky bifurcation curve appears (at least in part) as a green line.

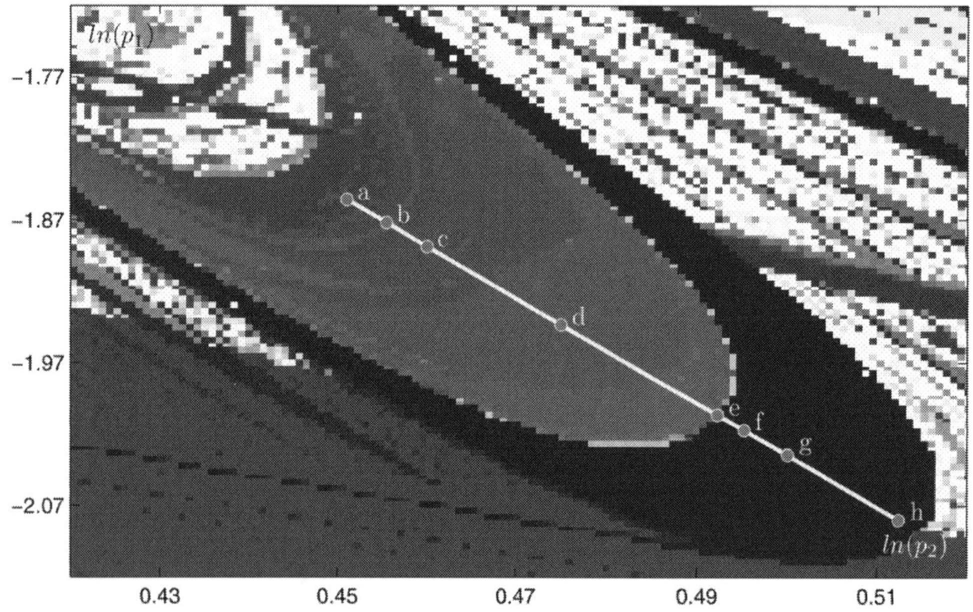

Figure 13. Detail of a one of the "boomerang"-like regions of the bifurcation diagram.

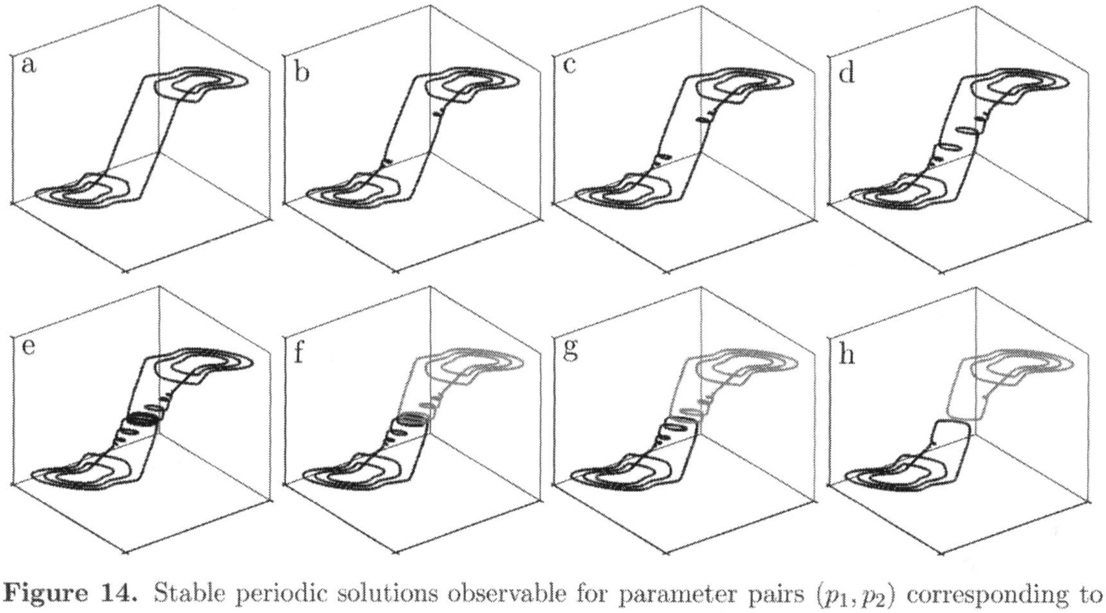

Figure 14. Stable periodic solutions observable for parameter pairs (p_1, p_2) corresponding to the points a to h in Fig. 13. The plot scales are the same for all figures.

Figure 14 shows the asymptotic (stable) solutions observable for parameter values corresponding to the points from a to h in Fig. 13.

By moving the parameter values along the yellow line of Fig. 13, from a to h, we observe a typical (symmetric) blue-sky scenario: an F-type periodic solution, shown in Fig. 14a, smoothly deforms winding around another small F-type limit cycle close to the origin (see Fig. 14b-e); correspondingly, its period and length grew asymptotically approaching the bifurcation boundary, as shown in Fig. 12; at the bifurcation point boundary (blue to gray transition in Fig. 13), the F-type periodic solution breaks up into a pair of S-type periodic solutions, as displayed in Fig. 14f; initially, these S-type solutions make many turns around the small F-type limit cycle; then, by further moving the parameter values along the yellow line, they unroll as shown in Fig. 14g-h, and their periods and lengths decrease correspondingly.

5. Overall bifurcation scenario

Figure 15 reports the bifurcation curves obtained by continuation superimposed to the brute-force simulations bifurcation diagram.

The colouring and labeling code of the curves is as in the sketches of Figs. 6 and 8, here summarized for the reader's convenience:

Colour	Type of curve	Example
blue	pitchfork of equilibria	P_E $(E_0 \to E_0,\ E_\pm)$
blue	pitchfork of cycles	P_C $(C_0 \to C_0,\ C_\pm)$
green	Hopf	$H_\pm$ $(E_\pm \to E_\pm,\ C_\pm)$
red	fold of limit cycles	F_0 $(C_0,\ C_h)$
cyan	period doubling	PD $(C_\pm)$
black	heteroclinic (for F-cycles)	Ht
grey	homoclinic (for S-cycles)	Hm

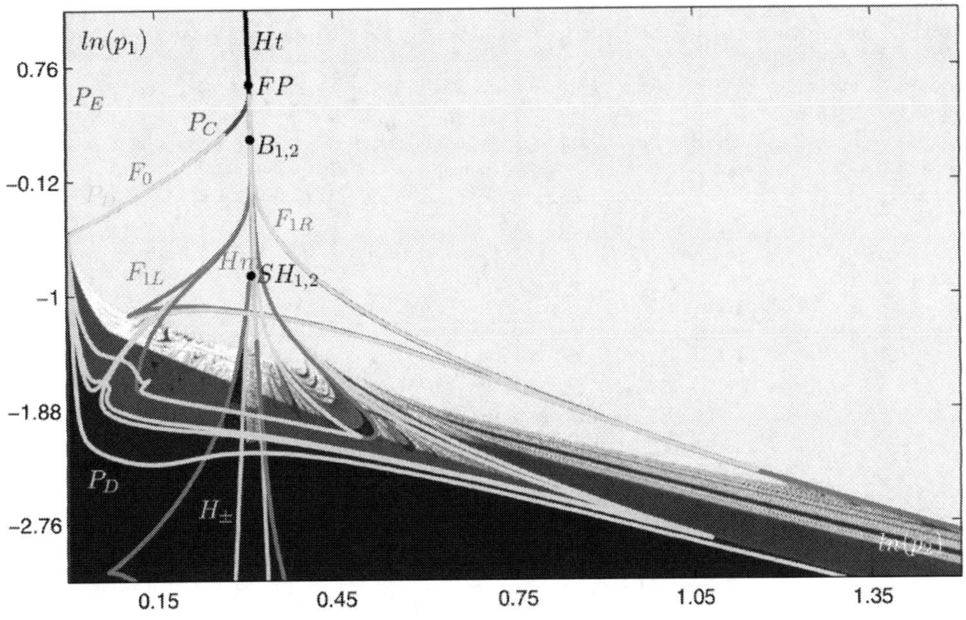

Figure 15. Overall bifurcation scenario ($ln(p_1)$ $vs.$ $ln(p_2)$).

As mentioned at the outset, for the limit cycles considered (with low values of n^+, n^0, and n^-), the continued bifurcation curves, and corresponding codimension-2 points organizing the scenario, lie, at least partially, in the region where no bounded dynamics can be observed. Nonetheless, they outline the prototype of the bifurcation diagram organizing the existence of one family of periodic solutions. The knowledge of this prototype diagram, combined with the fractal (self-similar) structure of the brute-force simulation bifurcation diagram highlights that the very same bifurcation scenario exists for all the families with different numbers n^+, n^0, and n^- of turns. In this respect, it is interesting to notice that only the solutions with rather high numbers of loops appear to be stable.

6. Conclusions

Simulations of the hysteresis-based oscillator model show the existence of several families of periodic and chaotic solutions characterized by different numbers of turns in the upper, central, and lower part of the state-space. The detailed bifurcation analysis reported here shows how their existence is justified, as well as organized in the parameter space, by a few codimension-2 bifurcation points.

This analysis is prelude to the definition of criteria for the oscillator design.

Acknowledgments

This work was supported by MIUR, within the FIRB framework (project RBNE012NSW), and by University of Genoa.

References

[1] Nakagawa S and Saito T 1996 An RC OTA hysteresis chaos generator *IEEE Trans. Circ. and Syst. I* **43** 1019

[2] Varrientos J and Sanchez-Sinencio E 1998 A 4D chaotic oscillator based on a differential hysteresis comparator *IEEE Trans. Circ. and Syst. I* **45** 3

[3] Elwakil A and Kennedy M P 1999 Chaotic oscillators derived from Saito's double-screw hysteresis oscillator *IEICE Trans. Fund.* **82** 1769

[4] Elwakil A and Kennedy M P 2000 Systematic realization of a class of hysteresis chaotic oscillators *Int. J. Circ. Th. Appl.* **28** 319

[5] Kataoka M and Saito T A two-port VCCS chaotic oscillator and quad screw attractor *IEEE Trans. Circ. and Syst. I* **48** 221

[6] Krasnosel'skii M A and Pokrovskii A V 1989 *Systems with Hysteresis* (Berlin: Springer)

[7] Mayergoyz I D 1991 *Mathematical Models of Hysteresis* (New York: Springer)

[8] Kennedy M P and Chua L O 1991 Hysteresis in electronic circuits: a circuit theorist's perspective *Int. J. Circ. Th. Appl.* **19** 471

[9] Parodi M, Storace M and Cincotti S 1994 A PWL ladder circuit which exhibits hysteresis *Int. J. Circ. Th. Appl.* **22** 513

[10] Bizzarri F, Storace M, Gardini L and Lupini R 2001 Bifurcation analysis of a PWL chaotic circuit based on hysteresis through a one-dimensional map *Int. J. Bif. Chaos* **11** 1911

[11] Lupini R, Bizzarri F and Storace M, 2001 Discontinuities in a one-dimensional map describing a hysteretic chaotic circuit *Nonlinear Analysis* **47** 5253

[12] Storace M and Bizzarri F 2001 Basic bifurcation analysis of a hysteresis oscillator *Int. J. Circ. Th. Appl.* **29** 343

[13] Bizzarri F and Storace M 2002 Two-dimensional bifurcation diagrams of a chaotic circuit based on hysteresis *Int. J. Bif. Chaos* **12** 43

[14] Bonnin M, Gilli M and Civalleri P P 2005 *A mixed time-frequency domain approach for the analysis of a hysteretic oscillator IEEE Trans. Circ. and Syst. II* **52** 525

[15] Bizzarri F, Stellardo D, Storace M and De Feo O 2005 *Some significant bifurcation curves in a hysteresis oscillator Proc. NOLTA'2005, Bruges, Belgium, 18-21 October 2005* 19

[16] Bizzarri F, Stellardo D and Storace M 2001 Bifurcation analysis and its experimental validation for a hysteresis circuit oscillator *IEEE Trans. Circ. and Syst. II* **53** in press

[17] Doedel E, Paffenroth R, Champneys A, Fairgrieve T, Kuznetsov Yu A, Sandstede B and Wang X 2001 AUTO 2000: Continuation and Bifurcation Software for Ordinary Differential Equations (with HomCont) *Computer Science Department, Concordia University, Montreal, Quebec, Canada*

[18] Kuznetsov Yu A 2004 *Elements of Applied Bifurcation Theory* (New York: Springer-Verlag 3rd ed)

[19] Belyakov L 1981 The bifurcation set in a system with a homoclinic saddle curve *Mat. Zametki* **28** 910

[20] Champneys A R and Kuznetsov Yu A 1994 Numerical detection and continuation of codimension-two homoclinic bifurcations *Int. J. Bif. Chaos* **4** 795

[21] Glendinning P and Sparrow C 1984 Local and global behavior near homoclinic orbits *J. of Statistical Phys.* **35** 645

Institute of Physics Publishing
doi:10.1088/1742-6596/55/1/003
Journal of Physics: Conference Series **55** (2006) 28–44
International Workshop on Multi-Rate Processes and Hysteresis

Singularly perturbed vector fields

V Bykov[1], I Goldfarb[2] and V Gol'dshtein[3]
[1] Institute of Technical Thermodynamics, Karlsruhe University, Karlsruhe, Germany
[2,3] Department of Mathematics, Ben Gurion University, Beer Sheva, Israel
E-mail: [1]bykov@itt.uka.de, [3]vladimir@bgu.ac.il

Abstract. A geometrically invariant concept of singularly perturbed systems of ordinary differential equations (singularly perturbed vector fields) is proposed in this paper. Singularly perturbed vector fields can be represented locally as singularly perturbed systems (for corresponding coordinate system choice. The paper focuses on possible ways of fast and slow directions/manifolds evaluations. A special algorithm for the evaluation is proposed. The algorithm is called as a global quasi-linearization procedure. A practical application of the proposed algorithm for numerical simulations is the main issue of the paper.

1. Introduction

The decomposition of complex systems into simpler subsystems using different rates of change (time scales) for different subsystems is widely used in physical and engineering models. The main difficulty for complex realistic models is the "hidden", implicit nature of the time scales of the original systems of governing equations. This gap between realistic models and the corresponding mathematical theory of singularly perturbed systems (SPS) represents a major problem for application of the mathematical theory based on an explicit knowledge about a "fast-slow" subdivision. For relatively simple models such a subdivision can be found by using various non-dimensionalization procedures and/or *a priori* physical or engineering knowledge about the processes.

For complicated models it might be reasonable to develop mathematical and/or algorithmic tools that can permit us to evaluate a possible "fast-slow" subdivision and this is a main issue of the current work. There are a number of numerical methods developed to deal with an implicit fast-slow structure [8, 11, 3, 10, 1]. These methods are based on a division of eigenvalues of the corresponding Jacobian matrix according to the values of their real parts into two groups: large and small. If such a division does exist then the first group of "large" eigenvectors are declared as the fast directions and the second group is declared as the slow directions. This decomposition permits us to evaluate slow invariant manifolds (see, for example, [1, 10, 13, 12, 8, 9]). Nevertheless, the procedure produces a number of so-called "ghost" manifolds that have no relation to the true invariant manifolds and the true system dynamics. From our previous research we can state that the main reason for the appearence of "ghost" manifolds' is an incorrect evaluation of fast directions through the Jacobi matrix eigenvalues.

In this paper we focus mainly on analytical and/or numerical algorithms that permit us to evaluate correctly the fast directions. Since we need a formal mathematical basis for such procedures we will start with the introduction and notion of singularly perturbed vector fields.

To the best of our knowledge this notion has not been discussed previously in any analytical framework.

Roughly speaking a singularly perturbed vector field (SPVF) $F(x, \varepsilon)$ is a vector field defined in a domain G of Euclidian space R^n that depends on a small parameter $\varepsilon \geq 0$ such that, for any point x, $F(x, 0)$ belongs to an a priori fixed fast subspace $L_f(x)$ of smaller dimension - $\dim L_f(x) < n$. Moreover, the dimension of $L_f(x)$ does not depend on the choice of the point x. Thus, the vector field $F(x, \varepsilon)$ can be decomposed into a fast subfield that belongs to the fast subspace $L_f(x)$ and its complement represents a slow subfield. Of course this is not a formal description, which is more sophisticated. Note, if $L_f(x)$ does not depend on x then the vector field $F(x, \varepsilon)$ represents (by definition) a linearly decomposed singularly perturbed vector field. Accordingly, the notion of the linearly decomposed singularly perturbed vector field is a geometrical analog of a singularly perturbed system of ordinary differential equations.

A formal concept (a theory of SPVFs) can be useful for practical applications if it is supported by an identification algorithm for fast subfields. In this paper we construct an algorithm for linearly decomposed singularly perturbed vector fields. This algorithm is based on a global linear interpolation procedure for an original vector field that we call a global quasi-linearization (GQL) (see e.g. [23]). As a reader will see, the procedure is flexible and permits us to adopt an appropriate version of the global quasi-linearization for different applications. A number of applications are discussed at the end of the paper. Some of these are purely illustrative and demonstrate the main properties of the proposed framework.

2. General Notion of Singularly Perturbed Vector Fields

In this section we introduce an abstract general notation for a singularly perturbed vector field based on the concept of vector bundles. On the one hand this general construction will be useful for our future investigations, on the other hand it permits us to understand better the corresponding simplest possible "linear" situation.

2.1. Motivation

Let us consider a standard singularly perturbed system (SPS). It can be written in two different equivalent forms [19]

$$\varepsilon \frac{dx}{dt} = F(x, y, \varepsilon) \tag{1}$$

$$\frac{dy}{dt} = G(x, y, \varepsilon) \tag{2}$$

for the so called fast time t or

$$\frac{dx}{d\tau} = F(x, y, \varepsilon) \tag{3}$$

$$\frac{dy}{d\tau} = \varepsilon G(x, y, \varepsilon) \tag{4}$$

for the so called slow time ($\tau : \varepsilon \, d\tau = dt$).

Here $x \in R^{n_f}$, $y \in R^{n_s}$, $n_f + n_s = n$, smooth vector functions $F(x, y, \varepsilon)$, $G(x, y, \varepsilon)$ are defined in a domain $V \subset R^n$ and $F(x, y, \varepsilon)$, $G(x, y, \varepsilon) \sim O(1)$. The vector field corresponding to the system (1)-(2) is

$$\Phi_\varepsilon(x, y) = \left(\frac{1}{\varepsilon} F(x, y, \varepsilon),\, G(x, y, \varepsilon)\right); \tag{5}$$

the vector field corresponding to the system (3)-(4) is

$$\Phi^{\varepsilon}(x,y) = (F(x,y,\varepsilon),\, \varepsilon G(x,y,\varepsilon)). \tag{6}$$

Both vector fields have the same directions but different asymptotic behavior for the limit $\varepsilon \to 0$:

1) $\Phi_{\varepsilon}(x,y) \to (\infty, G(x,y,0))$ for any point $(x,y) \in V$ such that $F(x,y,0) \neq 0$, and $\Phi_{\varepsilon}(x,y) = (0, G(x,y,0))$ for any point (x,y) such that $F(x,y,0) = 0$, i.e. for slow manifolds;

2) $\Phi^{\varepsilon}(x,y) \to (F(x,y,0),0)$ for any point $(x,y) \in V$.

Hence the first asymptotic representation is more convenient for describing the slow motion $G(x,y,\varepsilon)$ and the second one is better for the fast motion $F(x,y,\varepsilon)$.

In this work we focus on the fast motion's subspace R^{n_f} and, therefore, the second asymptotic representation is more important for our study. Using the language of geometric invariance the second asymptotic representation can be described in the following way: $\lim_{\varepsilon\to 0} \Phi^{\varepsilon}(x,y) \subset R^{n_f}$ for any $(x,y) \in V$. This remark is a motivation for the definition of asymptotic singularly perturbed vector fields.

2.2. Asymptotic singularly perturbed vector fields

We will start from a standard definition of vector bundles and will use vector bundles as a formal substitute for so-called nonlinear coordinate systems.

Definition 2.1 *A vector bundle ξ over a connected manifold $N \subset R^m$ consists of a set $E \subset R^m$ (the total set), a smooth map $p : E \to N$ (the projection) which is onto, and each fiber $F_x^{\xi} = p^{-1}(x)$ is a finite dimensional affine subspace. These objects are required to satisfy the following condition: for each $x \in N$, there is a neighborhood U of x in N, an integer k and a diffeomorphism $\varphi : p^{-1}(U) \to U \times R^k$ such that is an isomorphism of vector spaces on each fiber φ.*

All fibers have the same dimension k.

Definition 2.2 *Call a domain $V \subset R^n$ a structured domain (or a domain structured by a vector bundle) if there exists a vector bundle ξ and a diffeomorphism $\psi : V \to U$ onto an open subset $U \subset E$, where E is the total set of ξ.*

Fix a parametric family of smooth vector fields $\Phi(z,\delta)$ defined in a domain $V \subset R^n$ for any $0 < \delta < \delta_0$. Here δ_0 is a fixed positive number and δ is a small positive parameter, an explicit form of the small parameter in the system is needed at least in initial stage.

A corresponding system of ODEs is

$$\frac{dz}{dt} = \Phi(z,\delta). \tag{7}$$

Definition 2.3 *Suppose that V is a domain structured by a vector bundle ξ and a diffeomorphism ψ. For any point $z \in G$ call $M_z := \psi^{-1}(p^{-1}(\psi(z) \cap U)$ a fast manifold associated with the point z. Call the set of all fast manifolds M_z a family of fast manifolds of V.*

By construction any point $z \in G$ belongs to only one fast manifold. If $z \neq z_1$ either $M_z \cap M_{z_1} = \emptyset$ or $M_z = M_{z_1}$. The dimension of any manifold M_z is the same. Denote this dimension by n_f and call it the fast dimension of G. For instance, a singularly perturbed system (1-2) has $M_z = R^{n_f}$ and $\dim M_z = n_f$.

Definition 2.4 *A family of fast manifolds M_z is linear if there exists a linear subspace L_f of R^n such that $M_z = \{z\} + L_f$ for any $z \subset V$.*

Call L_f a fast subspace.

This is the simplest possible "linear" situation. Using a corresponding linear change of variables it is possible to move L_f to a coordinate subspace like in the case (1-2).

Denote by TM_z a tangent space to M_z at the point z.

Definition 2.5 *A parametric family $\Phi(z, \delta) : V \to R^n$ of vector fields defined in a domain V structured by a vector bundle ξ and a diffeomorphism ψ is an asymptotic singularly perturbed vector field if $\lim_{\delta \to 0} \Phi(z, \delta) \in TM_z$ for any $z \in V$ and the structure of the domain G is minimal for the vector field $\Phi(z, \delta) : G \to R^n$ in the following sense.*

There is no proper vector subbundle ξ_1 of the vector bundle ξ such that $\Phi(z, \delta) : V \to R^n$ is an asymptotic singularly perturbed vector field in a domain V structured by the vector subbundle ξ_1 and the same diffeomorphism ψ.

Remark 1 *This property of minimality means that it is not possible to reduce the dimension of fast manifolds $\{M_z\}$ using subbundles.*

From this point onwards, without loss of generality, we suppose that a family of fast manifolds $\{M_z\}$ associated with a singularly perturbed vector field $\Phi(z, \delta)$ is minimal.

For a linear family of fast manifolds associated with a singularly perturbed vector field $\Phi(z, \delta)$ the property of minimality can be written in a rather simple way. If L_f is a minimal fast linear subspace associated with a singularly perturbed vector field $\Phi(z, \delta)$ then dimension $n_f = \dim L_f$ cannot be reduced.

Call this minimal subspace L_f a linear subspace of fast motions of $\Phi(z, \delta)$.

2.3. Fast-slow decomposition of singularly perturbed vector fields.

Fix an asymptotic singularly perturbed vector field $\Phi(z, \delta)$. Suppose $\{M_z\}$ is a fast family associated with $\Phi(z, \delta)$ and the fast dimension of $\{M_z\}$ is n_f. Then the vector field $\Phi(z, \delta)$ is a sum of two vector fields $\Phi_f(z, \delta) := Pr_f\Phi(z, \delta)$ and $\Phi_s(z, \delta) := \Phi(z, \delta) - Pr_f\Phi(z, \delta)$. Here $Pr_f\Phi(z, \delta)$ is a projection of $\Phi(z, \delta)$ onto the tangent space TM_z of the fast manifold M_z, and $\Phi_s(z, \delta)$ is a projection of $\Phi(z, \delta)$ onto the linear subspace TM^z of slow motions that is orthogonal to TM_z.

Call an asymptotic singularly perturbed vector field $\Phi(z, \delta)$ a uniformly asymptotic singularly perturbed vector field (or simply a uniform vector field) if

$$\lim_{\delta \to 0} \sup_{z \in V} |Pr_s\Phi(z, \delta)| = 0.$$

Define $\varepsilon := \sup_{z \in V} |Pr_s\Phi(z, \delta)|$ which is a new small parameter, $\varepsilon < \varepsilon_0 = \sup_{z \in V} |Pr_s\Phi(z, \delta_0)|$; $F(z, \delta) := Pr_f\Phi(z, \delta)$ is the fast subfield and $G(z, \delta) := \frac{Pr_s\Phi(z,\delta)}{\sup_{z \in V}|Pr_s\Phi(z,\delta)|}$ is a slow subfield of $\Phi(z, \delta)$. Then the vector field $\Phi(z, \delta)$ is a linear combination of its fast and slow subfields i.e.

$$\Phi(z, \delta) = F(z, \delta) + \varepsilon G(z, \delta). \tag{8}$$

Remark: the small parameter ϵ is a function of the small parameter δ. If $\delta \to 0$ then $\epsilon \to 0$.

For any practical implementation of the proposed construction of singularly perturbed vector fields we have to find a way to determine the fast manifolds. For the moment we have this for the linear case i.e. for the case where all fast manifolds are parallel to a fixed linear subspace L_f.

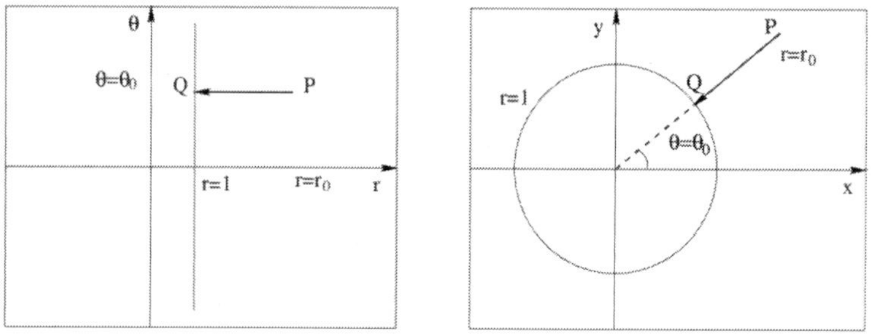

Figure 1. The left figure shows the system dynamics in a polar coordinates system; the right figure presents the system behavior in a Cartesian coordinate system after transformation.

2.4. Analytical example of a singularly perturbed vector field - Polar coordinates
This example was constructed by V.A Sobolev and it demonstrates that a fast subfield can depend on a point even in very simple cases.

Consider a simple SPS in polar coordinates

$$\begin{cases} \varepsilon \frac{dr}{dt} = (1-r) \\ \frac{d\theta}{dt} = 1 \end{cases}, \tag{9}$$

with initial conditions $r(t=0) = r_0$, $\theta(t=0) = \theta_0$. Now, we change the polar coordinates to Cartesian

$$\begin{cases} x = r\cos(\theta) \\ y = r\sin(\theta) \end{cases}. \tag{10}$$

The system (9) is singularly perturbed and its dynamics in both $(r,\ \theta)$ and $(x,\ y)$ planes are schematically presented in Figure 1.

After the substitution (10) the system (9) has the following representation

$$\begin{cases} \frac{dx}{dt} = -\frac{y}{(x^2+y^2)} + \frac{1}{\varepsilon}\frac{x\left(1-\sqrt{x^2+y^2}\right)}{(x^2+y^2)^{3/2}} \\ \frac{dy}{dt} = \frac{x}{(x^2+y^2)} + \frac{1}{\varepsilon}\frac{y\left(1-\sqrt{x^2+y^2}\right)}{(x^2+y^2)^{3/2}} \end{cases}, \quad \Rightarrow \quad z = \begin{pmatrix} x \\ y \end{pmatrix}, \quad \frac{dz}{dt} = \Phi(z), \tag{11}$$

with initial conditions $x_0 = r_0\cos(\theta_0)$, $y_0 = r_0\sin(\theta_0)$. Assume that the original form (9) of the system is unknown. Then a natural question is how to find (redefine) new global coordinates similar to (9) in such a way that the system goes into the standard form of SPS. Remark: despite the existence a small parameter and the multi-scale (singularly perturbed) nature of the system, system (11) is not written as SPS in Cartesian coordinates.

The proposed framework allows us treat and redefine the global transformation. An initial step in this direction is a decomposition of the given vector field (11) and it might be performed according to

$$\Phi = F + G, \tag{12}$$

where the fast subfield F and the slow subfield G can easily be found due to the explicit appearance of the small parameter ε:

$$F = \begin{pmatrix} F_1 \\ F_2 \end{pmatrix} = \frac{1}{\varepsilon} \begin{pmatrix} \frac{x\left(1-\sqrt{x^2+y^2}\right)}{(x^2+y^2)^{3/2}} \\ \frac{y\left(1-\sqrt{x^2+y^2}\right)}{(x^2+y^2)^{3/2}} \end{pmatrix} \sim O(\tfrac{1}{\varepsilon}),$$
$$G = \begin{pmatrix} G_1 \\ G_2 \end{pmatrix} = \begin{pmatrix} -\frac{y}{(x^2+y^2)} \\ \frac{x}{(x^2+y^2)} \end{pmatrix} \sim O(1). \tag{13}$$

Now, in order to represent (11) as SPS we can analyze the fast and slow subfields separately. This approach represents a considerable simplification which is important in an analytical study (it reduces the system dimension and leads to simple subsystems). For example, if we consider the original system (11), it is clear that to find the solution in any form is not a simple task. Nevertheless, the fast subfield (subsystem) - $\frac{dz}{dt} = F$ can be integrated separately and it has the general solution in implicit form - $u = \ln(\frac{x}{y}) + C_u = 0$. This means that the fast motion is performed in the (x, y) plane along straight lines $y = Cx$, which represents the fast manifold $\{M_{(x,y)}\}$ of the singularly perturbed vector field F. Moreover, it would be natural to expect that the new variable u is slow. In fact, the variable u describes the slow system dynamics of the original system

$$\frac{du}{dt} = u_x \frac{dx}{dt} + u_y \frac{dy}{dt} = u_x(F_1 + G_1) + u_y(F_2 + G_2) = u_x G_1 + u_y G_2 \sim O(1).$$

At the same time the slow subsystem $\frac{dz}{dt} = G$ has the general solution $v = x^2 + y^2 + C_v = 0$ and it is expected to be fast according to

$$\frac{dv}{dt} = v_x \frac{dx}{dt} + v_y \frac{dy}{dt} = v_x(F_1 + G_1) + v_y(F_2 + G_2) = v_x F_1 + v_y F_2 \sim O(\frac{1}{\varepsilon}).$$

Hence, the standard SPS form of the system (11) becomes

$$\frac{du}{dt} = -\frac{y}{x\,(x^2+y^2)} - \frac{x}{y\,(x^2+y^2)} = -\frac{(\exp(-u) + \exp(u))}{v},$$

$$\frac{dv}{dt} = \frac{1}{\varepsilon}\left(\frac{2x^2\left(1-\sqrt{x^2+y^2}\right)}{(x^2+y^2)^{3/2}} + \frac{2y^2\left(1-\sqrt{x^2+y^2}\right)}{(x^2+y^2)^{3/2}}\right) = \frac{2(1-\sqrt{v})}{\varepsilon\sqrt{v}}.$$

Here the constants - C_u, C_v were chosen equal to zero. This finalizes the task of the redefinition of a global coordinate system. Future simplifications and redefinition of the polar coordinates can be done by taking $\theta = \arctan(\frac{y}{x}) = \arctan(\exp(u))$ as the slow variable instead of u, because, as was mentioned during the fast stage, the angle between the x - axis and direction of the fast motion is constant. Additionally one can use $r = \sqrt{v} = \sqrt{x^2+y^2}$ instead of v.

It should be noted that this type of the analysis might be considered as a preliminary step (converting to the standard SPS form (1)-(2)) of asymptotic analysis and it is possible only if a decomposition of a singularly perturbed vector field is known and/or a small system parameter is explicitly specified as in the example considered.

2.5. Pressure driven flames (model system)

Let us now present another example representing an application of the suggested theoretical framework to a 'hot' combustion problem. Namely, we consider the phenomenon of flame propagation through an inert porous media filled with a gaseous combustible mixture. Such flame waves are often referred to as pressure driven or barodiffusion flames [25]. It should be mentioned that our attempts to use a conventional approach (the method of inner and outer asymptotic expansions, well accepted in the field of the mathematical theory of combustion) to solve the problem under consideration failed. In the authors' opinion this failure was caused by the failure of the standard assumption of a constant pressure which is obviously not correct for pressure driven flames.

The machinery based on the SPVF approach permits us to overcome the main problems of conventional asymptotics involved in the analysis of combustion problems, and allows us to investigate the flame structure and to obtain an asymptotic estimation of the flame propagation velocity.

The main physical assumptions required to construct this simplified model are the following: the heat transfer mechanism is considered negligible (to single out an impact of the local pressure elevation), the inertia effects are negligible and the friction force is proportional to the gas velocity (Darcy law) (see e.g. [26]). Non-dimensionalization and suitable integration allow us to re-write the original system in the form of the three ODEs [5, 6]

$$\frac{d\theta}{d\xi} = \Lambda\frac{\Pi-\theta}{1+\beta\theta} + \varepsilon_1\eta\frac{1+\beta\Pi}{1+\beta\theta}\exp\left(\frac{\theta}{1+\beta\theta}\right), \tag{14}$$

$$\sigma\frac{d\Pi}{d\xi} = \Lambda\frac{\Pi-\theta}{1+\beta\theta}, \tag{15}$$

$$\frac{d\eta}{d\xi} = -\varepsilon_2\eta\frac{1+\beta\Pi}{1+\beta\theta}\exp\left(\frac{\theta}{1+\beta\theta}\right). \tag{16}$$

The system is written in the moving frame coordinate system using the so called auto-model variable ξ. Here Π, θ, η are dimensionless quantities of pressure, temperature and deficient reactant concentration and Λ is a dimensionless flame velocity that corresponds to the quasi-stationary approach.

To demonstrate the essence of the new approach, let us additionally apply the Frank-Kamenetskii approximation and put $\beta = 0$ in the set of the governing equations (14)-(16). The simplified model system reads

$$\frac{d\theta}{d\xi} = \Lambda(\Pi-\theta) + \varepsilon_1\eta\exp(\theta), \tag{17}$$

$$\sigma\frac{d\Pi}{d\xi} = \Lambda(\Pi-\theta), \tag{18}$$

$$\frac{d\eta}{d\xi} = -\varepsilon_2\eta\exp(\theta). \tag{19}$$

Further simplifications are possible due to the fact that the systems (17)-(19) and (14)-(16) have an energy integral (the adiabatic approach was used)

$$\eta - 1 + \frac{\varepsilon_2}{\varepsilon_1}(\theta - \sigma\Pi) = 0, \tag{20}$$

which permits us to determine the parameters of the burnt mixture (behind the flame front):

$$\eta_b = 0;\ \theta_b = \Pi_b = \frac{\varepsilon_1}{\varepsilon_2(1-\sigma)}. \tag{21}$$

In addition, the existence of the energy integral allows to reduce a number of governing equations by one. The reduced system reads

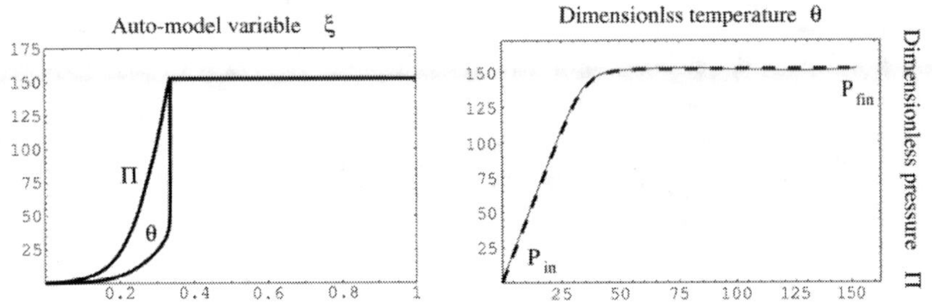

Figure 2. Profiles of quasi-stationary solution (left) and phase plane of reduced system with solution trajectory (right figure). Dashed line - approximate analytical solution, solid line - numerical solution.

$$\frac{d\theta}{d\xi} = \Lambda\left(\Pi - \theta\right) + \varepsilon_1\left(1 - \frac{\varepsilon_2}{\varepsilon_1}(\theta - \sigma\Pi)\right)\exp\left(\theta\right), \tag{22}$$

$$\sigma\frac{d\Pi}{d\xi} = \Lambda\left(\Pi - \theta\right), \tag{23}$$

and is subject to the initial conditions

$$\theta(0) = \varepsilon \ll 1;\ \Pi(0) = \frac{\varepsilon}{\sigma} \ll 1, \tag{24}$$

whose choice in the form (24) eliminates the so-called cold-boundary problem for initial data (a well known problem for initial data in combustion models). The boundary conditions (24) and (21) permit us to solve the problem of the evaluation of the wave velocity Λ since the system is over-determined. It becomes a boundary value problem.

Before starting the asymptotic investigation, let us mention typical values of the parameters for combustion systems: $0.1 \leq \sigma \leq 0.3$, $10^{-6} \leq \varepsilon_2 \ll \varepsilon_1 \leq 10^{-4}$. Accordingly, a characteristic value of the adiabatic temperature and pressure has the order of magnitude: $O(10^2)$.

Additionally, the singular/adiabatic point of the system (22)-(23) is of saddle type. It means in particular, that this point is reachable only along the unique direction (separatrix) which is found from

$$\frac{d\theta}{d\Pi} = \sigma\left(\frac{1}{\sigma} + \frac{\varepsilon_2}{\Lambda}\exp\left(\theta_b\right)\right) \tag{25}$$

corresponding to the negative eigenvalue $\lambda = -\varepsilon_2\exp\left(\theta_b\right)$ and, therefore, if a trajectory of the system reaches this singular point it must do so along this direction.

Now, consider the system (22)-(23) and construct an approximate solution in the vicinity of the singular (adiabatic) saddle point P_{fin} of the system (see Fig. 2, where profiles and the trajectories of the system in the phase plane are presented). Then using the analysis of the approximate solution we transform the system into standard SPS form and resolve the boundary value problem for the system eigenvalue Λ.

The behavior of the trajectory of the system in the phase plane (θ, Π) is determined by the solution of the following equation (plane system)

$$\frac{d\theta}{d\Pi} = \sigma\left(1 + \frac{\varepsilon_1\left(1 - \frac{\varepsilon_2}{\varepsilon_1}(\theta - \sigma\Pi)\right)}{\Lambda\left(\Pi - \theta\right)}\right)\exp\left(\theta\right). \tag{26}$$

In order to find an approximate solution let us simplify this equation using the information regarding the direction of the system trajectory in the vicinity of the singular point. This direction can be readily found from (26)

$$\left.\frac{d\theta}{d\Pi}\right|_{\substack{\theta\to\theta_b\\ \Pi\to\Pi_b}} = \sigma\left(1 + A\,\exp\left(\theta_b\right)\right), \tag{27}$$

where the limit of the pre-exponential term A is evaluated using (25) in the following way (an exponentially small term (i.e. $\exp(-\theta_b) \ll 1$, $\theta_b = O(10^2)$) is omitted)

$$\begin{aligned}\sigma\left(\tfrac{1}{\sigma} + \tfrac{\varepsilon_2}{\Lambda}\,\exp\left(\theta_b\right)\right) &\equiv \sigma\left(1 + A\,\exp\left(\theta_b\right)\right),\\ A = \left(\left(\tfrac{1}{\sigma} - 1\right)\exp\left(-\theta_b\right)\right) &+ \tfrac{\varepsilon_2}{\Lambda} \approx \tfrac{\varepsilon_2}{\Lambda}.\end{aligned} \tag{28}$$

Thus, the simplified solution, which defines the fast motion zero order approximation as well as the transformation into SPS form, can be easily found from (26)

$$\int \frac{d\theta}{\left(1 + \frac{\varepsilon_2}{\Lambda}\exp\left(\theta\right)\right)} = \sigma \int d\Pi, \tag{29}$$

which, after integration, yields

$$\theta - \ln\left(1 + \frac{\varepsilon_2}{\Lambda}\exp\left(\theta\right)\right) = \sigma\Pi + C. \tag{30}$$

This corresponds to the solution of the fast subsystem where the decomposition of the type (8), (12) is made by using the special system behavior near the singular point (21). Since the fast motion is realized along the integral curve (30) (at least near the stationary point) a candidate for the slow variable can be extracted easily as

$$u = \sigma\Pi - \theta + \ln\left(1 + \frac{\varepsilon_2}{\Lambda}\exp\left(\theta\right)\right). \tag{31}$$

We now transform the model system considered to the standard SPS form. Following the suggested framework we apply a new coordinate system (u, θ) in the form

$$\frac{d\theta}{d\xi} = \Lambda\left(\Pi\left(u,\theta\right) - \theta\right) + \varepsilon_1\left(1 - \frac{\varepsilon_2}{\varepsilon_1}\left(\theta - \sigma\Pi\left(u,\theta\right)\right)\right)\exp\left(\theta\right), \tag{32}$$

$$\frac{du}{d\xi} = \varepsilon_1 \frac{\left(1 - \frac{\varepsilon_2}{\varepsilon_1}(1-\sigma)\Pi\left(u,\theta\right)\right)\exp\left(\theta\right)}{\left(1 + \frac{\varepsilon_2}{\Lambda}\exp\left(\theta\right)\right)}, \tag{33}$$

where $\Pi\left(u,\theta\right)$ is given by (31). We remark that the RHS of (33) is $O(\varepsilon_1)$ everywhere in the phase plane, at the initial point as well as near the stationary point. Therefore θ is the fast variable. The slow variable u does not change much along the system solution and has to be close to a constant defined at the stationary point $u = u_b$.

To find the zero order approximation for $\varepsilon_1 \ll 1$ we have to find the constant of integration C in (30) which is determined by the values at the stationary point. After substitution we have

$$\theta - \theta_b - \ln\left(\frac{1 + \frac{\varepsilon_2}{\Lambda}\exp\left(\theta\right)}{1 + \frac{\varepsilon_2}{\Lambda}\exp\left(\theta_b\right)}\right) = \sigma(\Pi - \Pi_b). \tag{34}$$

Now, expand the approximate solution of the fast subsystem up to the initial point taking the limit $\theta(0) = \varepsilon \to 0$

$$-\ln\left(\frac{1+\frac{\varepsilon_2}{\Lambda}}{1+\frac{\varepsilon_2}{\Lambda}\exp(\theta_b)}\right)=\theta_b(1-\sigma). \tag{35}$$

An estimate of the parameter Λ is obtained as the solution of this equation

$$\Lambda=\varepsilon_2\exp(\sigma\theta_b)\frac{(1-\exp(-\sigma\theta_b))}{(1-\exp(-(1-\sigma)\theta_b))}\approx\varepsilon_2\exp(\sigma\theta_b). \tag{36}$$

It has to be mentioned here that further improvements and increasing accuracy is possible by the standard SPS analysis of the decomposed system (32)-(33). Comparison with previous results [5, 7, 23] shows that these estimates coincide with exponentially small order of magnitude ($\theta_b \gg 1$)

$$\Lambda=\frac{\varepsilon_1}{(1-\sigma)\theta_b}\exp(\sigma\theta_b)=\varepsilon_2\exp(\sigma\theta_b). \tag{37}$$

The formula (37) finalizes the solution of the problem (17)-(19). Thus, the well-known property of the saddle point together with the proposed framework allowed us to construct the required decomposition in the whole phase plane. Moreover, it justified our confidence that there are asymptotic tools which are able to provide us with more accurate asymptotic solutions of the combustion problems than the existing ones.

3. Slow Invariant Manifolds

A few words about slow manifolds. Suppose $\Phi(z,\delta)=F(z,\delta)+\varepsilon G(z,\delta)$ is a singularly perturbed vector field. The equation $F(z,\delta)=0$ represents a fast isocline of the vector field $\Phi(z,\delta)$. If its solution is a manifold S_0 with tangent spaces transverse to fast manifolds then this manifold has dimension n_s and using local charts of the corresponding vector bundle and a standard local representation of manifold S_0 as a graph of a vector function it is possible to prove the existence of the slow invariant manifold S. Similarly to the standard theory of singularly perturbed systems, S_0 is a zero approximation (for $\delta=0$) of the slow invariant manifold S.

Of course this restriction on S_0 does not hold for all points of S_0 in any realistic situation. Even if S_0 is a manifold it can happen that for some submanifold its tangent space is not transverse to the fast manifolds. This submanifold represents a manifold of turning (jumping) points of S_0 where slow dynamics change to fast ones (for realistic situations). In this paper we will avoid any discussion about these delicate submanifolds and will leave it for future work.

Dynamics on S_0 is dynamics of the slow subfield $G(z,\delta)$ which belongs to the tangent bundle of S_0. To obtain the first approximation of the slow invariant manifold let us differentiate the fast vector field $F(z,\delta)$ w.r.t. time (using the dynamical system $dz/dt=\Phi(z,\delta)$). Because $\Phi(z,\delta)=F(z,\delta)+\varepsilon G(z,\delta)$ we obtain an implicit expression

$$F_z(z,\delta)(F(z,\delta)+\varepsilon G(z,\delta))=0$$

as the first approximation, where terms of order $o(\delta)$ are neglected. By repeated differentiation, approximations of high orders can be obtained. This procedure coincides with the evaluation of implicit invariant slow manifolds (see e.g. [15], [16]). Since this paper is focused on the evaluation of fast manifolds, detailed analysis of slow manifolds will be given in future works.

4. Singularly Perturbed Vector Fields with Linear Fast Subspace

For any realistic complex model the small parameter δ is unknown and this fact restricts possible applications of the proposed asymptotic theory. In this section we will try to explain how to adapt the asymptotic theory developed for a practical problem in the simplest possible case of linear fast manifolds.

Thus any fast manifold at any point z is parallel to a linear subspace $L_f(z)$ with fixed dimension n_f. Note that for many applications an assumption that L_f does not depend on z is natural.

4.1. Fast-slow decomposition

Fix a uniformly asymptotic singularly perturbed vector field $\Phi(z, \delta)$. Suppose that the fast subspace L_f does not depend on z and $\dim L_f = n_f$. The vector field $\Phi(z, \delta)$ is a sum of two vector fields $\Phi_f(z, \delta) := Pr_f\Phi(z, \delta)$ and $\Phi_s(z, \delta) := \Phi(z, \delta) - Pr_f\Phi(z, \delta)$. Here $Pr_f\Phi(z, \delta)$ is a projection of $\Phi(z, \delta)$ onto the fast subspace L_f, and $\Phi_s(z, \delta)$ is a projection of $\Phi(z, \delta)$ onto the linear slow subspace L_s that is orthogonal to L_f.

The uniformity condition permits us to represent a uniformly singularly perturbed vector field and a corresponding dynamical system (7) as a standard singularly perturbed system (SPS). A corresponding construction is now possible. Suppose $x := Pr_f z$ and $y := Pr_s z$ are fast and slow variables that represent a new coordinate system with n_f fast variables x and $n_s = n - n_f$ slow variables y; $\varepsilon := \sup_{z \in V} |Pr_s\Phi(z, \delta)|$ is a small parameter, $\varepsilon < \varepsilon_0 = \sup_{z \in V} |Pr_s\Phi(z, \delta_0)|$; $F(x, y, \delta)$ is a representation of $F(z, \delta) := Pr_f\Phi(z, \delta)$ in the new coordinate system (x, y) and $G(x, y, \delta)$ is a representation of $G(z, \delta) := \frac{Pr_s\Phi(z,\delta)}{\sup_{z \in V}|Pr_s\Phi(z,\delta)|}$ in the new coordinate system (x, y).

Hence the system (7) has the standard SPS form

$$\frac{dx}{d\tau} = F(x, y, \delta)$$

$$\frac{dy}{d\tau} = \varepsilon G(x, y, \delta).$$

in the new coordinate system (x, y).

Remember that the small parameter ϵ is a function of the small parameter δ. If $\delta \to 0$ then $\epsilon \to 0$.

4.2. Singularly perturbed vector fields: non asymptotic definition.

In many practical situations the previous definition of an asymptotic singularly perturbed vector field $\Phi(z, \delta)$ cannot be useful because the small parameter δ is unknown. Meanwhile the main geometrical idea is still useful if some previous knowledge about a scaling is known. This means that some "small" number ε_0 is fixed for corresponding processes (models) and any parameter $\varepsilon < \varepsilon_0$ can be considered as a small parameter.

Suppose a smooth vector field $\Phi(z)$ is defined in a structured domain $V \subset R^n$ and L_f is the fast subspace. Moreover $\sup_{z \in V} |Pr_s\Phi(z)| < \varepsilon_0$.

Suppose as well, as in the previous subsection, that $x := Pr_f z$ and $y := Pr_s z$ are fast and slow variables that represent a new coordinate system with n_f fast variables x and $n_s = n - n_f$ slow variables y; $\varepsilon := \sup_{z \in V} |Pr_s\Phi(z)|$ is a small system parameter; $F(x, y)$ is a representation of $Pr_f\Phi(z)$ and $G(x, y)$ is a representation of $\frac{Pr_s\Phi(z)}{\sup_{z \in V}|Pr_s\Phi(z)|}$ in the new coordinate system (x, y).

Hence the system (7) has a standard SPS form

$$\frac{dx}{d\tau} = F(x, y) \tag{38}$$

$$\frac{dy}{d\tau} = \varepsilon G(x, y). \tag{39}$$

in the new coordinate system (x, y) and $\Psi(x, y, \varepsilon) = F(x, y) + \varepsilon G(x, y)$ is a singularly perturbed vector field.

We remark that previous formal descriptions are useless without an efficient procedure permitting us to evaluate a fast subspace L_f for an arbitrary singularly perturbed vector field $\Phi(z)$.

5. Evaluation of fast subspaces

In this section some procedures for the evaluation of the dimension and structure of fast subspaces is proposed. We call such a type of procedure quasi-linearizations of vector fields. Different versions of quasi-linearization were discussed in our previous paper [23]. In what follows we present a general construction with the help of the approach developed here.

5.1. Asymptotic global quasi-linearization procedure

Suppose $\Phi(z, \delta)$ is a singularly perturbed vector field defined in a structured domain V. This means that $\Phi(z, \delta)$ can be represented as a sum $\Phi(z, \delta) := F(z, \delta) + \varepsilon G(z, \delta)$ where $F(z, \delta)$ is a fast part of the vector field $\Phi(z, \delta)$ and $\varepsilon G(z, \delta)$ is a slow part of the vector field $\Phi(z, \delta)$. Exact descriptions of $F(z, \delta)$, $G(z, \delta)$ and ε can be found as in the previous section.

Suppose there exist such n points $z_1, ..., z_n \in V$ such that vectors $\Phi(z_1, \delta), \ldots, \Phi(z_n, \delta)$ are linearly independent and its fast parts $F(z_1, \delta), \ldots, F(z_n, \delta)$ represent a basis of L_f for any $0 \leq \delta \leq \delta_0$. Call an $n \times n$-matrix $T := [\Phi(z_1, \delta), ..., \Phi(z_n, \delta)]$ a quasi-linearization matrix of the vector field $\Phi(z, \delta)$ associated with points $z_1, ..., z_n$. As usual, $\Phi(z_1, \delta), \ldots, \Phi(z_n, \delta)$ are columns of the matrix T. Using the representation $\Phi(z, \delta) := F(z, \delta) + \varepsilon G(z, \delta)$, the matrix T can be represented as a sum

$$T = T_f + \varepsilon T_s = [F(z_1, \delta), ..., F(z_n, \delta)] + \varepsilon [G(z_1, \delta), ..., G(z_n, \delta)]$$

of the fast and the slow matrices T_f and εT_s.

Denote by A^* a matrix conjugate to a matrix A. For the evaluation of the fast subspace L_f we use the matrix TT^*, because the eigenvalue problem (possible complex eigenvalues, non stable numerics etc.) for T that can lead to non coincidence of the "fast" subspace of T with L_f ([4], [24]). Recall that all eigenvalues of the matrix TT^* are nonnegative (even positive) and the corresponding eigenspaces are orthogonal. By the construction of matrices T_f and T_s the matrix $TT^* = (T_f + \varepsilon T_s)(T_f + \varepsilon T_s)^* = T_f T_f^* + \varepsilon \left(T_s T_f^* + T_f T_s^* \right) + \varepsilon^2 T_s T_s^*$. If $\varepsilon = 0$ then $TT^* = T_f T_f^*$ has no more then n_f positive eigenvalues. Additionally all these eigenvalues are positive (non negative) for small enough ε and for $\varepsilon \to 0$ another n_s eigenvalues tend to zero. Eigenvectors that correspond to n_f positive eigenvalues belong to L_f and form a basis of L_f.

Therefore for small enough ε (asymptotically) the matrix TT^* has two groups of eigenvalues: n_f "large" eigenvalues, whose eigenvectors asymptotically form a basis of the fast subspace L_f and n_s "small" eigenvalues.

5.2. Practical implementation

Here we will discuss a variation of the global quasi-linearization procedure (GQL) that was motivated by numerical simulations. A reason for the modification is that in many practical situations a system has dimensional form and therefore cannot be used without proper non-dimensionalization beforehand. Otherwise the system hierarchy can be perturbed by different scales of the system variables. The suggested modification overcomes this problem and can be applied to the system in dimensional form as well.

Suppose now $\Phi(z)$ is a vector field and our assumption is that this vector field is singularly perturbed and it depends on a "hidden" small parameter δ. We will not even try to look for δ because only the parameter $\varepsilon = \varepsilon(\delta)$ has practical importance.

The proposed modification of GQL is the following procedure. Choose n linearly independent points $z_1, ..., z_n$ in such a way that vectors $\Phi(z_1), ..., \Phi(z_n)$ are also linearly independent. It is impossible to check linear independence for any relatively small δ because the small parameter δ is unknown. Using $z_1, ..., z_n$ and $\Phi(z_1), ..., \Phi(z_n)$ as columns for matrices we define a new matrix T as follows

$$T := [\Phi(z_1), ..., \Phi(z_n)] [z_1, ..., z_n]^{-1}.$$

This matrix has a simple geometrical interpretation. It is the matrix of the linear mapping that transforms points $z_1, ..., z_n$ to $\Phi(z_1), ..., \Phi(z_n)$. Of course the efficiency of this procedure depends essentially on the choice of $z_1, ..., z_n$. Practical recommendations for the choice are the following. Points $z_1, ..., z_n$ cannot be close to another because it can lead to a degeneracy of the matrix $[z_1, ..., z_n]^{-1}$; also the vector field $\Phi(z)$ should have essentially "different" behavior for different points $z_1, ..., z_n$. It is obvious that this is a crucial point of the algorithm and must be adopted in every particular model.

For the evaluation of the fast subspace L_f we will use the matrix TT^* in the previous subsection.

6. Examples

Let us finally demonstrate the global quasi-linearization procedure on the basis of comparatively simple examples.

6.1. Linear example

The idea to use principal subspaces of matrix TT^* instead of the principal subspaces of T is illustrated by a simple linear system

$$\begin{cases} \frac{du}{dt} = au + bv \\ \frac{dv}{dt} = cv \end{cases} \tag{40}$$

Because this system is linear any quasi-linearization procedure gives the same system. Suppose that $|b| \gg |c|, |b| \gg |a|$ and $|c| \sim |a|$. Using the previous procedure for the evaluation of the fast direction it is necessary to calculate first the eigenvalues of TT^*; where $T = \begin{bmatrix} a & b \\ o & c \end{bmatrix}$ is the matrix of the system (40) (it is the Jacobian matrix of the system). It is obvious that any method based on the eigenvectors of $(J = T)$ will fail to produce any decomposition. In contrast consider the matrix TT^*, which has two positive eigenvalues

$$\mu_{1,2} = \frac{a^2 + b^2 + c^2}{2} \pm \sqrt{\left(\frac{a^2 + b^2 + c^2}{2}\right)^2 - a^2c^2}.$$

We can conclude that $\mu_1 \approx a^2 + b^2 + c^2$ is a "big" eigenvalue and μ_2 (which is close to zero) is a "small" eigenvalue because $|b| \gg |a|, |b| \gg |c|$. A "fast" eigenvector is $e_f = \left(1, \frac{\mu_1 - a^2 - b^2}{bc}\right) \approx \left(1, \frac{c}{b}\right)$. Therefore we obtain a reasonably good fast direction whose deviation from $e_1 = (1, 0)$ is of order of c/b.

6.2. Numerical example - Rotation of the original coordinates

The second type of example illustrates an application of the linear version of the singularly perturbed vector field framework and the global quasi-linearization procedure. It has been developed to treat a system having a relatively simple inner hidden hierarchy. To construct a model example let us take the following systems in the standard SPS form

$$\begin{aligned} \varepsilon \frac{dx}{dt} &= x^3 - 2xy + y - 1 \\ \frac{dy}{dt} &= -(x + y - 1) \end{aligned} \tag{41}$$

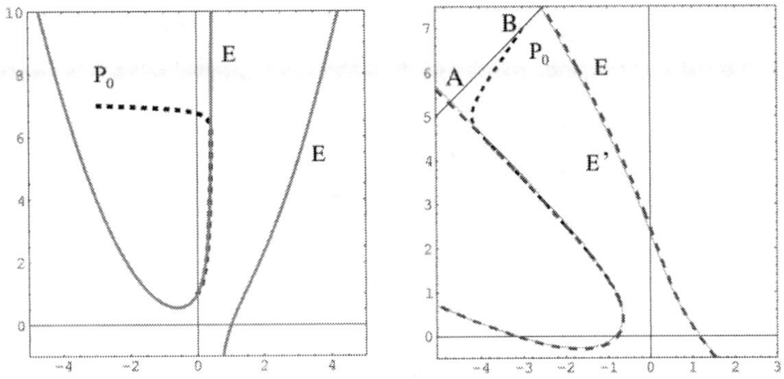

Figure 3. The left figure shows the system dynamics of (41) in the (x, y) state plane of original variables with $\varepsilon = 0.3$; the right figure - the same as in the left, but in the new system coordinates (u, v) after rotation with $\theta = -\pi/4$.

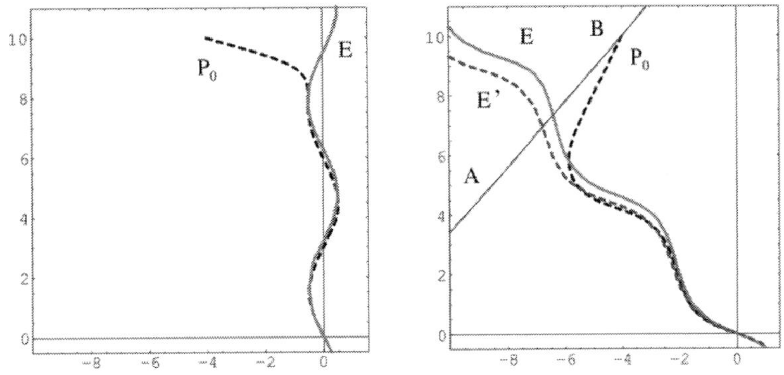

Figure 4. The left figure shows the system dynamics of (42) in the (x, y) state plane of the original variables with $\varepsilon = 0.1$; the right figure - the same as in the left, but in the new system coordinates (u, v) after rotation with $\theta = -\pi/4$.

and

$$\begin{array}{r} \varepsilon\frac{dx}{dt} = -\left(x + sin\left(x\right) + sin\left(y\right)\right) \\ \frac{dy}{dt} = -y \end{array} \tag{42}$$

Rotate the system coordinates according to

$$\begin{cases} u = cos(\theta)x + sin(\theta)y \\ v = -sin(\theta)x + cos(\theta)y \end{cases}$$

and apply now the suggested global quasi-linear approximation (GQL) to redefine the decomposition in the (x, y) phase plane. Reference points were chosen uniformly distributed in the cube $(0, x_0) \times (0, y_0)$ where $(x_0, y_0) = P_0$ is an initial point. Results for $\varepsilon = 0.3$ and 0.1, $\theta = -\pi/4$ are shown in figures 3 and 4, where the slow curves and the system trajectories are drawn in both (x, y) and (u, v) phase planes. Solid curves E represent the slow curves of the original system which are based on standard SPS theory and defined by the RHS of the fast subsystem (first equations) in (41) and (42). The solid line AB shows the direction of the fast motion redefined by the GQL approximation, while the dashed curves - E' are slow curves obtained with the help of the approximation (GQL). Dashed black lines are system trajectories starting at the same P_0, but in different phase planes. It is not hard to see that the proposed method produces a very good approximation for the decomposition, especially for the fast manifolds.

6.3. Numerical example - Michaelis-Menten enzyme model
The last example is the Michaelis-Menten model [17, 18]. A simplified mathematical model is the following

$$\begin{cases} \frac{dx}{dt} = -x\,(1-y) + L_1 y \\ \frac{dy}{dt} = -\frac{1}{L_2}\,(x\,(1-y) - y) \end{cases} \tag{43}$$

This example has been extensively studied in [17] to show, in particular, that the standard method of quasi-stationary concentrations leads to a significant inaccuracy in defining the slow system manifold. It was shown that the zero order approximation for the slow invariant manifold given by the RHS of the second equation in (43), when the value of the system parameter L_2 is small, is not good enough (see Fig. 5 line E).

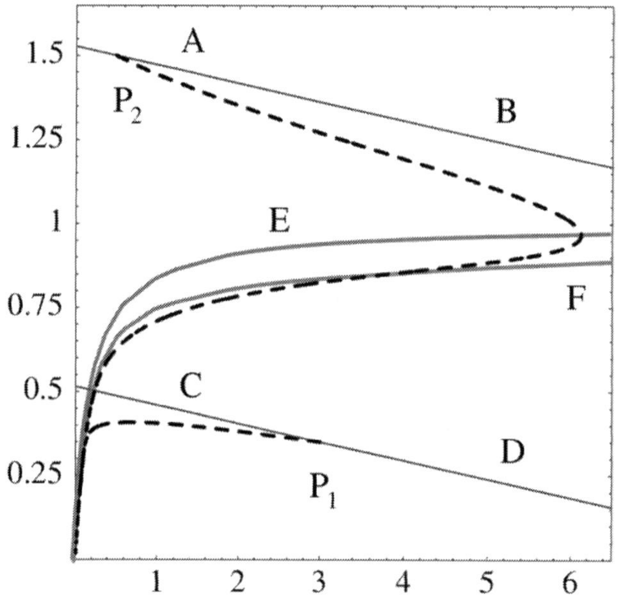

Figure 5. State plane with two system trajectories - dashed black lines, fast directions (AB, CD) - solid blue lines and the slow curve E - solid green line. The first order approximation - F is shown by solid red curve. The system parameters equal $L_1 = 25$, $L_2 = 0.2$.

Indeed, for relatively large but still finite values of the system parameter, one gets only a rough approximation at least to zero order.

Let us demonstrate now how the suggested method (GQL) overcomes this problem. Figure 5 shows typical trajectories of system solutions (starting at some initial points P_1 and P_2) by dashed black lines. Straight lines AB and CD represent the linear approximation to the fast system manifold.

Curve F corresponds to the first order approximation of the system slow invariant manifold based on the new SPS form generated by GQL and the corrected direction of the fast motion AB. It has to be emphasized that the zero order approximation (slow curve) based on GQL is also not so good and actually close to E since the assumption about the asymptotic smallness of L_2 is not fully satisfied. Nevertheless, the first order approximation, even for this value of the system's small parameter, approximates quite well (line F) the slow invariant manifold of the system.

7. Conclusions

The theoretical framework of Singular Perturbed Vector Fields (SPVF) has been proposed in this work. To some extent it can be considered as a generalization of a standard approach of the SPS. It was shown that the proposed approach can be successfully applied in both analytical and numerical versions. While the original system of ODEs is treated analytically, SPVF provides us with an accurate description of the fast and slow system manifolds and allows to transform the original system into standard SPS form so that a powerful method of singular perturbations can be applied afterward. Additionally, within the framework of SPVF a numerical technique for the construction of a linear approximation of the system's fast manifold(s) has been suggested.

A number of examples were considered. They clearly illustrate that the proposed method allows us to overcome a number of difficulties, which other analyzes face. It was demonstrated that the proposed theoretical framework of SPVF has potential as a possible theoretical background for the construction of numerical schemes to automatically identify the decomposition of the system into the "slow" and "fast" subsystems.

The proposed technique can be useful for researchers interested in analytical or numerical evaluation of a 'hidden' decomposition of the system dynamics into so called "slow"and "fast" motions.

References

[1] Maas U and Pope S B 1992 Simplifying chemical kinetics: intrinsic low-dimensional manifolds in composition space *Combustion and Flame* **117** 99

[2] Rhodes C, Morari M and Wiggins S 1999 Identification of the low order manifolds: validating the algorithm of Maas and Pope *Chaos* **9** 108

[3] Kaper H G and Kaper T J 2001 Asymptotic analysis of two reduction methods for systems of chemical reactions *Preprint* Argonne National Lab, ANL/MCS-P912-1001

[4] Goldfarb I, Gol'dshtein V and Maas U 2004 Comparative analysis of two asymptotic approaches based on integral manifolds *IMA J. of Appl. Math.* **69** 353

[5] Bykov V, Goldfarb I and Gol'dshtein V 2003 On one asymptotical approach to pressure driven flames in porous media, *Int. J. of Pure and Appl. Math.* **9** no 4 p 403

[6] Bykov V, Goldfarb I, Gol'dshtein V and Sivashinsky G 2004 Effect of hydraulic resistance and heat losses on detonability and flammability limits *Combustion Theory and Modeling* **8** 1

[7] Bykov V, Goldfarb I and Gol'dshtein V 2004 On the integral manifold approach in flame propagation problem: pressure driven flame in porous media *IMA J. of Appl. Math.* **69** 335

[8] Gorban A and Karlin I 2002 Methods of invariant manifolds for kinetic problems *J. of Chemical Kinetics* **396** 197

[9] Gorban A, Karlin I and Zinoviev A Constructive methods of invariant manifolds for kinetic problems *Phys. Reports* **396** 197

[10] Lam S H and Goussis D M 1994 The GSP method for simplifying kinetics, *Intl. J. of Chemical Kinetics* **26** 461

[11] Hadjinicolaou M and Goussis D M 1999 Asymptotic solutions of stiff PDEs with the CSP method: the reaction diffusion equation *SIAM Journal of Scientific Computing* **20** 781

[12] Masias A, Diamantis D, Mastorakos E and Goussis D A 1999 An algorithm for the construction of global reduced mechanisms with CSP data *Combustion and Flame* **117** 685

[13] Valorani M and Goussis D M 2001 Explicit time-scale splitting algorithm for stiff problems: auto-ignition of gaseous mixtures behind a steady shock *J. of Computational Phys.* **169** 44

[14] Neophytou M K, Goussis D A, Loon M M and Mastorakos E 2004 Reduced chemical mechanism for atmospheric pollution using computational singular perturbation analysis *Atmospheric Environment* **38** 3661

[15] Gol'dshtein V and Sobolev V 1988 *Qualitative Analysis of Singularly Perturbed Systems* (Novosibirsk: Institute of Mathematics, Siberian Branch of USSR Academy of Science)

[16] Gol'dshtein V and Sobolev V 1992 Integral manifolds in chemical kinetics and combustion *Singularity Theory and Some Problems of Functional Analysis* American Mathematical Society 73

[17] Roussel M R and Fraser S J 1993 Global analysis of enzyme inhibition kinetics *J. Phys. Chem* **97** 8316-8327; errata, ibid. **98** 5174

[18] Roussel M R and Fraser S J 2001 Invariant manifold methods for metabolic model reduction *Chaos* **11** 196

[19] Fenichel N 1979 Geometric singular perturbation theory for ordinary differential equations *J. Differential Equations* **31** 53

[20] Mitropolskiy Yu A and Lykova O B 1968 *Lectures on the Methods of Integral Manifolds* (Kiev: Institute of Mathematics Ukrainean Akademy of Science) (in Russian)

[21] Bykov V, Goldfarb I, Gol'dshtein V, Sazhin S S and Sazhina E M 2005 An asymptotic approach to numerical modelling of spray autoignition in hot gas *Proceedings of the Fourth Mediterranean Combustion Symposium* MCS4 Paper **6** Lisbon (Portugal) 6

[22] Sazhina E M, Bykov V, Goldfarb I, Gol'dshtein V, Sazhin S S and Heikal M R 2005 Modelling of spray autoignition by the ODE system decomposition technique *Proceedings of HEFAT2005 — 4th International Conference on Heat Transfer, Fluid Mechanics and Thermodynamics* Cairo (Egypt) Paper number SE2

[23] Bykov V, Goldfarb I and Gol'dshtein V 2005 Novel numerical decomposition approaches for multiscale combustion and kinetic models *J. Phys.: Conf. Series* **22** 1

[24] Bykov V, Goldfarb I, Gol'dshtein V and Maas U On a modified version of ILDM approach: asymptotic analysis based on integral manifolds *IMA J. of Appl. Math.* to appear

[25] Babkin V S 1993 Filtration combustion of gases. Present state of affairs and prospects *Pure and Applied Chemistry* **64** 335

[26] Gordon P, Kagan L and Sivashinsky G 2003 Fast subsonic combustion as a free-interface problem *Interfaces and Free Boundaries* **5** no 1 p 47

Institute of Physics Publishing
doi:10.1088/1742-6596/55/1/004
Journal of Physics: Conference Series **55** (2006) 45–54
International Workshop on Multi-Rate Processes and Hysteresis

A multi-scale problem arising in a model of avian flu virus in a seabird colony

C F Clancy[1], M J A O'Callaghan[2*] and T C Kelly[3]

[1,2]Department of Applied Mathematics, Aras na Laoi, University College Cork, Ireland
[3]Department of Zoology, Ecology and Plant Science, Distillery Fields, North Mall, University College Cork, Ireland
*Corresponding author

E-mail: [2]`mja.ocallaghan@ucc.ie`

Abstract. Understanding the dynamics of epidemics of novel pathogens such as the H5N1 strain of avian influenza is of crucial importance to public and veterinary health as well as wildlife ecology. We model the effect of a new virus on a seabird colony, where no pre-existing Herd Immunity exists. The seabirds in question are so-called K-strategists, i.e. they have a relatively long life expectancy and very low reproductive output. They live in isolated colonies which typically contain tens of thousands of birds. These densely populated colonies, with so many birds competing for nesting space, would seem to provide perfect conditions for the entry and spread of an infection. Yet there are relatively few reported cases of epidemics among these seabirds. We develop a SEIR model which incorporates some of the unusual features of seabird population biology and examine the effects of introducing a pathogen into the colony.

1. Introduction

Epidemics continue to be a major focus of modern science. Epidemic disease poses a threat to human kind, farm animals, crops and to wildlife and therefore to the stability of ecological systems and to the global economy (Millennium Ecosystems Synthesis Report 2005) [15]. Seabirds are an unusual element in the earth's fauna. They are on average larger in size than land birds but there are many fewer species ($n = 300$ approximately — only 3% of the total). However, seabirds foregather and breed in very large inaccessible colonies which may number in excess of one million individuals (Croxall, 1987 [8]). It is therefore remarkable that there have been relatively few reported incidences of epidemics among marine birds at their densely populated colonies e.g. Newton (1998) [16].

Most directly transmitted infectious disease travels from host to host in a chain type reaction. Therefore the more densely packed the hosts the greater the opportunity for both the entry and spread of an infectious pathogen. In this paper we attempt to model the introduction and transmission dynamics of a novel avian pathogen, i.e. the bird flu H5N1, within the receiving population which is believed to have had no previous exposure to the virus and thus there is no Herd Immunity. More specifically we aim to model the epidemiology of a novel pathogen in a dense population of homeothermic vertebrates with exceptional life history strategies. Marine birds are so-called K-strategists. This means that they have a long life expectancy but in any one year have a very low — often minimal — reproductive output. Since the reproductive rate

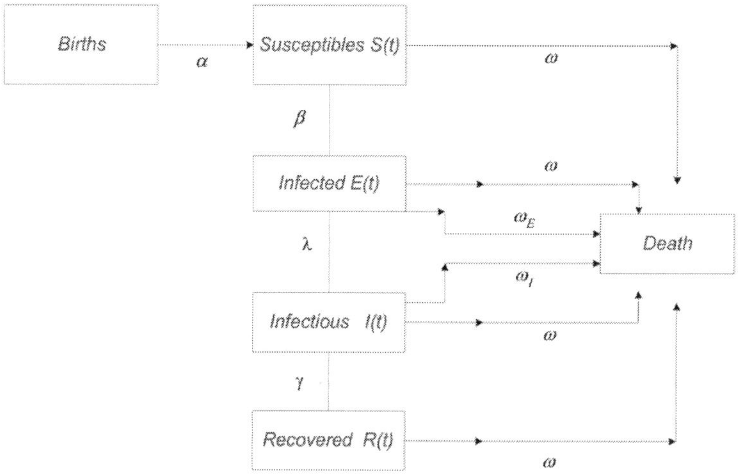

Figure 1. Dividing the population into subclasses

is directly proportional to the number of susceptibles in the host population it is of interest to model the spread of a pathogen in this somewhat unusual potential disease focus.

Clearly it is vital to understand the dynamics of such an outbreak so that sufficient preventative measures can be taken. Mathematical modelling is a useful tool for this. Van der Goot *et al* [21], for instance, apply a model to identify the most effective vaccination strategy in chickens. The aim of this project is to analyze the spread of the H5N1 virus in a colony of seabirds. In particular, parameters such as transmission rates and lethality will be examined in order to assess their importance in the propagation and maintenance of the virus. It should be noted that the global stability properties of SEIR epidemic models with constant recruitment (Li and Wang 2000 [11]) have not been addressed here.

2. Basic Model

2.1. SEIR Approach

Many different methods are used to analyze virus transmission dynamics, including cellular automata (Keeling and Gilligan [13]) and forest-fire lattices (Rhodes *et al* [19]), and even field theory from physics (Adamek *et al* [1]).

In 1927, Kermack and McKendrick approached the problem by dividing the population under consideration into disjoint classes of those who are susceptible to an infection, those who are infected and those who have had the disease and are recovered. This is a very popular and much studied model (Anderson and May [3], James and Steele [12]). It has been used to describe the dynamics of numerous infectious diseases, including Bubonic plague (Keeling and Gilligan [13]), Severe Acute Respiratory Syndrome, SARS, (Anderson *et al* [2]) and West Nile virus (Wonham *et al* [23]).

In order to model the effect of the H5N1 influenza virus on a seabird colony, we divide the total population $N(t)$, a function of time, into those birds which are susceptible $S(t)$; those which have been infected but are not yet infectious $E(t)$, i.e. cannot yet transmit the virus to others; those which are infectious, $I(t)$, and those which have recovered and are immune, $R(t)$. These subclasses are clearly disjoint so that at any time t, the total population is $N(t) = S(t) + E(t) + I(t) + R(t)$. This division is shown in Figure 1.

The following assumptions are made for the purpose of modelling:

- The colony is initially assumed to be isolated, so no birds arrive or leave.
- The population is assumed to be a function of time only. Spatial variation is not taken into account. This is reasonable due to the dense nature of seabird colonies.
- The birth rate, α, is set to be constant. While this may seem unrealistic, there are seabird species in the tropics which breed continuously. Furthermore, it is assumed that only healthy birds, i.e. the susceptibles and recovered, will breed.
- A natural death rate of ω applies to all birds. This corresponds to an average life span of $1/\omega$ years. Infected birds in the E class also die at a rate ω_E due to the lethality of the virus, while the lethality rate for the infectious class I is ω_I.
- The roles of the rates β, λ and γ are as shown in Figure 1. The latency period of the virus (the time between the point of infection and the point when the bird becomes infectious) is $1/\lambda$, while the recovery period is $1/\gamma$. Time is measured in years. The β parameter, which measures the transmission rate, will be further discussed later on.

The system can thus be described with the following set of ordinary differential equations, which are simply derived from an examination of Figure 1:

$$\begin{aligned} \dot{S} &= \alpha(S+R) - \beta SI - \omega S, \\ \dot{E} &= \beta SI - \lambda E - \omega E - \omega_E E, \\ \dot{I} &= \lambda E - \gamma I - \omega I - \omega_I I, \\ \dot{R} &= \gamma I - \omega R, \end{aligned} \tag{1}$$

where dot denotes the time derivative.

2.2. Initial Conditions

It is important to examine what happens when a new pathogen is introduced to a population which has no existing herd immunity, i.e. none of the population is immune. This can be a devastating event (Kelly [14]). In 1950, for example, myxomatosis was introduced to control the rabbit population in Australia. The rabbits were completely defenseless to this highly virulent disease and 99.8% of all those infected died (Shigesada and Kawasaki [20]).

In order to solve this system, we need one initial condition for each equation. In our model we denote the initial population $N(0)$ by N_0. There is just a single infectious bird and so $I(0) = 1$. All other birds are susceptible, so that $S(0) = N_0 - 1$. The other initial conditions must then be $E(0) = R(0) = 0$ so as to satisfy $N(0) = S(0) + E(0) + I(0) + R(0)$.

2.3. The R_0 Parameter

The basic reproductive rate, R_0, is one of the key concepts in epidemiological modelling. It is defined as the average number of further infections that a single virus can produce. Clearly, a virus can only establish itself and remain in a population if $R_0 > 1$.

There is no definite way of calculating R_0. Heffernan *et al* [10] describe a number of commonly used methods. Anderson and May [3] extensively use the formula

$$R_0 = 1/s^*, \tag{2}$$

where s^* is the fraction of the population of susceptibles at equilibrium, that is, when the rates of change of the infected and infectious populations are zero. This will be the form of R_0 which we adopt here.

3. Modelling Adults and Chicks

3.1. The Seabird Colony

We now adapt the basic model described in the previous section to define the colony as accurately as possible.

Seabirds live in colonies on islands and cliff faces containing tens of thousands of birds. These colonies are very densely populated. As such, they would seem to provide ideal conditions for virus epidemics to occur. Yet, surprisingly, outbreaks are, in fact, quite rare.

The reproductive output of seabirds tends to be quite low. When a guillemot chick is born, it remains in the colony for about 22 days. It then leaves and does not return for 4–5 years (Crespin *et al* [7]).

Each year then, there are many birds returning to find a nesting site. However, as previously mentioned, the colony is very densely inhabited and space is scarce. Thus, a 'queue' is formed, whereby there may considerable numbers of birds waiting nearby for an empty site to appear. A key assumption in analyzing the adult population, which seems to reasonably model the reality for seabird colonies, is that as a bird dies, thereby vacating a space in the colony, a bird from the queue fills the vacated space.

To consider these factors, we treat the adults and chicks as separate, albeit connected, populations, and apply the SEIR idea as before to each population. The extra assumptions for this separation are as follows:

- The adult susceptible population, $S(t)$, is maintained by a 'return' rate which matches the total number of adults dying, both naturally and from the virus. This is done to model the queue for space that was mentioned. As soon as a bird dies, the site it vacates is immediately filled.
- Chicks are assumed to be born susceptible.
- All chicks leave at a rate δ. A value of $365/22$ will be used for δ, i.e. they leave after 22 days.
- We consider the special case in which adults infect adults and chicks infect chicks. The more likely scenario of cross infection between both groups is not addressed at this time.
- All returning birds are assumed to return as susceptibles, that is, any immunity from earlier infection would be lost in the years in which it was away from the colony.
- The birth rate is still assumed constant.

The separation is as shown in Figure 2.

3.2. Equations

The following equations are readily established:

$$\begin{aligned}
\dot{S} &= \omega(S+E+I+R) + \omega_E E + \omega_I I - \beta S I - \omega S, \\
\dot{E} &= \beta S I - \lambda E - \omega E - \omega_E E, \\
\dot{I} &= \lambda E - \gamma I - \omega I - \omega_I I, \\
\dot{R} &= \gamma I - \omega R, \\
\dot{S}_c &= \alpha(S+R) - \beta S_c I_c - \omega S_c - \delta S_c, \\
\dot{E}_c &= \beta S_c I_c - \lambda E_c - \omega E_c - \omega_E E_c - \delta E_c, \\
\dot{I}_c &= \lambda E_c - \gamma I_c - \omega I_c - \omega_I I_c - \delta I_c, \\
\dot{R}_c &= \gamma I_c - \omega R_c - \delta R_c.
\end{aligned} \tag{3}$$

The initial conditions, as discussed earlier, are given by:

$$S(0) = N_0 - 1, \quad E(0) = 0, \quad I(0) = 1, \quad R(0) = 0, \quad S_c(0) = E_c(0) = I_c(0) = R_c(0) = 0. \tag{4}$$

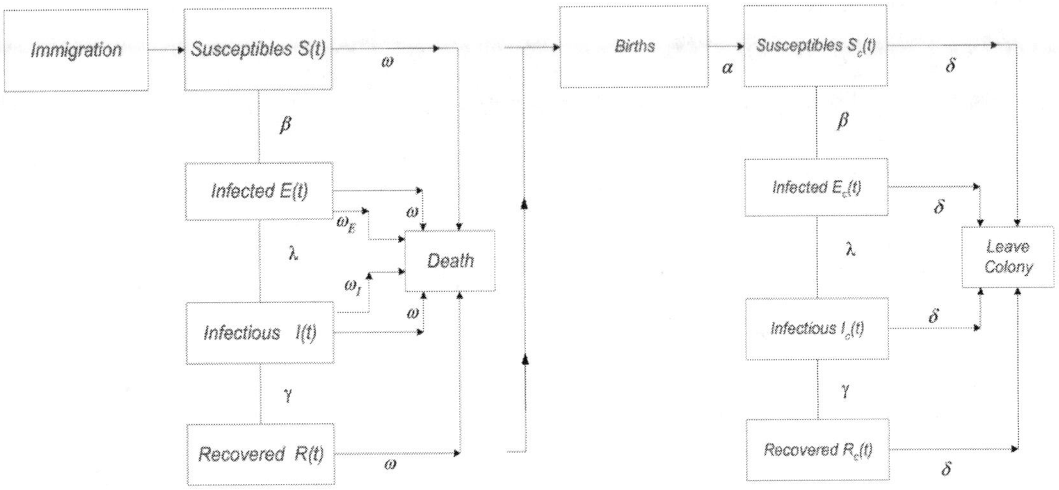

Figure 2. Adult and chick populations separated

3.3. Scaling

We introduce the following scaling:

$$S = N_0 A, \quad E = N_0 B, \quad I = N_0 C, \quad R = N_0 D,$$
$$S_c = N_0 A_c, \quad E_c = N_0 B_c, \quad I_c = N_0 C_c, \quad R_c = N_0 D_c.$$

in which, now, each function is a fraction of the total initial population; for example, $A(t)$ represents the fraction of N_0 who are adult susceptibles. On substituting these into the equations in (3), we obtain the following system:

$$\begin{aligned}
\dot{A} &= \omega(B + C + D) + \omega_E B + \omega_I C - \beta N_0 AC, \\
\dot{B} &= \beta N_0 AC - \lambda B - \omega B - \omega_E B, \\
\dot{C} &= \lambda B - \gamma C - \omega C - \omega_I C, \\
\dot{D} &= \gamma C - \omega D, \\
\dot{A}_c &= \alpha(A + D) - \beta N_0 A_c C_c - \omega A_c - \delta A_c, \\
\dot{B}_c &= \beta N_0 A_c C_c - \lambda B_c - \omega B_c - \omega_E B_c - \delta B_c, \\
\dot{C}_c &= \lambda B_c - \gamma C_c - \omega C_c - \omega_I C_c - \delta C_c, \\
\dot{D}_c &= \gamma C_c - \omega D_c - \delta D_c.
\end{aligned} \tag{5}$$

The initial equations given in (4) now become

$$\begin{aligned}
A(0) = 1 - N_0^{-1}, \quad B(0) = 0, \quad C(0) = N_0^{-1}, \quad D(0) = 0, \\
A_c(0) = B_c(0) = C_c(0) = D_c(0) = 0.
\end{aligned} \tag{6}$$

With this scaling the total population of adult birds remains constant, i.e.

$$\dot{A} + \dot{B} + \dot{C} + \dot{D} = 0.$$

This satisfies the assumption that when a bird dies, another immediately fills place.

4. Analysis and Results

4.1. The basic reproductive rate R_0

We use equation (2) to derive an expression for R_0. For this we need the fraction of adult susceptibles at equilibrium. Thus we set $\dot{B} = \dot{C} = 0$ and solve for A in the system of equations (5), thereby obtaining a value

$$s^* = A = (\lambda + \omega + \omega_E)(\gamma + \omega + \omega_I)/(\beta\lambda N_0)$$

and so, from $R_0 = 1/s^*$, we get

$$R_0 = \beta\lambda N_0/((\lambda + \omega + \omega_E)(\gamma + \omega + \omega_I)). \tag{7}$$

4.2. Numerical Analysis

Together with the initial conditions (6), the system (5) was solved numerically using the Runge–Kutta method, as outlined in Burden and Faires [5], implemented in a C program. The outputted results were then plotted in Mathematica.

4.3. Parameter Values

For the numerical simulations, values for each of the equation parameters were needed. These were taken to accurately model the birds and the H5N1 virus. As well as this, the parameters were varied to test how strongly the system depended on them.

Colonies of seabirds tend to be quite large. Crespin *et al* [7] record up to 20,000 breeding pairs of common guillemots *Uria aalge* on the Isle of May in Scotland. With this information, the initial bird population N_0 was chosen, conservatively, to be 10,000. The birth rate, α was chosen to be 0.5. As previously mentioned, newly born chicks leave the colony after around 22 days, or 22/365 years, so their rate of leaving δ was set at 365/22.

Seabirds are known to have a long lifespan and high survival rates. Oesterblom *et al* [17] show that the annual survival rate of the guillemot in the Baltic seas is over 90%, while Wanless *et al* [22] give a similar rate for the gannet *Morus bassanus*. We thus take the death rate ω to be 0.1 in this analysis.

In their model of an influenza pandemic, Gani *et al* [9] take the latent period to be 2 days, with an infectious period of 4 days, that is, $\lambda = 365/2$ and $\gamma = 365/4$. Van der Goot *et al* [21] also quote a latent period of 2 days, while the infectious period is found experimentally to range from 1 to 7 days.

It is difficult to know exactly the rate of lethality induced by the H5N1 strain. Rajgopal [18] reports that when avian influenza was first discovered, the mortality was approaching 100%. The lethality parameters ω_E and ω_I will initially be taken to be 0.5 and 0.75 but will be varied to study their influence.

The final parameter needed to numerically solve the system is β, which controls the transmission of the virus. From equation (7) we have

$$\beta = (\lambda + \omega + \omega_E)(\gamma + \omega + \omega_I)R_0/(\lambda N_0). \tag{8}$$

So a value for β can be found by inputting a value for R_0. The following table shows estimated R_0 values for various infections, taken from Heffernan *et al* [10], Chung [6], Brockman *et al* [4] and Anderson *et al* [2]:

Infection	R_0
Pandemic Influenza	2–4
Avian Influenza, Netherlands 2003	up to 6.5
SARS	2–3
BSE	10–12
Foot and Mouth Disease	about 4.5

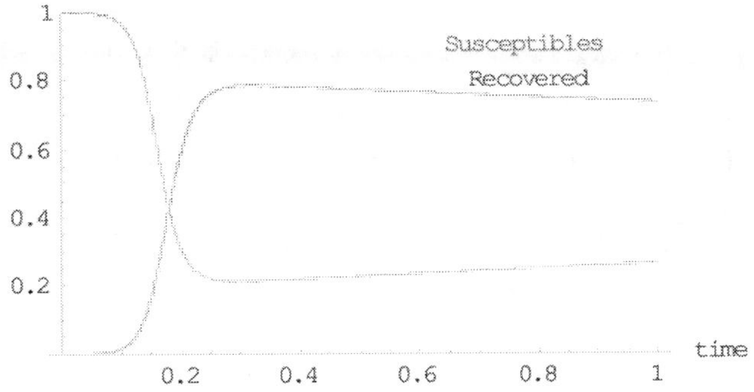

Figure 3. Fraction of susceptibles and recovered over the first year for $R_0 = 2$

As mentioned earlier there is no one "correct" way to calculate R_0, so a precise value for the H5N1 strain of avian influenza isn't known. Here we vary the R_0 value from 2 up to 10 to examine the resulting dynamics.

4.4. Results

In all of the simulations, the following parameter values were kept fixed, as described in the previous section, chosen to accurately reflect seabird profiles:

Parameter	Value
N_0	10,000
α	0.5
ω	0.1
δ	16.59

Equation (8) gives a relationship between R_0 and β, and so the values of γ, λ, ω_E, ω_I and R_0 were varied at each simulation.

It was found that, regardless of changing these values, the chicks tended to make up only around 3% of the total population, and so had negligible impact on the general dynamics of the infection. Instead, we focused on the adult populations.

An R_0 value of 10, representing a highly infectious disease, resulted in the general behaviour, similar for all other parameter values, which is shown in Figure 4.

The fraction of susceptibles begins close to 1. Over the first two to three months or so, (similar to that seen in Figure 3, even though the latter is for the case for $R_0 = 2$) there is a devastating outbreak, affecting almost all of the birds. After this, the number of susceptibles grows slowly, due to the low birthrate. Despite further small tremors, the majority of birds are recovered and thus immune. It is unlikely that another catastrophic epidemic involving this particular virus will occur.

For R_0 equal to 2, we obtain the pseudo-periodic behaviour portrayed in Figure 5.

Again, there is an initial population crash as in Figure 3. This time, however, it is less severe and the population of susceptibles has a greater chance of recovery. Subsequent outbreaks occur, represented by the jumps in the graph. We varied the values of the lethality parameters ω_E and ω_I from the probable values assumed above to unreasonably high values such as 80 and 100 respectively. It was found that these lethality parameters govern the length of time between the outbreaks — the higher the values, the closer the occurrences. We further investigated the results of changing the recovery rate γ and the latency period $1/\lambda$ from the estimated values

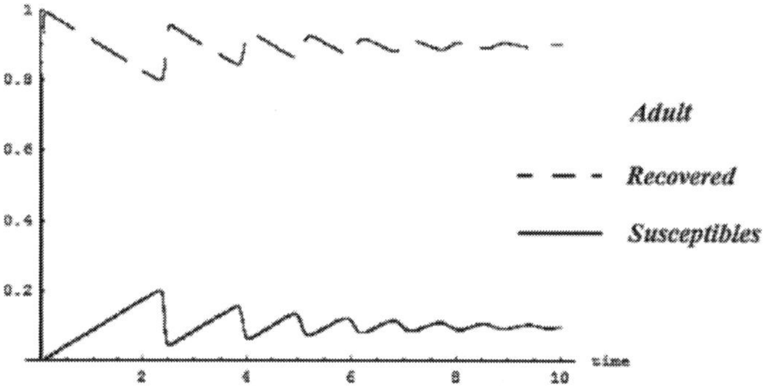

Figure 4. Fraction of adult susceptibles and recovered for $R_0 = 10$; time in years (note that the initial population crash follows the trajectory shown in Figure 3, which refers to a time interval of one year)

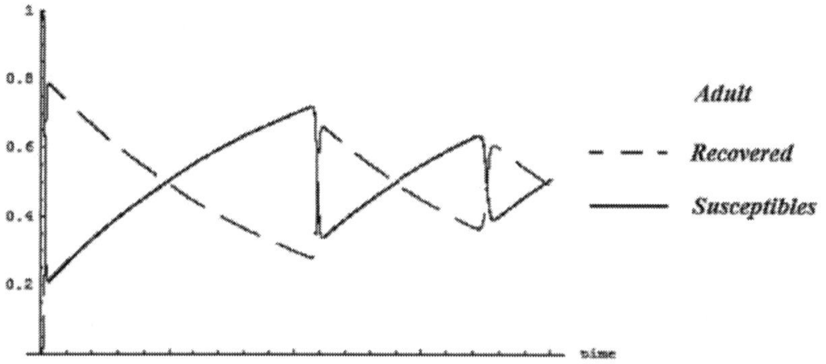

Figure 5. Pseudo-periodic behaviour in the system

as above to values as low as 20 and 50 respectively. The steepness of the jumps, that is, the duration of the epidemics, seems to be controlled by the rate of recovery and the latent period of the virus.

5. Discussion and Conclusion

5.1. Dynamics

When a novel pathogen, such as the H5N1 strain of avian influenza, is introduced into a seabird colony, the resulting dynamics depend largely on the virus's basic reproductive rate R_0. For a high R_0 value of 10, a devastating epidemic occurs very soon after the infected bird is introduced. Almost all of the adult birds become sick. Once the epidemic has passed however, the colony consists almost entirely of recovered birds which are immune to the virus and the low birth rate of the birds means that this situation remains largely unchanged. Therefore, while small outbreaks may frequently occur, the virus does not have enough susceptible hosts to begin another significant epidemic.

The dynamics of the virus are quite different for a lower R_0 value of 2.

The initial epidemic is not as severe in this case and fewer birds are affected. There remains a significant population of susceptible birds and this facilitates more future outbreaks. The general system behaviour looks like that in Figure 4, with the jumps corresponding to outbreaks.

The lethality of the virus governs the length scales involved in these 'cycles'. The higher the value of the lethality, the closer together the outbreaks will occur. The severity of the population changes are governed by the latent period and recovery period of the virus. For longer periods the changes are less sudden.

It is interesting that, regardless of how devastating the virus, the overall population size, in the long term, remains largely unaffected which appears to be the case for seabird colonies in general. This may be explained by the combination of a low reproductive rate and a long life span of seabirds. The proportion of adults in the population is so high that the colony dynamics are undisturbed by the chicks. The colony tends to be largely inhabited by adult birds who have been sick at some point but who have recovered and gained immunity. This makes it difficult for the virus to continually establish itself.

Acknowledgments
The authors wish to gratefully acknowledge Prof. Vladimir Sobolev, Samara State University, Russia, and Prof. Alexei Pokrovskii, University College Cork for helpful discussions and advice. The helpful assistance in rewriting the paper in LaTeX by Dr. Dmitrii Rachinskii of University College Cork is also gratefully acknowledged.

References

[1] Adamek J, Keller M, Senftleben A and Hinrichsen H 2005 Epidemic spreading with long-range infections and incubation times *J. of Stat. Mech.* P09002

[2] Anderson R M, Fraser C, Ghani A C, Donnelly C A, Riley S, Ferguson N M, Leung G M, Lam T H and Hedley A J 2005 Epidemiology, transmission dynamics, and control of SARS: the 2002–2003 epidemic *SARS: A Case Study In Emerging Infections* ed A R McLean, R M May, J Pattison and R A Weiss (Oxford: Oxford University Press)

[3] Anderson R M and May R M 1992 *Infectious Diseases of Humans: Dynamics and Control* (Oxford: Oxford University Press)

[4] Brockmann D, Hufnagel L and Geisel T 2005 Dynamics of modern epidemics *SARS: A Case Study In Emerging Infections* ed A R McLean, R M May, J Pattison and R A Weiss (Oxford: Oxford University Press)

[5] Burden R L and Faires J D 1997 *Numerical Analysis* (California: Brooks/Cole Publishing)

[6] Chung P H A 2005 Preparing for the worst-case scenario *Science* **310** 1117

[7] Crespin L, Harris M P, Lebreton J D, Frederiksen M and Wanless S 2006 Recruitment to a seabird population depends on environmental factors and on population size *J. of Animal Ecology* **75** 228–38

[8] Croxall J P (ed) 1987 *Seabirds: Feeding Ecology and Role in Marine Ecosystems* (Cambridge University Press)

[9] Gani R, Hughes H, Fleming D, Grin T, Medlock J and Leach S 2005 Potential impact of antiviral drug use during influenza pandemic *Emerging Infectious Diseases* www.cdc.gov/eid **11** no 9

[10] Heffernan J M, Smith R J and Wahl L M 2005 Perspectives on the basic reproductive ratio *J. R. Soc. Interface* **2** 281–93

[11] Li M Y and Wang L 2000 Global stability in some SEIR models *Preprint* http://mat.ualberta.ca

[12] James G and Steele N 1990 Epidemics and the spread of diseases *Mathematical Modelling: a Source Book of Case Studies* ed I D Huntley and D J G James (Oxford: Oxford University Press)

[13] Keeling M J and Gilligan C A 2000 Bubonic plague: a metapopulation model of a zoonosis *Proc. R. Soc. Lond.* B **267** 2219–30

[14] Kelly T 2004 Globetrotting Germs — and the transport of their carriers *The Effects of Human Transport on Ecosystems: Cars and Planes, Boats and Trains* ed J Davenport and L J Davenport (Royal Irish Academy) 227–43

[15] *Millennium Ecosystems Synthesis Report 2005* (Washington DC: Island Press) http://www.millenniumassessment.org/en/Products.Synthesis.aspx

[16] Newton Ian 1998 *Population Limitation in Birds* (London and New York: Academic Press)

[17] Oesterblom H, Van Der Jeugd H P and Olsson O 2004 Adult survival and avian cholera in common guillemots Uria aalge in the Baltic Sea (Ibis, British Ornithologists' Union) 146

[18] Rajgopal T 2005 Coping with pandemic avian influenza *Indian J. of Occup. Environ. Med.* **9** 99–102

[19] Rhodes C J, Jensen H J and Anderson R M 1997 On the critical behaviour of simple epidemics *Proc. R. Soc. Lond.* B **264** 1639–46

[20] Shigesada N and Kawasaki K 1997 *Biological Invasions: Theory and Practise* (Tokyo: Oxford University Press)

[21] Van der Goot J A, Koch G, de Jong M C M and van Boven M 2005 Quantification of the effect of vaccination on transmission of avian in-fluenza (H7N7) in chickens *Proc. Natl. Acad. Sci. USA* **102** no 50, 18141–6

[22] Wanless S, Frederiksen M, Harris M P and Freeman S N 2006 Survival of Gannets Morus bassanus in Britain and Ireland, 1959–2002 *Bird Study* **53** 79–85

[23] Wonham M J, de-Camino-Beck T and Lewis M A 2004 An epidemiological model for West Nile virus: invasion analysis and control applications *Proc. R. Soc. Lond.* B **271** 501–7

Institute of Physics Publishing
doi:10.1088/1742-6596/55/1/005
Journal of Physics: Conference Series **55** (2006) 55–62
International Workshop on Multi-Rate Processes and Hysteresis

A mean-field model of investor behaviour

R Cross[1], **M Grinfeld**[2] **and H Lamba**[3]

[1]Department of Economics, University of Strathclyde, Sir William Duncan Building, 130 Rottenrow Glasgow G4 0GE, Scotland, UK

[2]Department of Mathematics, University of Strathclyde, Livingstone Tower, 26 Richmond Street, Glasgow G1 1XH, Scotland, UK

[3]Department of Mathematical Sciences, George Mason University, MS 3F2, 4400 University Drive, Fairfax, VA 22030 USA

E-mail: `rod.cross@strath.ac.uk`, `caas05@maths.strath.ac.uk`, `hlamba@gmu.edu`

Abstract. In this note we investigate the ability of a mean-field model distilled from a heterogeneous agent model to simulate stylized facts of financial times series.

Keywords: heterogeneous agent models, mean-field approximations, economic modeling, market dynamics, herding.

1. Introduction

Since the early statistical studies of Bachelier [3], the degree of randomness of stock market prices has stimulated debates about the extent of stock market efficiency and about the underlying behaviour of investors, or their agents, that could explain the features of stock market price statistics. The efficient markets hypothesis (EMH) has, until recently, dominated the mainstream finance literature [9]. A typical Brownian motion formulation of the EMH has the stock log-price differences being identically and independently distributed (IID) across time, no matter what the time interval, and obeying a Gaussian distribution. The underlying behaviour is taken to be akin to a fair game in which investors exploit all the information relevant to the determination of stock prices so that the expected gain or loss is zero. The assumption is either that investors are homogeneous in the sense that they use the same information sets; or that informed investors quickly arbitrage away the gains to be had from the behaviour of uninformed investors [16]. The random arrival of new information that leads actual stock prices to differ from those expected by investors is translated into a Brownian motion in, and a Gaussian distribution of, stock returns.

A major problem with the EMH is that the returns on stocks, the log-price differences, are not well approximated by a Gaussian distribution [6]. The serial autocorrelation of returns is close to zero for returns defined over periods longer than about 15-30 minutes, but the Gaussian is not a good approximation for even the first moment of the returns distribution for difference periods of less than a month [10]. Two key departures from a Gaussian distribution are the fat tails and clustered volatility observed in actual returns distributions. The fat tails indicate that extreme returns occur more frequently than in a Gaussian world, with the tails displaying signs of scaling that are possibly consistent with power laws [10]. Clustered volatility indicates that the returns are not identically distributed, the variance changing over time. Volatility indices such as the US VIX chart the positive serial correlation observed in the second moments or absolute values of returns and thus reveal periods of quiescence punctuated sporadically by episodes of high volatility.

The implications of these, and other [17], non-Gaussian features of stock returns have been the subject of intense debate. From proponents of the EMH there is a transactions cost defence: "While the stock market may not be a mathematically perfect random walk, it is important to distinguish statistical from economic significance—the statistical dependencies are not likely to permit investors to realise excess returns—anyone who pays transactions costs is unlikely to fashion a trading strategy based on the kinds of momentum found in these studies that will beat a buy-and-hold strategy" [15, p. 62]. The decline in trading costs arising from electronic trading [20] suggests that this defence has become less plausible.

It was perhaps the excess volatility observed in stock returns [18] that most stimulated interest in alternatives to the EMH account of investor behaviour. The point of departure in the behavioural finance (BF) approach was to point out the limits of arbitrage [19]. Stocks as a whole do not have close substitute portfolios, so if they are mispriced there is no riskless hedge for the arbitrageur. The risk that new information will be such that the stocks become even more mispriced exacerbates the problem, even assuming that investors can identify 'fundamental' values. In this world the behaviour of 'noise' investors is driven by changes in sentiment, as well as by the arrival of new information relevant to 'fundamental' stock values. If the changes in sentiment are correlated to some degree across investors, the 'noise' trades will not cancel each other out, and so there will be an impact on market prices. This framework raises problems with regard to 'fundamentalist' investors. Do they stick to 'fundamentals', or do they also take account of the impact on market prices of 'noise' investors? In this world, Keynes' beauty contest metaphor [14] becomes not completely implausible.

An implication of the BF approach is that affective as well as cognitive processes can influence investor behaviour [5]. In a previous paper [7], we introduced affective responses in the form of 'inaction' and 'herding' thresholds to model investor psychology and market sentiment. The idea was to simulate an heterogeneous agent model (HAM) of the stock market in which investors are, in part, motivated to switch positions when certain psychological pressures become unbearable. These will be described in more detail in Section 2. The model was largely successful in replicating the main non-Gaussian features, but so have been other HAMs (see the review in [12]). The parsimonious representation of investor psychology in the HAM of [7] arguably has an edge over more complicated behavioural specifications in alternative HAMs. This still leaves unresolved the question of whether there is an underlying common mechanism for the generation of the non-Gaussian features of stock returns. As Farmer [10, p. 34] puts it, "the problem is not 'how do we get realistic deviations from normality and IID behaviour?' but rather, 'how do we determine the necessary and sufficient conditions, and how do we identify which factors drive such effects in real markets?' ".

It is difficult to do much with HAMs by way of analysis due to their high dimensionality. Hence it makes sense to distill the salient features of HAMs, into a simpler model based on easily understandable principles and check whether the time series of price returns it generates would retain desirable properties. This process in general requires a significant dimension reduction, and can only be justified in a small number of situations.

One such situation is considered, among other places, in [7], where all the traders in a market are only coupled globally, through the price of the asset and through their (not necessarily complete) knowledge of market sentiment. In such a situation, one is allowed to pass to the mean field limit, as in Horst [13], where such a process is considered in detail; we will return to consider this model in detail in the next section.

In this paper we show that a distillation of psychological facts of [7] into statements of probabilities of transitions between market positions leads to a system of a stochastic difference equation coupled to a deterministic one, the statistics of which retain many of the observed features of financial time series. The resulting framework can in principle be analysed by tools of random dynamical systems theory [2]; we leave this analysis for a later publication.

2. A mean-field model

A good starting point for our derivation is the model of Horst [13]. Even though this model cannot reproduce the key features of financial time series (see below), it is the most thorough and rigorous derivation of a mean-field model from a Markov chain one known to us; it is also very useful for comparison with our approach.

Consider a market for an asset with price p_t, at time t populated by (an infinity of) traders. Let the proportion of long at time t traders be denoted by m_t^+, while that of short traders by m_t^-. Clearly, $m_t^+ + m_t^- = 1$, and we define an order parameter m_t to be $m_t := m_t^+ - m_t^-$. Thus m_t is a measure of the sentiment of the market; below we will refer to it as *the* sentiment.

Briefly, in [13] it is assumed that the price p_t is slaved to m_t,

$$p_t = p_F + \alpha m_t, \tag{1}$$

where p_F is "the" fundamental price of the asset and α is some positive constant (a similar assumption is made in [1]), while the sentiment $m_t \in [-1, 1]$ evolves via an equation of the form

$$m_{t+1} = f(e_{t+1}, h_t). \tag{2}$$

Here $h_{t+1} = h_t + \eta_t$, $\eta_t \sim \mathcal{N}(0, \sigma_h)$, and corresponds to exogenous 'rational' market-changing information entering the system. The quantity $e_{t+1} = m_t + \epsilon_t$, with $\epsilon_t \sim \mathcal{N}(0, \sigma_\epsilon)$, represents uncertainty in the estimation of the market sentiment. The function f in (2) is of the same type as in (6) below. The resulting dynamics of m_t for reasonable choices of distributions of η_t and ϵ_t are roughly bimodal, which by (1) means that the price returns implied by this model cannot correspond to those observed in real markets.

We now derive a new mean-field model based partly upon (1) and (2) but also using ideas from previous HAMs, in particular, from the one of [7]. Firstly we note that in (1) the price p_t is a deterministic function of the sentiment m_t. This occurs because all the random effects, including the exogenous information h_t entering the marketplace, affect the sentiment directly in (2). Thus new information affects the sentiment which then affects the price in a deterministic way. However, we shall argue that new information entering the market affects the price directly. This is much closer to the spirit of rational market theory and the EMH where the concept of sentiment does not even exist. We leave a detailed justification of (3) to a subsequent publication, but note for now that it is plausible to suggest that as new information arrives, it is liable, for example, to result in changes of both bids and offer prices on limit orders (see [8] for more details on this terminology and a discussion of the underlying trade mechanism). Hence one can effect a change in the stock price virtually without any increase in transactions. Another way of saying this is that we shall model the exogenous noise as directly affecting the fundamental price p_F, rather than suppose p_F to be a constant. We also assume that the new updated price will also depend in a linear manner upon the change in sentiment over the last timestep (via the laws of supply and demand) resulting in the alternative pricing formula

$$p_{t+1} = p_t + \eta_t + \kappa \Delta m_t, \tag{3}$$

where $\Delta m_t := m_t - m_{t-1}$ and $\eta_t \sim \mathcal{N}(0, \sigma)$. This price updating model was employed in the HAM of [7] (note that if a geometric pricing model is being used then the price variable p_t is in fact the log-price but, for brevity, we shall simply refer to p_t as the price). Here the variance σ is chosen to reflect the magnitude of the external fluctuations occurring over one time period (whether that corresponds to days or months etc). The parameter κ is a measure of the liquidity of the market and the extent to which pricing changes are affected by the arrival of external information as opposed to internally generated market dynamics. Finally we point out that setting $\kappa = 0$ leads to EMH dynamics of returns.

We now suggest an updating formula for the sentiment, similar in form to (2), but where the change in sentiment now depends upon the change in market price. In particular, the rules determining how each agent decides whether or not to switch are as follows. Firstly each agent has an 'inaction' threshold whereby when the price moves far enough away from the price at which she last traded, she will switch. This mimics many different (rational and irrational) market effects such as taking profits, cutting losses, or just an irrational need to do something. Intervals of this kind also arise naturally when one considers the effects of transaction/sunk costs that prevent agents switching arbitrarily often. The second threshold affecting each agent is a 'herding' threshold. Here agents can only tolerate being in the minority for so long, after which point they have to switch. This herding tendency is strongly suspected, via its incorporation into many different successful HAMs, as being a major contributor to the 'boom-and-bust' dynamics of financial markets. The reader is directed towards [7] for further details of the modeling and its justification. The numerical results presented there are interesting in that the herding tension is shown to be directly responsible for the fat-tails and excess kurtosis.

The master equation for the sentiment is

$$m_{t+1}^{+} = m_t^{+} P_{(+,+)} + m_t^{-} P_{(-,+)}, \tag{4}$$

where $P_{(+,+)}$, for example, is the probability of a long trader staying long.

Now, $m_t^{+} = (m_t + 1)/2$ and $m_t^{-} = (1 - m_t)/2$, so we get

$$m_{t+1} = 2\left(\frac{m_t + 1}{2} P_{(+,+)} + \frac{1 - m_t}{2} P_{(-,+)}\right) - 1. \tag{5}$$

Note that $1 = (m_t + 1)/2 + (1 - m_t)/2$. Now let us consider the transition probabilities. The range of the function $P_{(+,+)}$ should be $[0, 1]$. We also want this function to be monotone in m_t; this is in agreement with the specification of the herding tendency of [7]. In addition, we want to allow a dependence on Δp_t. Similar considerations for $P_{(-,+)}$ lead to the following possible form of the two transition probabilities:

$$P_{(+,+)} = \frac{1}{2}(1 - \tanh\beta(-Jm_t + f_1(\Delta p_t))) \text{ and } P_{(-,+)} = \frac{1}{2}(1 - \tanh\beta(-Jm_t + f_2(\Delta p_t)),)$$

where as in [4], J is a measure of the strength of social interactions. We shall leave f_1 and f_2 undetermined for now. Opening up the brackets in (5) and using the fact that the tanh function is odd, gives

$$m_{t+1} = \frac{m_t + 1}{2}\tanh\beta(Jm_t - f_1(\Delta p_t)) + \frac{1 - m_t}{2}\tanh\beta(Jm_t - f_2(\Delta p_t)) \tag{6}$$

Next we need to determine what the dependence of f_1 and f_2 on Δp_t should be in order to incorporate the inactivity tension described above. A simple and plausible choice is $f_1(x) = |x|$. This would mean that if there is a large jump in price, the probability $P_{(+,+)}$ is small: the traders will prefer to modify their market positions. Similarly, a good choice for f_2 is $f_2(x) = -|x|$: if a jump in price is large, so will be $P_{(-,+)}$. With these assumptions, we have

$$m_{t+1} = \frac{m_t + 1}{2}\tanh\beta(Jm_t - |\Delta p_t|) + \frac{1 - m_t}{2}\tanh\beta(Jm_t + |\Delta p_t|). \tag{7}$$

Remarks. Since it is always the case that $P_{(+,+)} \geq P_{(-,+)}$, in the parlance of [11] the model (7) is antihysteretic. Furthermore note that the model no longer has the skew-symmetric structure as in (1)–(2). In other words, the price p_t is not slaved to the sentiment m_t and so, for example, bimodal dynamics of m_t no longer imposes a bimodal distribution on p_t.

The resulting model (3),(7) has rich dynamics and requires some parameter tuning to exhibit stylized market facts. In particular two highly unrealistic effects occur for large parameter regimes. Firstly, highly oscillatory behaviour can occur where the sentiment displays period-two behaviour with a large fraction of the agents switching every time period. Such a jittery population of agents would quickly be wiped out by transaction costs. The other is the presence of a fixed point at $m_t \equiv 0$ which can be stable for a wide range of parameters. The stability analysis for $\kappa = 0$ is straightforward (see also [4]). Briefly, if we consider

$$m_{t+1} = \frac{m_t + 1}{2} \tanh \beta(Jm_t - s) + \frac{1 - m_t}{2} \tanh \beta(Jm_t + s) \tag{8}$$

with $s > 0$, it is not hard to check that $m_t = 0$ is asymptotically stable if

$$s > \frac{1}{\beta} \tanh^{-1}\left(1 - \frac{1}{\beta J}\right).$$

Now suppose $s = s_t = |\eta_t|$. Then using Maple we obtain

$$\mathbb{E}(s) = \frac{2^{1/2}}{\pi^{1/2}}\sigma.$$

Assuming that it is the expectation of h_t that plays the important rôle, we thus have the following rough statement: $m_t = 0$ is unstable as long as

$$\sigma < \frac{\pi^{1/2}}{2^{1/2}} \frac{1}{\beta} \tanh^{-1}\left(1 - \frac{1}{\beta J}\right).$$

In the next section we shall see that in parameter regimes where both the period-2 cycles and the trivial fixed point are unstable, price/sentiment dynamics can be compared with the stylized facts.

3. Numerics

First we choose the variance of the external information stream η_t since this determines the timescale corresponding to each timestep. For many real-world financial markets the daily standard deviation in prices is approximately 0.5–0.6% and this corresponds to $\text{var}(\eta) = 0.00004$. The liquidity parameter κ is chosen to be 0.12 (similar to the value used in [7]), $J = 0.05$ and $\beta = 40$. A typical simulation is shown in Figure 1 over 10000 timesteps which equates to approximately 40 years of trading.

Figure 1a) plots the price p_t (solid line) and the 'fundamental' price p_t^F (dashed line) which is defined by either setting $\kappa = 0$ in (3) or, equivalently, by supposing that the sentiment is always 0. In other words it is the price obtained when the market sentiment is always neutral and the exogenous noise is correctly incorporated into the market price. The fundamental price does not appear directly in the model but when plotted with p_t in Figure 1a), it allows us to see the effect of the sentiment changes on the price dynamics. As expected, the actual price is more variable than the fundamental one and significant deviations can last several years.

In Figure 1b), the sentiment is plotted. Polarized markets, both positive and negative, can remain for significant periods before the sentiment suddenly switches. There are also long intervals where the sentiment fluctuates about 0 without major price booms or crashes forming. Thus the two variables in the model appear to be behaving reasonably, at least in a crude

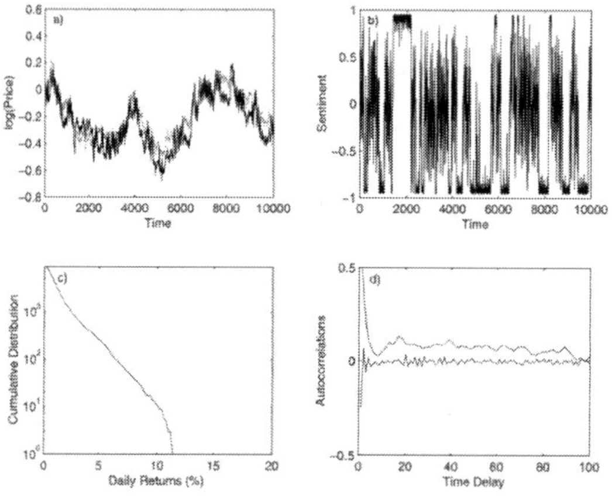

Figure 1. Numerical simulation over a 40 year period. See the text for full details.

qualitative sense, and so we now look more closely at the stylized facts and the ability of the model to reproduce them.

In Figure 1c) we plot the cumulative distribution of the absolute price changes (the number of days on which the absolute percentage price change exceeds a given value). On a semi-log scale a Gaussian distribution of returns would be approximately a straight line. However, there is a clear excess of days on which the return is greater than approximately 3% (in good agreement with statistical data and HAMs). Thus the fat-tail phenomenon is present and this is also reflected in the kurtosis which fluctuates significantly between runs but averages at approximately 9, which is also consistent with measurements from real markets if a little on the low side.

We now examine the autocorrelations present in the price data. Figure 1d) shows two sets of data. The upper curve, exhibiting a very slow decay, is the autocorrelation of the volatility which is defined as the absolute value of the daily price returns. This is indicative of volatility clustering and a time series of the price returns (Figure 2) confirms this. That the model can reproduce this effect is perhaps the most surprising aspect of the simulation. In real markets volatility clustering is often assumed to be due to 'memory effects', especially since the Hurst exponent of the volatility time-series is usually much greater than 0.5, suggesting persistence in the data. The model being simulated here has minimal memory, since the sentiment equation can be written as $m_{t+1} = F(m_t, m_{t-1}, \eta_t)$. The volatility autocorrelations arise because the average absolute change in sentiment is itself a function of the sentiment. Close to $m_t = 0$ the absolute change in sentiment over any timestep is very small (because of the multiplying factor m_t on the RHS of (7)) but this is not the case away from $m_t = 0$. Since the price changes themselves depend in part upon the sentiment changes, this induces volatility clustering in the price data. A more general observation [7] is that volatility clustering will occur whenever the average magnitude of price changes is dependent upon a more-slowly varying quantity such as the sentiment. Furthermore there are good economic arguments for supposing that this is indeed the case. For example, in a highly-polarized market, especially a bull market, the presence of extra market participants such as day-traders will increase the volatility until the bubble deflates and sentiment returns to more normal levels. However, to date there appears to have been very little research into the relationship between volatility and sentiment. Such investigations should help clarify the origins of volatility clustering in real markets.

Finally, the lower data-set in Figure 1d) shows the autocorrelation of the daily price-changes themselves (rather than their absolute value). Both the EMH and the stylized facts conclude that autocorrelations in price data vanish over all but the very shortest timescales. However

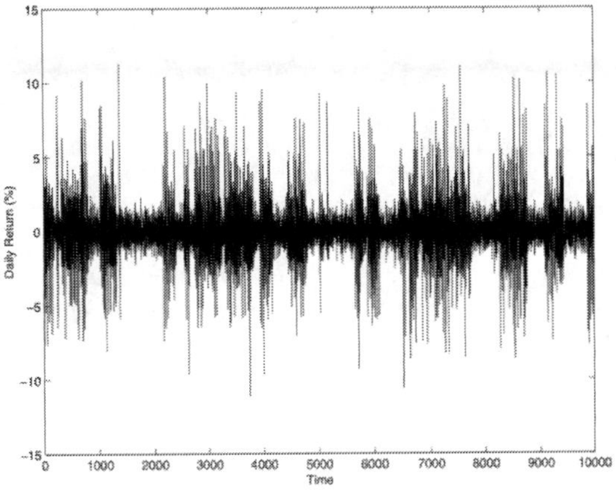

Figure 2. The price return data for the simulation in Figure 1.

we see a noticeable (and therefore exploitable) negative correlation between price changes on adjacent days. This appears to be an artifact of the discretized modeling process and could be mitigated by, for example, choosing the variance of the external noise to correspond to a time-period smaller than the one of interest and then recovering the daily/weekly/monthly price data as required.

4. Conclusions

There are many heterogeneous agent models in the economics and econophysics literature that seem to be capable of reproducing the most important stylized facts of real market data. It has been claimed that the heterogeneity of the agents is necessary for realistic model behaviour.

The model introduced in this paper is not intended to be a replacement for HAMs. The questions concerning when and how trading rules at the micro-level percolate through to global statistics and structures are profoundly important ones. Furthermore, HAMs appear to be more capable of accurately preserving important but short-scale properties (such as the non-correlation of price changes over all but very short timescales) over a very wide range of parameters. However, small, fast models that can quickly produce plausible market-pricing data for long simulation times may be sufficient for many financial purposes and yield important analytical insights.

Furthermore, the model (3), (7) clearly shows the necessity of the two psychological mechanisms postulated in [7]: both the herding (cowardice) effects expressed in the relationships $\partial P_{(+,+)}(x,y)/\partial x > 0$ and $\partial P_{(-,+)}(x,y)/\partial x > 0$ and the inaction effects ($\partial P_{(+,+)}(x,y)/\partial y < 0$ and $\partial P_{(-,+)}(x,y)/\partial y > 0$) are required for the correct prediction of financial times series properties. Thus, it makes sense to consider the model introduced here as in some sense a minimal model capable of generating stylized market facts.

It is also worth noting that the model (3), (7) can be considered as a return of representative agent economics, with m_t being then interpreted as the agent's expected market position. The lesson here is that in order to obtain realistic price return statistics, such an agent must be based on plausible psychological principles and has to follow stochastic behavioral rules.

Acknowledgments

MG would like to thank Prof. O. Penrose for clarifying the legitimacy of passing to the mean field limit and Dr. G. Berkolaiko for stimulating discussions.

References

[1] Alfarano S and Lux T 2003 A minimal noise trader model with realistic time series properties *Economics Working Paper No. 2003–15, Department of Economics, Christian-Albrechts-Universität, Kiel*

[2] Arnold L 1998 *Random Dynamical Systems* (Berlin: Springer)

[3] Bachelier L 1900 *The Theory of Speculation* reprinted in 1964 *The Random Character of Stock Market Prices* ed P H Cootner (Cambridge: MIT Press)

[4] Brock W A and Durlauf S N 2001 Discrete choice with social interactions *Rev. of Economic Studies* **68** pp 235–60

[5] Camerer C, Loewenstein G and Prelec D 2005 Neuroeconomics *J. of Economic Literature* **43** pp 9–64

[6] Campbell J Y, Lo A W and MacKinlay A C 1997 *The Econometrics of Financial Markets* (Princeton: UP)

[7] Cross R, Grinfeld M, Lamba H and Seaman T 2005 A threshold model of investor psychology *Physica* A **354** pp 463–78

[8] Daniels M G, Farmer J D, Gillemot L, Iori G and Smith E 2003 Quantitative model of price diffusion and market friction based on trading as a mechanistic random process *Phys. Rev. Letters* **90** pp 108102-1–4

[9] Fama E F 1970 Efficient capital markets: a review of theory and empirical work *J. of Finance* **25** pp 283–417

[10] Farmer J D 1999 Physicists attempt to scale the ivory towers of finance *Computing in Science and Engineering* **1** pp 26–39

[11] Grinfeld M, Piscitelli L and Cross R 2000 A probabilistic framework for hysteresis *Physica* A **287** pp 577–86

[12] Hommes C H 2006 Heterogeneous agent models in economics and finance, in: *Handbook of Computational Economics, vol 2: Agent-Based Computational Economics* ed L Tesfatsion and K L Judd (Amsterdam: Elsevier) pp 1–66

[13] Horst U 1999 Ergodic fluctuations in a stock market model with interacting agents: the mean field case *Discussion paper No. 106, Sonderforschungbereich 373, Humboldt Universität, Berlin*

[14] Keynes J M 1936 *The General Theory of Employment, Interest, and Money* (London: Macmillan)

[15] Malkiel B G 2003 The efficient markets hypothesis and its critics *J. of Economic Perspectives* **17** pp 59–82

[16] Merton R and Brodie Z 2005 Design of financial systems *J. of Investment Management* **3** pp 6–28

[17] Shiller R J 2003 From efficient markets to behavioural finance *J. of Economic Perspectives* **17** pp 83–104

[18] Shiller R J 1981 Do stock prices move too much to be justified by subsequent changes in dividends? *American Economic Rev.* **71** pp 421–36

[19] Shleifer A and Summers L H 1990 The noise trader approach to finance *J. of Economic Perspectives* **4** pp 19–33

[20] Stoll H R 2006 Electronic trading in stock markets *J. of Economic Perspectives* **20** pp 153–74

Institute of Physics Publishing
doi:10.1088/1742-6596/55/1/006

Journal of Physics: Conference Series **55** (2006) 63–73
International Workshop on Multi-Rate Processes and Hysteresis

Numerical solution of ODEs involving the derivative of a Preisach operator and with discontinuous RHS

D Flynn[1], **J P O'Kane**[2] **and A Zhezherun**[3]

[1,2,3] Dept of Civil & Environmental Engineering, University College Cork, Ireland

E-mail: [1]`d.flynn@mars.ucc.ie`, [2]`p.okane@ucc.ie`, [3]`a.zhezherun@mars.ucc.ie`

Abstract. The purpose of this paper is to numerically solve a special class of ordinary differential equations with discontinuities on the right-hand side and with the Preisach operator under the derivative. We present an algorithm that we devised to numerically integrate this equation making use of a specific approximation from our companion paper, which provided an approximate solution where singularities were expected to occur. We found when the discontinuities forced the right-hand side closer to zero, that the graph of the solution had more parabolic-like curvature similar to the graphs of the solutions using our approximation.

1. Introduction

In this paper we study a model for the discontinuous flow of water through porous media. Such models are of interest in cases where there are almost instantaneous changes in the flow rate, such as when there is switching between precipitation and evaporation in a hydrological system [1]. We can model this using an operator differential equation based on Darcy's law [2], where the operator models the hysteresis in the system and is known as the Preisach operator, see the companion paper [3] for an introduction. The particular form of this operator ODE is unusual in that the Preisach operator is on the left hand side of the ODE and is under the derivative. These types of Preisach operator ODEs have been the subject of research in [4, 5].

A porous medium is a substance which consists of two or more phases of matter such as solids, liquids and gases. These form a network of pore spaces interconnected by pore throats. Within this network, liquids and gases can move throughout it, but not freely as there is some resistance depending on the type of solid phase. Some examples of porous media are soils, sand, and even biological tissue such as bones. In this paper the porous media of interest to us are soils and the three phases of matter that constitutes a soil are: particles of ground rock or clay, water and air.

Three phase systems of this sort generally exhibit hysteresis in the relationship between moisture content and the capillary pressure. Until recently it was believed that the main cause of this hysteresis was due to the Haines-jumps [6] but nowadays the literature indicates that the contact angle between a liquid and a solid is a more fundamental mechanism [7].

Hysteresis in hydrology has been modeled since the 1930's by Haines [6], who discovered that the relationship between the soil-moisture content and the capillary pressure was in fact hysteretic. Subsequently Childs and Poulovassilis [8, 9] applied the independent domain model developed by Preisach [10], Néel [11, 12], Everett [13] and others to capillary hysteresis. This was improved upon by Mualem [14, 15, 16] who simplified the model such that only two outer boundary hysteresis loops were required to determine the inner loops. Parlange improved upon this conceptual model further by reducing the number of required boundary curves to one [17] and this was the approach used by Haverkamp et. al. [18] to fit a series of data from his GRIZZLY database of soils. More recently Flynn et. al. [19] used a one parameter Preisach model to fit this same GRIZZLY data with better accuracy.

The Preisach model [10] was originally conceived for hysteresis in magnetic materials. It was not until the 1970's that it became a formal mathematical discipline, when Mark Krasnoselskii [20], one of the founders of nonlinear analysis, led an investigation into hysteresis phenomena using the theory of nonlinear operators. This has led to a fertile research area with many papers and monographs being published such as [21, 22, 23]. Not only has the Preisach model been used to model electromagnetism, but it is now used for models in disparate disciplines such as Hydrology [24, 25, 26, 27, 19], Macroeconomics [29, 30], Neural networks [31, 32] and Shape memory alloys [22, 33].

Standard integration algorithms (e.g., Runge–Kutta) are not applicable to operator ODEs because of specific singularities introduced by hysteresis operators. In the previous paper [4] these difficulties have been overcome for systems with a Preisach operator under the derivative and a smooth right-hand side. It was suggested that in addition to standard integration steps, special "linear" steps should be made at certain moments. In this paper we consider equations with a right-hand side discontinuous in time, and we suggest a new technique, which we call "square-root" step, to adapt the previously published algorithm to the discontinuous case. In the companion paper [3] we discuss the questions of existence and uniqueness of a solution of the equation in question, and we also prove a theorem which describes the behaviour of the solution at the points of discontinuities of right-hand side. This theorem provides reasoning for the "square-root step" method. For a simple explanation of the "square-root" step refer to section 5 of the companion paper [3].

The paper is organised as follows. In section 2 we describe the equations that we are interested in. Section 3 describes the physical problem that we wish to model namely the flow of liquids through porous media. Then in section 5 we present an algorithm to numerically solve this type of equation and in section 4 results of numerical simulations will be shown. Finally some conclusions will be drawn in section 6.

2. Equation with discontinuity

Consider the following differential-operator equation

$$\frac{dP(x(t))}{dt} = f(t, x(t)), \tag{1}$$

where f is a smooth function and P is the Preisach operator, for a description see [4] or [3]. Now we consider the addition of an extra term to the right hand side of equation (1) as follows.

$$\frac{dP(x(t))}{dt} = f(t, x(t)) + g(t) = F(t, x(t)), \tag{2}$$

This extra term, $g(t)$ is a discontinuous function in time only. From now on we will work with a simple discontinuous function where the points of discontinuity are known in advance, such as

$$g(t) = \operatorname{sgn}(\cos(n\pi t)), \tag{3}$$

where n determines the frequency of the jump discontinuities.

Even though this is a very simple type of discontinuity (see figure 1), it can be used to model a number of physical systems, e.g. the switching between raining and drying in a hydrological system. This trivial function still presents a nontrivial problem to solve (2) numerically due to the discontinuity.

In [4] discontinuities in the right hand side were avoided, now a means to numerically solve equations such as equation (2) will be presented.

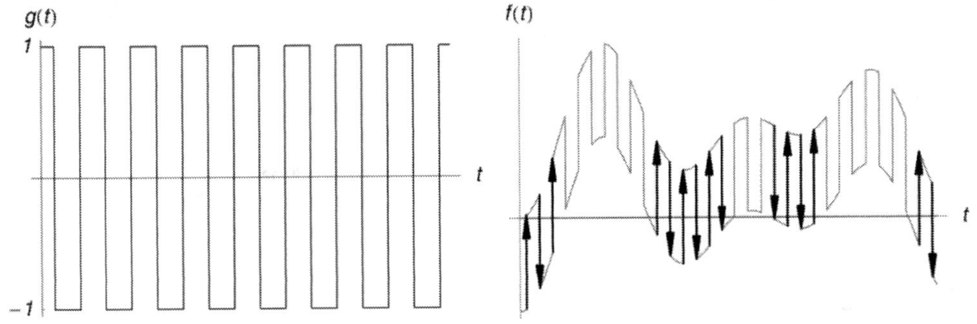

Figure 1. The figure on the left is an example of a simple discontinuity $g(t)$. The figure on the right is a modulated sinusoid with the jumps from the left figure included as described by the function $F(t, x(t))$. The arrows indicate the location and direction of the discontinuous jumps.

3. Physical system to be modelled

Flows through unsaturated porous media are usually assumed to satisfy three conditions

1. conservation of water mass, or of water volume when the density is constant,
2. a non-linear potential flow obeying a generalised Darcy Law,
3. a rate-independent relationship between water content and potential which closes the system of equations that express the first two conditions.

It is widely known from experiment that such unsaturated flows are hysteretic [18, 19]. In the past, hysteretic flows have either been avoided by considering non-reversing flows, or by treating flow-reversal in an ad hoc manner. The integration of this hysteretic system is explored in this paper using a simplified case where there is no spatial variation—the zero-dimensional case.

The dimensionless 'volumetric water content', θ, is defined to be the volume of water per volume of block of "soil-solids, water and air". θ is zero when the soil is completely dry; θ is maximum, less than one, when the block is saturated with water and no air is present. The specific volumetric flow rate of water q into or out of the block is the volumetric flow per volume of block. Flow into the block is positive; flow out of the block is negative.

Conservation of the volume of water in the block requires the rate of change of the volumetric water content to equal the specific flow at all times. It is expressed as follows:

$$\dot{\theta} = q, \quad \theta(0) = \theta_0, \quad 0 \leq \theta \leq \theta_s < 1. \tag{4}$$

Since spatial variation is excluded, the generalised Darcy Law says in this case that the instantaneous specific flow depends on the difference between the potential energy per unit mass of water in the block and a reference value outside the block. When there is no difference in the vertical position of water inside and outside the block, the gravitational potential plays no role. Only the difference in the matric, or capillary, potential drives the flow. Consequently,

$$q = -k\left(\psi(\theta) - \psi_{ref}(t)\right), \tag{5}$$

where k is the hydraulic conductivity of the medium, $\psi_{ref}(t)$ is a time dependent reference potential, $\psi(\theta)$ is the matric potential at a given moisture content θ and is defined by Campbell [34] as: the amount of work, per unit mass of water, required to transport an infinitesimal quantity of soil solution from the soil matrix to a reference pool of the same soil solution at the same elevation, pressure and temperature.

Both the hydraulic conductivity and the matric potential may exhibit hysteresis. There are a number of possible choices for further simplification of expression (5). In the first and most simple case, the rate-independent relationship $q = -k\theta$ (the so-called linear reservoir of hydrological theory) can be made hysteretic by replacing k with a Preisach operator. An examination of the hysteretic linear reservoir is reported in [26, 27]. In this paper, k is taken as a constant and a realistic Preisach operator models the relationship between matric potential and volumetric water content.

The combined ODE and Preisach operators exhibit non-local memory. See [28] for a discussion of non-local memory and rate independence, which are the key characteristics of dynamic hysteretic systems. The existence of two return-points in the non-local memory, completely dry and saturated respectively, ensure predictability.

For the reference potential we will choose the same reference potential $\psi_{ref}(t)$ as in [4], but this time we will include an extra term $g(t)$ for the discontinuity. Thus the reference potential will be a time-dependent sinusoidal driving function with an additional discontinuous time dependent function

$$\psi_{ref}(t) = -A\left(1 + \cos\left(\frac{2\pi t}{T}\right) + \cos\left(\frac{\pi t}{\sqrt{2}T}\right) + g(t)\right), \tag{6}$$

where $A > 0$ is the amplitude, T is the period and $g(t) = \operatorname{sgn}(\cos(t))$. Note that this function satisfies the condition of always being non-positive.

If we combine equations (6) and (5) and substitute them into equation (4), we then have the following:

$$\begin{aligned} \dot{\theta}(\psi) &= -k\left(\psi(\theta) + A\left(1 + \cos\left(\frac{2\pi t}{T}\right)\right) + \cos\left(\frac{\pi t}{\sqrt{2}T}\right) + g(t)\right), \\ \theta(\psi) &= P[\psi], \end{aligned} \tag{7}$$

where P is the Preisach operator [3].

4. Numerical Simulation

In the numerical experiments we solved equation (7) with the following parameter values: $k = 0.001, A = 20$ and $T = 1$. These were chosen such to cover all possible situations which the algorithm tests for.

The algorithm presented in section 5 was implemented in the C++ programming language. The simulations were performed on a 1.7GHz Pentium Xeon© dual-processor PC.

In figure 2 we show two overlayed plots. The solid line is the right-hand side of equation (7) and the dashed line is its solution $x(t)$, which has been numerically solved by the algorithm. In this figure there are two regions of interest where different parts of the algorithm were dominant. These regions have been magnified in figures 3 and 4 and are discussed below.

- Discontinuous jump in $F(t, x(t))$ not crossing t-axis Shown in figure 3 is a magnified view of the region 1 in figure 2. The dashed graph is a plot of the solution $x(t)$ against t and the solid graph is a plot of the right hand side, $F(t, x(t))$ against t . At the point **a**, $F(t, x(t))$ shows a jump discontinuity, but this does not cross the t-axis. The solution $x(t)$ at point **a**, shows a sudden increase in its slope. Likewise a similar situation occurs at point **b**, where the direction of the jump is reversed and the slope of the solution $x(t)$ suddenly decreases.

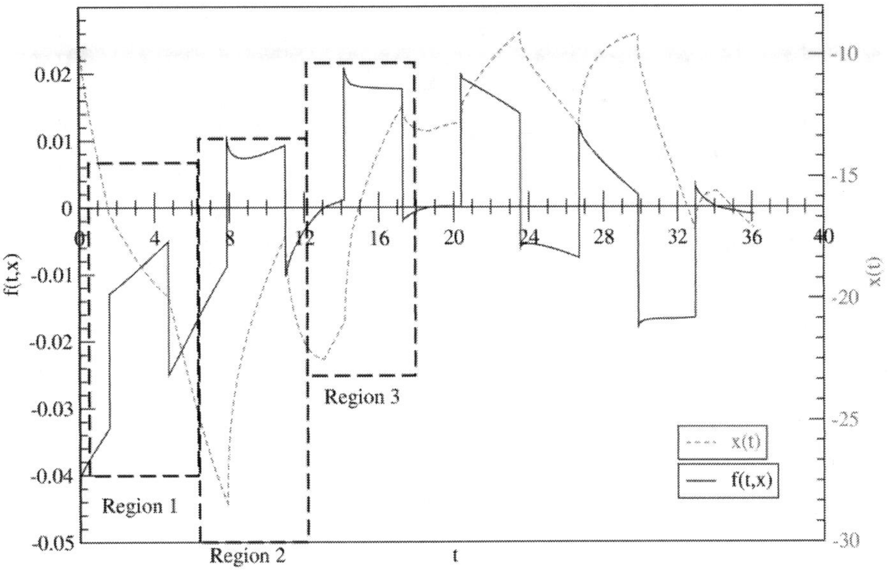

Figure 2. Plot of $F(t, x(t))$ superimposed on plot of $x(t)$against t. Indicated on graph are three regions of particular interest, which are magnified in the following figures.

- Discontinuous jump in $F(t, x(t))$ crossing the t-axis Shown in figure 4 is a magnified view of region 2 in figure 2. The dashed graph is a plot of the solution $x(t)$ against t and the solid graph is a plot of the right hand side, $F(t, x(t))$ against t. At point **a** the right-hand side of the ODE, $F(t, x(t))$ shows a jump discontinuity, which *does* cross the t-axis. At this point the normal methods of numerical integration break down and a step known as the "square-root" step is implemented for one step of the program. After this the program starts with a new initial condition and proceeds as normal. The result of this special step, is a parabolic-like curve in the solution and in the right-hand side $F(t, x(t))$ since the solution $x(t)$ is placed back into $F(t, x(t))$. Again a similar situation occurs at point **b**.
- Discontinuous jump in $F(t, x(t))$ almost crossing the t-axis Shown in figure 5 is a magnified view of region 3 in figure 2. The dashed graph is a plot of the solution $x(t)$ against t and the solid graph is a plot of the right hand side, $F(t, x(t))$ against t. At point **a** $F(t, x(t))$ crosses the t-axis without any discontinuity. Here the linear-step from [4] is implemented as $F(t, x(t))$ becomes zero when crossing the t-axis. At point **b** the right-hand side of the ODE, $F(t, x(t))$ shows a jump discontinuity, which *almost* crosses the t-axis. It can be observed here that the behaviour of $F(t, x(t))$ and $x(t)$ are quite similar to the case where the discontinuity crosses the t-axis, i.e. it has parabolic-like curvature. Note that the "square-root" approximation was not implemented in generating the solution in this case.

5. Algorithm

5.1. General description

The algorithm presented below has been written to solve Preisach operator ODEs of the type shown in equation (2). This algorithm begins at a time $t = t_0$ and will cycle until t reaches a predetermined end time $t = t_{end}$ or if the solution is no longer unique according to proposition 3.1 in [4].

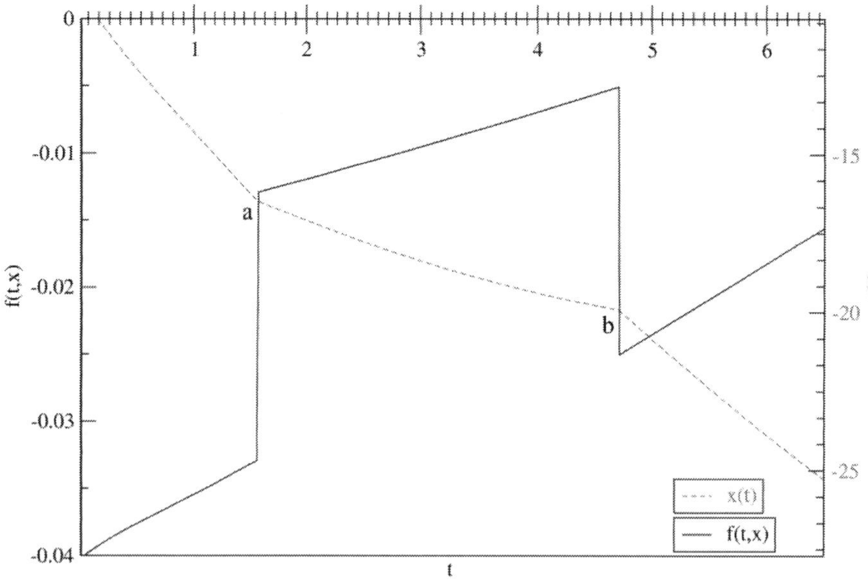

Figure 3. The magnified area which is denoted by Region 1 in figure 2.

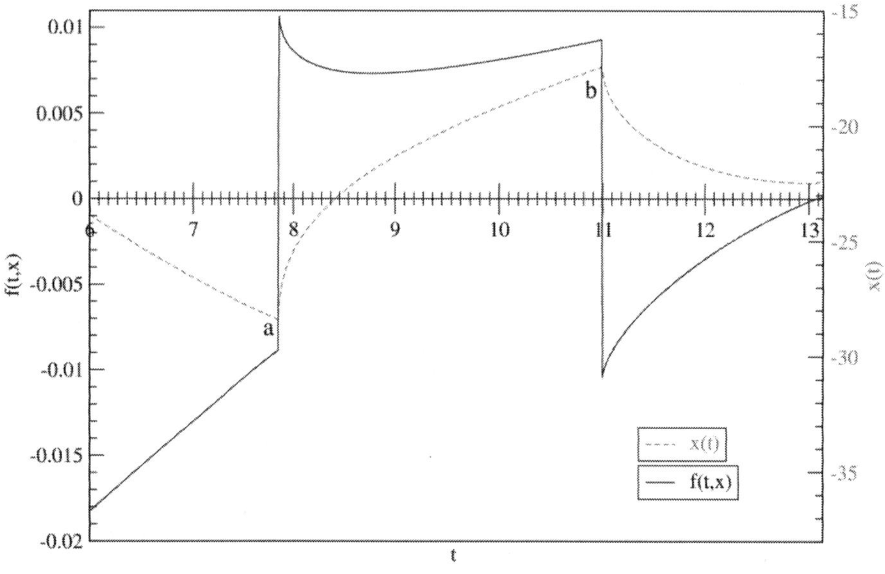

Figure 4. The magnified area which is denoted by Region 2 in figure 2.

5.1.1. Normal steps During each cycle a series of checks are performed to determine if the right-hand side of equation (2) changes sign. Special steps are taken when this occurs, these are described below. If there is no special step required, then equation (2) is solved by a numerical integration step $\nu(t_n, x_n, \eta_n; \tau_n)$ where t_n is the current time, x_n is the solution of the ODE as calculated in a previous step, or if there is no previous step it is the initial condition x_0. η_n is the current state of the Preisach operator and τ_n is the current time-step to be taken.

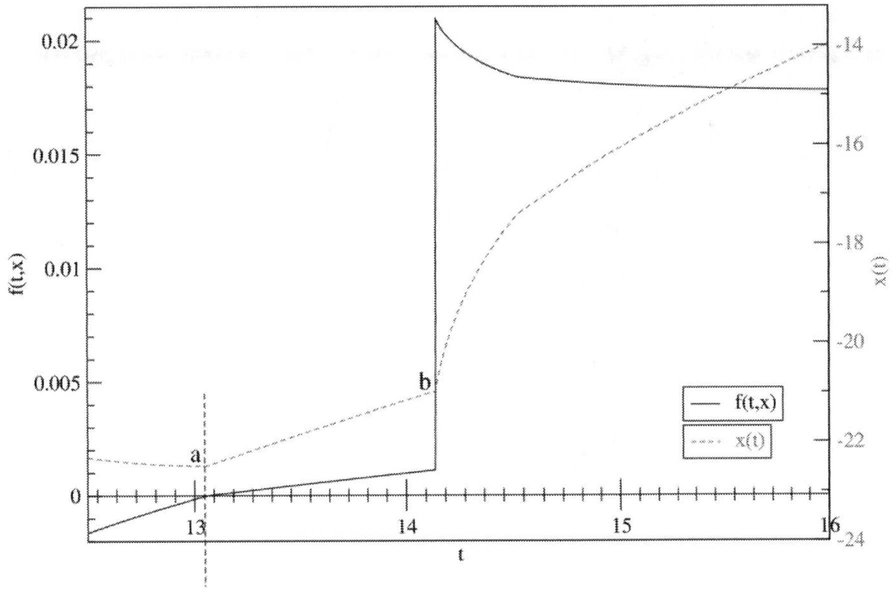

Figure 5. The magnified area which is denoted by Region 3 in figure 2 .

5.1.2. Special steps There are three cases, which can occur where the regular integration method cannot be used. These occur where there is either a change in sign of $F(t, x(t))$ and/or where $g(t)$ has a discontinuity at t. The cases which can occur are as follows

(i) $F(t, x)$ changing sign and $g(t)$ does not have a discontinuity.

(ii) $F(t, x)$ not changing sign and $g(t)$ has a discontinuity.

(iii) $F(t, x)$ is changing sign and $g(t)$ has a discontinuity at t.

Case i) is simply the case that was treated in [4]. There, a step known as the "linear step" was taken. It approximated a solution to the ODE, x_n, from $t = t_*$ to $t = t_* + h_{linear}$. It then returned to using the numerical integrator with the initial condition $x(t_0) = x_n$, where $t_0 = t_* + h_{linear}$.

Case ii) occurs when a discontinuity in $F(t, x)$ does not cause a change in its sign. This amounts to a sudden change in the slope of the solution. The algorithm determines where this point occurs i.e. when $t = t_*$. It then makes a step of size $t_* - t_n$.

Case iii) requires a special step, which was named the "square-root" step and was derived in a companion paper [3]. As with the other steps mentioned above, the time t_* at which the step is used is determined, in this case by a call to a function GetNextDJump(). The "square-root" step approximates the solution for x_n from $t = t_*$ to $t = t_* + h_{sqrt}$. As with the other special cases, the algorithm returns to using numerical integration with initial condition $x(t_0) = x_n$, where $t_0 = t + h_{sqrt}$.

5.2. Integration algorithm

INTEGRATE1()

Input: Initial values t_0, η_0, x_0; time step range $[h_{min}, h_{max}]$ and end time t_{end};

Output: Result of the integration is an array of $\{(t_i, x_i, \eta_i, \varsigma_i)\}_{i=0}^{no}$

(1) $n = 0$

(2) ς_0=YDIRECTION()

```
    ς = ς_0,        τ_n = h
    x_jump = GetNextQJump()
    t_discon = GetNextDJump()
    if |F(t_0, x_0)| == 0
        LinearStep()
        t_{n+1} = t_n + τ_n,        η_{n+1} = UpdatePreisach(η_n, x_{n+1}),        n++
    while t <= t_end
        τ_n = h
        t_discon = GetNextDJump()
        x_{n+1} = ν_1(t_n, x_n, η_n; τ_n)
        if (x_n − x_jump) · (x_{n+1} − x_jump) ≤ 0
            τ_n = JumpToY()
            x_jump = GetNextQJump()
            t_discon = GetNextDJump()
        else if (t_n − t_discon) · (t_{n+1} − t_discon) ≤ 0
            τ_n = t_discon − t_n
            x_{n+1} = ν_1(t_n, x_n, η_n; τ_n)
            if F(t_n, x_n) · F(t_{n+1} + ε, x_{n+1}) ≤ 0
                t_{n+1} = t_n + τ_n,        η_{n+1} = UpdatePreisach(η_n, x_{n+1}),        n++
                SqrtStep()
                η_{n+1} = UpdatePreisach(η_n, x_{n+1})
                ς_{n+1} = −ς_n,        t_{n+1} = t_n + h_sqrt,        n++
            x_jump = GetNextQJump()
            t_discon = GetNextDJump()
        else if F(t_n, x_n) · F(t_{n+1}, x_{n+1}) ≤ 0
            JumpToYP()
            η_{n+1} = UpdatePreisach(η_n, x_{n+1})
            ς_{n+1} = −ς_n,        t_{n+1} = t_n + τ_n,        n++
            x_jump = GetNextQJump()
            LinearStep()
        else
            ς_{n+1} = ς_n
        t_{n+1} = t_n + τ_n,        η_{n+1} = UpdatePreisach(η_n, x_{n+1}),        n++
```

GetNextQJump()

Input: t_n, x_n

Output: Return value gives the position x_{jump} of first discontinuity of Q in the direction of growth of the solution such that $Q(x_{jump}) \neq 0$.

```
    if ς_n == +1
        if K_↓ < 3 then return −∞
                else return w_n^(K_↓ − 2)
    else
        if K_↑ < 3 then return ∞
                else return w_n^(K_↑ − 2)
```

GetNextDJump()

Input: t_n, ε, where ε is the smallest possible time step before the discontinuity occurs.

Output: Since the discontinuous part of the input is based on the function $\mathrm{sgn}(\cos(\frac{t}{T}))$, where T is the period, then the jump points will occur at $t = \frac{m\pi T}{2}$, where $m \in \{\ldots, 1, 3, 5, \ldots\}$. The return value gives the time t_{discon} of the first discontinuity of Q in the direction of growth of the solution such that $Q(t_{discon}) \neq 0$.

(1) $t_{tmp} = 2t_n/\pi T$
(2) **if** $t_{tmp} \pmod 2 == 0$ **then** $t_{tmp}++$
(3) **return** $\pi T t_{tmp}/2 - \varepsilon$

YDIRECTION()
Input: Current step n
Output: $+1/-1$ if solution is going to increase/decrease
(1) **if** $F(t_n, x_n) == 0$ **and** $\frac{\partial F(t_n, x_n)}{\partial t} == 0$ **then error**
(2) **if** $F(t_n, x_n) < 0$ **or** ($F(t_n, x_n) == 0$ **and** $\frac{\partial F(t_n, x_n)}{\partial t} \neq 0$)
(3) **return** -1
(4) **else**
(5) **return** +1

LINEARSTEP()
Input: Current step n
Output:
(1)

$$a = \frac{1}{\rho(x_n, x_n)} \cdot \frac{\partial F(t_n, x_n)}{\partial x}, \qquad b = \frac{1}{\rho(x_n, x_n)} \cdot \frac{\partial F(t_n, x_n)}{\partial t}$$

(2) **if** $b < 0$ **and** $\varsigma_n < 0$
(3) $x_{n+1} = x_n + (-a - \sqrt{a^2 - 4b})h_{linear}/2$
(4) **else if** $b > 0$ **and** $\varsigma_n > 0$
(5) $x_{n+1} = x_n + (a + \sqrt{a^2 + 4b})h_{linear}/2$
(6) **else**
(7) **error**
(8) $\tau_n = h_{linear}$

SQRTSTEP()
Input: Current step n
Output:
(1)

$$c = \sqrt{\frac{2|F(t_n + \varepsilon, x_n) \cdot (t - t_0)|}{\rho(x_n, x_n)}}$$

(2) $x_{n+1} = x_n + \text{sgn}(F(t_n + \varepsilon, x_n)) \cdot ch_{sqrt}, \qquad \varepsilon << t$

JUMPTOY()
Input: Current step n
Output: Updated values of τ_n, x_{n+1}
(1) Using bisection modify τ_n so that new value $x_{n+1} = \nu_1(t_n, x_n, \eta_n; \tau_n)$ satisfies $|x_{n+1} - x_{jump}| < 10^{-8}$. $x_{n+1} = x_{jump} + \zeta \cdot 10^{-8} \cdot x_{jump}$
(2) $\eta_{n+1} =$ UPDATEPREISACH(η_n, x_{n+1})

JUMPTOYP()
Input: Current step n
Output: Updated values of τ_n, x_{n+1}
(1) Using bisection modify τ_n so that new value $x_{n+1} = \nu_1(t_n, x_n, \eta_n; \tau_n)$ satisfies $|F(t_n + \tau_n, x_{n+1})| < 10^{-8}$.

UPDATEPREISACH(η_{old}, x)
Input: Old state η_{old}, new input x
Output: Updated value $\eta = \{w^{(i)}\}$ and appropriate K

```
η = η_old,        K = K_old
if |x| >= γ
    K = 1        w^(1) = sign(x) · γ
    return η
if x > w^(K_old)
    if K mod2 == 0 then K++
    while K ≥ 3 and x > w^(K-2)
        K = K - 2
else if x < w^(K_old)
    if K mod2 == 1 then K++
    while K ≥ 3 and x < w^(K-2)
        K = K - 2
if K ≥ K_max then error
w^(K) = x
return η
```

6. Conclusions

In this paper a toolkit for the numerical solution of Preisach operator ODEs with discontinuities on the right-hand side was presented. It was used to model a simple system from hydrology based on Darcy's law and a number of interesting features were observed in solutions to this system. One of them was that the solution was more parabolic-like when the discontinuity either crossed or was very close to the t-axis. Although the physical model used to demonstrate the numerical techniques was based in hydrology, the numerical techniques are not limited to this. This algorithm can be used in physical models which satisfy the following conditions:

- The model must be of the general form of equation (2), i.e. with the derivative of the Preisach operator on the left of the ODE.
- When the right-hand side contains a discontinuity, the discontinuity can only be in the time variable [3] and [5].

Acknowledgments

The authors wish to acknowledge the many fruitful discussions with Alexei Pokrovskii, Oleg Rasskazov and Dmitrii Rachinskii. This research was partially supported by SFI grant 05/RFP/ENG062, by a private bequest to Cork University Foundation and by the Boole Centre For Research In Informatics (BCRI).

References

[1] O'Kane J P 2005 The FEST model—a test bed for hysteresis in hydrology and soil physics *J. Phys.: Conf. Series* **22**, ed M P Mortell, R E O'Malley, A Pokrovskiĭ and V Sobolev 148–63
URL `http://stacks.iop.org/1742-6596/22/148`

[2] Darcy H 1856 *Les Fontaines Publiques de la Ville de Dijon* (Dalmont, Paris)

[3] Zhezherun A and Flynn D 2006 ODE with Preisach operator under derivative and discontinuous in time right-hand side *J. Phys.: Conf. Series* (this issue)

[4] Flynn D and Rasskazov O 2005 On the integration of an ode involving the derivative of a preisach nonlinearity *J. Phys.: Conf. Series* **22**, ed M P Mortell, R E O'Malley, A Pokrovskiĭ and V Sobolev 43–55
URL `http://stacks.iop.org/1742-6596/22/43`

[5] Krejčí P, O'Kane J P, Pokrovskiĭ A and Rachinskiĭ D 2006 Mathematical models of hydrological systems with Preisach hysteresis *Preprint* http://www.bcri.ucc.ie/BCRI_58.pdf

[6] Haines W B 1930 Studies in the physical properties of soil. v. the hysteresis effect in capillary properties, and the modes of moisture distribution associated therewith. *J. of Agricultural Science* pp 97–116

[7] de Gennes P G, Brochard-Wyart F and Quere D 2003 *Capillarity and Wetting Phenomena: Drops, Bubbles, Pearls, Waves* (New York: Springer) ISBN 0387005927

[8] Poulovassilis A 1962 Hysteresis of pore water, an application of the concept of independent domains *Soil Science* **93** 405–12

[9] Poulovassilis A and Childs E C 1971 Hysteresis of pore water - non-independence of domains *Soil Science* **112** (5) 301–&
[10] Preisach P 1935 Über die magnetische nachwirkung *Zeitschrift für Physik* **94** 277–302
[11] Néel L 1943 Theories des lois d'aimantation de lord rayleigh, 2'. *Cahiers de physique* **13** 19–30
[12] Néel L 1942 Theories des lois d'aimantation de lord rayleigh, 1'. *Cahiers de physique* **12** 1–20
[13] Everett D H 1975 Capillary condensation and adsorption hysteresis *Berichte Der Bunsen-Gesellschaft-Physical Chemistry Chemical Physics* **79** (9) 732–4
[14] Mualem Y 1974 A conceptual model of hysteresis *Water Resources Research* **10**(3) 514–20
[15] Mualem Y 1973 Modified approach to capillary hysteresis based on a similarity hypothesis *Water Resources Research* **9**(5) 1324–31
[16] Mualem Y and Dagan G 1975 Dependent domain model of capillary hysteresis *Water Resources Research* **11**(3) 452–60
[17] Parlange J Y 1975 Solving flow equation in unsaturated soils by optimization - horizontal infiltration *Soil Science Society of America Journal* **39**(3) 415–8
[18] Haverkamp R, Reggiani P, Ross P and Parlange J 2002 *Enviromental Mechanics : Water, Mass and Energy Transfer in the Biosphere. The Philip Volume. Geophysical Monograph 129.* (The American Geophysical Union) chap Soil Water Hysteresis Prediction Model Based on Theory and Geometric Scaling, pp 213–46
[19] Flynn D, McNamara H, O'Kane J P and Pokrovskiĭ A 2005 Application of the Preisach model to soil-moisture hysteresis *Science of Hysteresis* vol 3 (Academic Press) ch 7, pp 689–744
[20] Krasnosel'skiĭ M and Pokrovskiĭ A 1989 *Systems with Hysteresis* (New York: Springer-Verlag) (translated from Russian by Marek Niezgódka)
[21] Visintin A 1994 *Differential Models of Hysteresis* (Berlin: Springer)
[22] Krejčí P and Sprekels J 1999 in *Variations of domain and free-boundary problems in solid mechanics (Paris, 1997)* vol 66 of *Solid Mech. Appl.* (Dordrecht: Kluwer Acad. Publ.) pp 237–44
[23] Mayergoyz I 2003 *Mathematical Models of Hysteresis And Their Applications* (Elsevier) ch 1
[24] O'Kane J P, Pokrovskii A, Krejčí P and Haverkamp R 2003 Hysteresis and terrestrial hydrology (Nice, France: EGS (European Geophyscial Society)-AGU (American Geophyscial Union)-EUG (European Union of Geosciences) Joint Assembly)
[25] O'Kane J P 2004 Hysteresis, the missing non-linearity *Hydrol. Earth Syst. Sci. Discuss* **1** 41–73 discussion of "Bringing it all together" by J.C.I. Dooge.
[26] O'Kane J P 2006 The hysteretic linear reservoir—a new Preisach model *Physica B: Condensed Matter* **372** (1-2) 388–92
[27] O'Kane J P 2005 Hysteresis in hydrology *Acta Geophys, Pol.* **53** 373–83
[28] O'Kane J P 2006 Thresholds, switches and hysteresis in hydrology from the pedon to the catchment scale—a non-linear systems theory *Hydrology and Earth System Sciences (X)* (submitted)
[29] Cross R and Allan A 1988 in R Cross, ed, *Unemployment, Hysteresis & Natural Rate Hypothesis* (Blackwell) pp 26–38
[30] Cross R 1995 *Hysteresis, The Handbook of Economic Methodology* (Edward Edgar)
[31] Cirrincione M, Miceli R and Galluzzo G 2002 Preisach function identification by neural networks *IEEE Transactions On Magnetics* **5**(38) 2421–3 part 1
[32] Kuczmann M and Ivanyi A 2001 Neural network based scalar hysteresis model *International journal of applied electromagnetics and mechanics* **15 (1-4)** 225–30
[33] Hughes D and Wen J T 1997 Preisach modeling of piezoceramic and shape memory alloy hysteresis *Smart Materials and Structures* **6**(3) 287–300 URL `http://stacks.iop.org/0964-1726/6/287`
[34] Campbell G 1985 *Soil Physics with BASIC–Transport Models for Soil-Plant Systems. Developments in Soil Science* vol 14 (New York: Elsevier)

Institute of Physics Publishing
doi:10.1088/1742-6596/55/1/007

Journal of Physics: Conference Series **55** (2006) 74–79
International Workshop on Multi-Rate Processes and Hysteresis

Canard solutions near a degenerate turning point

T Forget

Laboratoire de Mathématiques et Applications Pôle, Sciences et Technologies - Université de La Rochelle, Avenue Michel Crépeau, 17042 La Rochelle, France

E-mail: `thomas.forget@univ-lr.fr`

Abstract. We are interested in the study of canard solutions in a singularly perturbed real first order ODE of the form $\varepsilon u' = \Psi(x, u, a, \varepsilon)$, where $\varepsilon > 0$ is a small parameter, and $a \in \mathbb{R}$ is a control parameter. An existence result for such solutions is given, and the method used in the demonstration allow us to conjecture the existence of a generalized $\varepsilon^{1/(p+1)}$-asymptotic expansion for those solutions.

1. Introduction

In the theory of first order singularly perturbed real ODEs, equations of the form

$$\varepsilon u' = \Psi(x, u, a, \varepsilon) \tag{1}$$

are considered, where Ψ is $\mathcal{C}^\infty$, $x \in [-x_0, x_0]$ and u is real, a is a control parameter, and $\varepsilon > 0$ is a small real parameter.

As we are concerned with canard solutions, we have to make assumptions to be in the appropriate region for the appearance of canard solutions. For this reason, we have to suppose that the following two assumptions are satisfied by (1):

(H1) (1) has a slow curve.

This means that there exists (a_0, u_0) with a_0 real, and u_0 a real $\mathcal{C}^\infty$ function, such that

$$\forall x \in [x_0, -x_0],\ \Psi(x, u_0(x), a_0, 0) = 0.$$

Without loss of generality, we may suppose that $a_0 = 0$ and $u_0(x) = 0$ for all $x \in [-x_0, x_0]$.

(H2) $\frac{\partial}{\partial u}\Psi(x, 0, 0, 0)$ is $\begin{cases} < 0 \text{ if } x < 0 \\ > 0 \text{ if } x > 0. \end{cases}$

This assumption allows us to suppose that the slow curve has a turning point at $x = 0$, which means that the slow curve is attractive to the left, and repulsive to the right, which are the conditions for the appearance of a canard.

Moreover $\frac{\partial}{\partial u}\Psi(x, 0, 0, 0)$ has a zero at $x = 0$ of order p, where p is odd.

Solutions of interest for (1) are the so-called canard solutions [4][7][3]which are families of pairs $(a_\varepsilon, u_\varepsilon)$ which tends to the slow curve $(0, 0)$ as $\varepsilon \to 0$.

A similar study has been done in the case of complex variables, where solutions $(a_\varepsilon, u_\varepsilon)$ which tends to the slow curve $(0, 0)$ as $\varepsilon \to 0$ for all x in a complex neighborhood of 0 are the so-called overstable solutions [6][2].

In our study, the difference is that we are using a 1 dimensional real parameter a, where the complex study required p complex parameters, this difference being due to the real characteristic of the neighbourhood studied.
In the continuation of our study, we will restrict ourselves to the equations

$$\varepsilon u' = x^p u + ax^L + \sum_i a^{k_i} x^{l_i} + \varepsilon P(x, u, a, \varepsilon) \tag{2}$$

where $L < p$ is even and, for all i, $l_i > L$, $k_i \geq 1$.
By assuming some strong assumptions, it is possible to transform equation (1) to such a general equation.
From those equations, we will define an operator $\Xi : (b, v) \mapsto (a, u)$. It will be proved that it is a contraction, with its Lipschitz constant equal to $K\varepsilon^{1/(p+1)}$.
Then, a fixed point theorem will prove the existence of a canard solution by giving an iterative sequence $((a_{n,\varepsilon}, u_{n,\varepsilon}))_n$ which has the expected canard solution $(a^*_\varepsilon, u^*_\varepsilon)$ in the limit as $n \to +\infty$.
It has to be noted that this result has been proved by Peter de Maesschalck and Freddy Dumortier who have used a blowing up method [5], but the demonstration we propose allows us to assume that the canard solution has an asymptotic expansion, and this will be the subject of the second part of this paper:
By writing

$$u^*_\varepsilon = \sum_{n \geq 1} (u_{n,\varepsilon} - u_{n-1,\varepsilon})$$

we see that for all n, $u_{n,\varepsilon} - u_{n-1,\varepsilon} = \mathcal{O}(\varepsilon^{n/(p+1)})$. Thus, the study of the existence and the uniqueness of an $\varepsilon^{1/(p+1)}$-asymptotic expansion for the canard solution is a subject of interest.
We recall that an asymptotic expansion is a series $\sum_k u_k \varepsilon^{k/(p+1)}$ such that

$$\forall K, \left\| u^*_\varepsilon - \sum_{k=0}^{K-1} u_k \varepsilon^{k/(p+1)} \right\| = \mathcal{O}\left(\varepsilon^{K/(p+1)}\right)$$

where the properties of the coefficients u_k have to be defined.
We will show that intermediary functions have to be defined in order to seek coefficients u_k as $\mathcal{C}^\infty$ functions for the variable x and those intermediary functions.
Implementing such formulae in the case $p = 0$, to study limit layers in the case of an attractive slow curve, allows us to retrieve the so-called Combined Asymptotic Expansion [1].
When $p \geq 1$, implementing it implies the existence of the expansion, but the question of uniqueness is not solved, as interactions between those intermediary functions need to be better controlled.
In what follows, we will denote for convenience $\eta := \varepsilon^{1/(p+1)}$.

2. Existence of canard solutions in (2)

Let Ω be an open subset of $\mathbb{R}^4$ containing 0, and let $x_0 > 0$, $A > 0$, $B > 0$, $\tilde{\eta} > 0$, such that $[-x_0, x_0] \times [-A, A] \times [-B, B] \times [0, \tilde{\eta}] \subset \Omega$.
We will denote by E the set of real continuous functions over $[-x_0, x_0]$, and note that the spaces $(E, ||.||)_\infty$ and $(\mathbb{R} \times E, |.|_\infty)$ are two Banach spaces.
For all $M > 0$, $\mathcal{B}_M$ wil be the set $\{v \in E; ||v||_\infty \leq M\}$.
For convenience, the polynomial $\sum_i a^{k_i} x^{l_i}$ will be denoted by $s(x, a)$.

Theorem 1 *We suppose (b, v) and (a, u) are in $[-B, B] \times \mathcal{B}_A$.*
For all $\eta \in]0, \tilde{\eta}[$ and (b, v) bounded, the equation

$$\begin{cases} \eta^{p+1} u'(x) = x^p u(x) + a x^L + s(x, a) + \eta^{p+1} P(x, v(x), b, \eta) \\ u(x_0) = 0 = u(-x_0) \end{cases}$$

has an unique bounded solution (a, u), that will be denoted by $\Xi_\eta(b, v)$.
This will define an application Ξ_η.
Moreover, $\forall \eta_0 > 0, \exists K; \forall \eta \in]0, \eta_0[$: Ξ_η is a $(K\eta)$-Lipschitz function.
Finally, for all $\delta_1 \in]0, B[$ and $\delta_2 \in]0, A[$, there exists $\eta_0 > 0$ such that for all $\eta \in]0, \eta_0[$, Ξ_η is an operator mapping from $[-\delta_1, \delta_1] \times \mathcal{B}_{\delta_2}$ to itself.

For (b, v) bounded, the linear equation can be solved, and the matching conditions in 0 imply the uniqueness for a bounded parameter a, and so the uniqueness of a bounded solution u follows.
By writing Ξ_η as a composition of two operators $\Xi_\eta = \Theta \circ \wp$ defined as follows:
$\wp(b, v)(x) := P(x, v(x), b, \eta^{p+1})$
Θ: linear operator such that $\Theta(w)$ is the solution of

$$\begin{cases} \eta^{p+1} u' = x^p u + \alpha x^L + s(x, \alpha) + \eta^{p+1} w \\ u(-x_0) = 0 = u(x_0), \end{cases}$$

we prove, by estimating integrals like $e^{(x/\eta)^{p+1}} \int_0^x \xi^k e^{-(\xi/\eta)^{p+1}} d\xi$, that the operator Θ satisfies a Lipschitz condition, with Lipschitz constant equal to $C\eta$, with C independent of η.
As the operator $\wp$ is a L-Lipschitz function, we conclude that Ξ_η satisfies a Lipschitz condition, with Lipschitz constant equal to $K\eta$, where K is independent of η.
This result implies the last part of the theorem.

As $[-\delta_1, \delta_1] \times \mathcal{B}_{\delta_2}$ is a closed set of the Banach space $(\mathbb{R} \times E, |.|_\infty)$, the fixed point theorem implies that Ξ_η has a fixed point $(\alpha^*_\eta, u^*_\eta) \in [-\delta_1, \delta_1] \times \mathcal{B}_{\delta_2}$ which satisfies the equation

$$\begin{cases} \eta^{p+1} u^{*'}_\eta(x) = x^p u^*_\eta(x) + \alpha^*_\eta x^L + s(x, \alpha^*_\eta) + \eta^{p+1} P(x, u^*_\eta(x), \alpha^*_\eta, \eta) \\ u^*_\eta(x_0) = 0 = u^*_\eta(-x_0) \end{cases} \quad (3)$$

As a direct consequence, the existence of a canard solution for (2) follows:

Corollary 1 *For all $\delta_1 \in]0, B[$ and $\delta_2 \in]0, A[$, there exists $\eta_0 > 0$ such that for all $\eta \in]0, \eta_0[$, the equation (3) has an unique solution $(\alpha_\eta, u_\eta) \in [-\delta_1, \delta_1] \times \mathcal{B}_{\delta_2}$.*
Moreover, as $\delta := \min\{\delta_1, \delta_2\}$ is fixed, the family $(u_\eta)_{\eta \in]0, \eta_0[}$ is uniformly convergent to 0, over $[-x_0, x_0]$, as $\eta \to 0$.

3. Asymptotic expansion

The fixed point theorem allows us to define an iterative sequence $((a_{n,\eta}, u_{n,\varepsilon}))_n$ whose limit is the canard solution (a^*_η, u^*_η), and so we have that

$$a^*_\eta = \sum_{n \geq 1} (a_{n,\eta} - a_{n-1,\eta})$$

where, as the operator Ξ_η is a $(K\eta)$-Lipschitz function, we have for all n,

$$a_{n,\eta} - a_{n-1,\eta} = \mathcal{O}(\eta^n)$$

This implies that we have an η-asymptotic expansion for a^*_η : $a^*_\eta \simeq \sum_k \tilde{a}_k \eta^k$, i.e.:

$$\forall K, ||a^*_\eta - \sum_{k=0}^{K-1} \tilde{a}_k \eta^k|| = \mathcal{O}(\eta^K)$$

In what follows, we will try to propose a similar result for u^*_η but, as it is a function of x and η, some problems will appear:

3.1. Implementation of a general structure for u
Similarly to the result for a, we write

$$u^*_\eta = \sum_{n \geq 1} (u_{n,\eta} - u_{n-1,\eta}) .$$

where, as the operator Ξ_η is a $(K\eta)$-Lipschitz function, for all n,

$$u_{n,\eta} - u_{n-1,\eta} = \mathcal{O}(\eta^n)$$

This brings us to the question of the existence and the uniqueness of an η-asymptotic expansion for the function u^*_η, $u^*_\eta \simeq \sum_k u_k \eta^k$, i.e.:

$$\forall K, ||u^*_\eta - \sum_{k=0}^{K-1} u_k \eta^k|| = \mathcal{O}\left(\eta^K\right) .$$

Unfortunately, the natural η-asymptotic expansion $\sum_k u_k \eta^k$ where all coefficients u_k are analytic in the variable x is not sufficient:
A formal substitution shows that an η-asymptotic expansion $\sum_k u_k \eta^k$, where the coefficients u_k are independent of η, implies that the coefficients u_k may have a pole at $x = 0$.
So, we will seek coefficients u_k to be analytic in x and in some intermediary functions φ which have to be well defined.
Some possible first choices for such functions are $\varphi(x, \eta) = e^{-(x/\eta)^{p+1}}$ or
$\varphi(x, \eta) = (e^{(x/\eta)^{p+1}} \int_x^{+\infty} \xi e^{-(\xi/\eta)^{p+1}} d\xi, \cdots, e^{(x/\eta)^{p+1}} \int_x^{+\infty} \xi^{p-1} e^{-(\xi/\eta)^{p+1}} d\xi)$.
In the second case, the integrals considered are ones which appeared when we tried to make a decomposition of Θ. As they do not have an η-asymptotic expansion, it is natural to consider them as the expected intermediary functions, but as they are not bounded in $\mathbb{R}$, we have to use a modified version of those functions:
Seeing that those functions are the most practical ones we are willing to use them in our study, but the lack of control at 0 may cause problems and thus, by pointing out that $e^{(x/\eta)^{p+1}} \int_x^{+\infty} v(\xi) e^{-(\xi/\eta)^{p+1}} d\xi$ is a solution of the differential equation

$$\begin{cases} \eta^{p+1} u'(x) = x^p u(x) + \eta^{p+1} v \\ u(+\infty) = 0, \end{cases}$$

we will introduce a parameter a to give us a control at 0, and so, the function denoted by $\mathcal{I}(v)$ will be a solution of the differential equation

$$\begin{cases} \eta^{p+1} u'(x) = x^p u(x) + a + \eta^{p+1} v \\ u(-\infty) = 0 = u(+\infty), \end{cases}$$

where the parameter a is calculated such that $\mathcal{I}(v)$ is continuous at 0.
Another problem may appear since x and $x\varphi$ may not have the same estimation, For example

$$||e^{-(x/\eta)^{p+1}}||_{x\in[-1,1]} = 1 \text{ and } ||xe^{-(x/\eta)^{p+1}}||_{x\in[-1,1]} = \frac{e^{-1/(p+1)}}{(p+1)^{p+1}}\eta,$$

which implies that those two elements are not associated with the same coefficients u_k in the η-asymptotic expansion of the solution. Thus we have to order the monomials $x^i\varphi^j\eta^l$ with respect to their estimate in η.
For this purpose, we define an order

$$\maltese(x^i\varphi^j\eta^l) := \lim_{\eta\to 0} \frac{\ln ||x^i\varphi(x,\eta)^j\eta^l||_{x\in[-1,1]}}{\ln \eta}.$$

Note that the set

$$\mathcal{A}_k = \text{Vect}\{x^i\varphi^j\eta^l;\ \maltese(x^i\varphi^j\eta^l) \leq k\}$$

will be the set of the principal terms with order k for the η-asymptotic expansion of u^*_η and, by using the transformation $\pi_k : \mathcal{A} \to \mathcal{A}_k$, where $\mathcal{A} := \bigcup_k \mathcal{A}_k$, we are able to construct an iterative sequence defined as

$$\begin{cases} u_0 := 0 \\ u_k := \pi_k(\Theta \circ \wp(u_{k-1})), \end{cases}$$

We are able to prove that this sequence is well defined and convergent, by making a decomposition of Θ as a linear combination of monomials $x^i\varphi^j\eta^l$.
The limit will be the expected η-asymptotic expansion, which will prove the existence of the asymptotic expansion.
The current problem we are trying to solve is the uniqueness of such asymptotic expansions, because intermediary functions have some interactions.
For example, it can be proved that

$$\mathcal{I}(x^k\mathcal{I}(u)) = \frac{1}{k+1}\left(x^{k+1}\mathcal{I}(u) - \mathcal{I}(x^{k+1}u) - c_u\mathcal{I}(x^{k+1})\right).$$

We can find an answer to this question in the case $p = 1$ which is the non degenerate case, and also in the case $p = 0$, even if it is not related to a study of canard solutions:

3.2. Application to the case $p = 0$
We will apply our general structure for the study of an asymptotic expansion in the case of a limit layer for an singularly perturbed real ODE with an attractive slow curve.
Thus, we are studying the equation

$$\begin{cases} \varepsilon u' = -u + \varepsilon P(x, u, \varepsilon) \\ u(0) = u_0 \neq 0. \end{cases} \tag{4}$$

We note that the slow curve of (4) is $u = 0$.
For this study, we will choose for an intermediary function $\varphi(x, \varepsilon) = e^{-x/\varepsilon}$.
The application of this structure gives the existence and uniqueness of an ε-asymptotic expansion for (4) of the form

$$u^*_\varepsilon(x) \approx \sum_{i,l} a_{i,0,l}x^i\varepsilon^l + \sum_{i,l,j\geq 1} a_{i,j,l}\left(\frac{x}{\varepsilon}\right)^i \varphi(x,\varepsilon)^j\varepsilon^{l+i}.$$

This is a particular form of the so-called Combined Asymptotic Expansion, which has the following form

$$u_\varepsilon^*(x) \approx \sum_n \left(f_n(x) + g_n\left(\frac{x}{\varepsilon}\right) \right) \varepsilon^n,$$

where for all n, f_n is analytic, and g_n is an exponentially decreasing function.
Functions f_n are related to the asymptotic expansion of the slow curve, whereas the functions g_n are related to the solution in the fast layer.

3.3. Application to the case $p = 1$
In this case, the turning point is not degenerate, and no intermediary functions are needed to give the asymptotic expansion of the canard solution.
Indeed, in this case, we prove that $\Theta(x^k)$ can be written as a linear combination of monomials $x^i \varepsilon^l$, which directly implies the uniqueness of the asymptotic expansion.

Acknowledgments
I am grateful to Eric Benoît for the scientific framing of this PhD research, and to Guy Wallet for helpful discussions as co-advisor. This work was supported by the Région Poitou-Charentes under grant no 03/RPC-R-148.

References

[1] Benoît E El Hamidi A and Fruchard A 2002 *Electron. J. Differential Equations* **51**
[2] Benoît E Fruchard A Schäfke R and Wallet G 1998*C. R. Acad. Sci. Paris Sér. I Math.* 326 **7** 873-8
[3] Cartier P. 1982 *Bourbaki Seminar* **92** 21-44
[4] Diener F and Diener M 1981 *Collect. Math.* 32 **1** 37
[5] de Maesschalck P and Dumortier F 2006 *Trans. Amer. Math. Soc.* 358 **5** 2291
[6] Wallet G 1994 *Bull. Soc. Math. France* 122 **2** 185
[7] Zvonkin A K and Shubin M A 1984 *Uspekhi Mat. Nauk* 39 **2(236)** 77

Institute of Physics Publishing
doi:10.1088/1742-6596/55/1/008

Journal of Physics: Conference Series **55** (2006) 80–93
International Workshop on Multi-Rate Processes and Hysteresis

Dynamics of a Lotka-Volterra type model with applications to marine phage population dynamics

C Gavin[1], A Pokrovskii[1], M Prentice[2] and V Sobolev[3]

[1]School of Mathematical Sciences University College Cork, Cork, Ireland
[2]Department of Microbiology University College Cork, Cork, Ireland
[3]Department of Differential Equations and Control Theory Samara State University, Akademika Pavlova Street, 1, 443011, Russia

E-mail: a.pokrovskii@ucc.ie, M.Prentice@ucc.ie, hsablem@yahoo.com

Abstract. The famous Lotka-Volterra equations play a fundamental role in the mathematical modeling of various ecological and chemical systems. A new modification of these equations has been recently suggested to model the structure of marine phage populations, which are the most abundant biological entities in the biosphere. The purpose of the paper is: (i) to make some methodical remarks concerning this modification; (ii) to discuss new types of canards which arise naturally in this context; (iii) to present results of some numerical experiments.

"Bacteria don't die, they just phage away"
— Mark Müller, microbiologist

Introduction

A phage (short for bacteriophage) is a microscopic virus that infects bacteria. There is a huge number of varieties of phage, but in general phages only infect specific bacteria. Upon infecting bacteria, a phage will either begin to reproduce, or else it will integrate its DNA into the host bacteria. Phages are microscopic organisms. Their dimensions are typically between 24 to 200 nm, and their basic structure is shown in Figure 1.

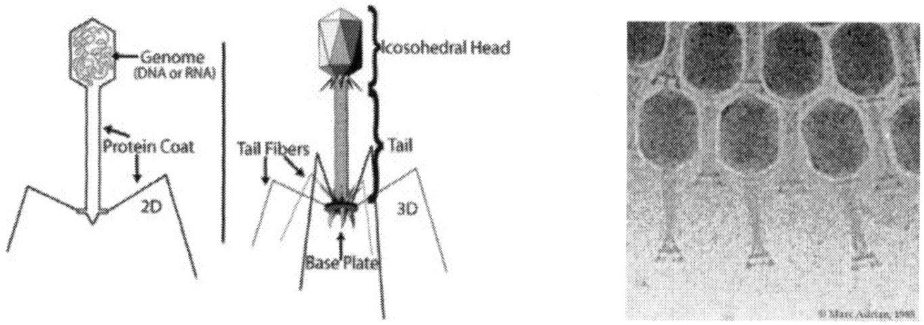

Figure 1. Structural overview (left), and actual picture (right) of phages

Once a phage encounters a bacteria, it attaches itself via its tail fibres, and injects its DNA down through the base plate and into the bacteria. Some phages, once they have infected a bacteria, begin to multiply immediately, and in a short time completely destroy the bacteria, releasing more phages. Other phages simply remain dormant once they are attached to bacteria. Their DNA incorporates itself into that of the bacteria, and remains attached to it until the bacteria is weak or close to death. Once they detect a decline in the bacteria, they then become active again, and reproduce before the bacteria's demise[2]. Phages are believed to be important in carbon cycling, with at least 25% of fixed carbon passing through them. They are also important in the absorption of CO_2 in the oceans.

A new modification of the classical Lotka–Volterra equations has been recently suggested to model the structure of marine phage populations. The purpose of the paper is: (i) to make some methodical remarks concerning this modification; (ii) to discuss new types of canards which arise naturally in this context; (iii) to present results of some numerical experiments. This paper addresses a wide range of readers, including biologists. Thus, we were trying to make the exposition as elementary and self contained as possible.

1. Methodical remarks on modified Lotka-Volterra equations

The classic Lotka-Volterra Equations were suggested to model population dynamics of a predator-prey system. They are [3, p.41]:

$$\dot{x} = x(\alpha - \beta y), \qquad \dot{y} = y(-\gamma + \delta x), \tag{1}$$

where x and y are functions of time representing the populations of the prey and predator respectively, and α, β, γ, and δ are positive constants which govern the strength of the interactions between the two species and amongst themselves. This system has a unique positive equilibrium $(\bar{x}, \bar{y})$ when $\dot{x} = \dot{y} = 0$, that is $\bar{x} = \gamma/\delta$, $\bar{y} = \alpha/\beta$.

From the analysis of experimental data, Hoffmann [1] proposed that for the phage-bacteria interaction the equations be modified such that a power law applies to the Lotka-Volterra equations. The modified Lotka-Volterra equations are therefore:

$$\dot{B} = B(\alpha - \beta\Phi^p), \qquad \dot{\Phi} = \Phi^p(-\gamma + \delta B). \tag{2}$$

We will assume the inequality

$$p > 1. \tag{3}$$

Loosely speaking the equations (2) mean that the "effective size" of the phage population differs from its "physical size", and this effective size is proportional to the power p of the physical size. A naive explanation may be as follows: in the traditional Lotka-Volterra equations (1) and (2), it is assumed that the important meetings are when one predator meets one prey. With the power of two, which was among the best fits in the modified equations, we can hypothesize that unlike the traditional system, it is the meeting of two phages and one bacteria that are important. (Hoffmann talks about the p=2 being due to attraction of more hosts to the area of phage lysis but there are some papers indicating 2-3 phages actually do infect each bacterium in laboratory tests.) Correspondingly, the power Φ^p in (2) indicated that that are "hunting-teams" of p phages which are effective. This sounds strange for a fractional p, but such ideas are not unusual in mathematics.

This system also has a the unique positive equilibrium:

$$B_* = \gamma/\delta, \qquad \Phi_* = (\alpha/\beta)^{1/p}. \tag{4}$$

Returning to equations (2), we can look for a constant of motion for the system. We do this by dividing the first equation (2) by the second to obtain

$$(-\gamma \log B(t) + \delta B(t)) + \left(\alpha\Phi(t)^{1-p}/(p-1) + \beta\Phi(t)\right) = const.$$

for any positive solution $(B(t), \Phi(t))$ of (1). Thus

$$W(B, \Phi) = (-\gamma \log B + \delta B(t)) + \left(\alpha \Phi^{1-p}/(p-1) + \beta \Phi\right) \tag{5}$$

is a conserved quantity, in other words, the first integral, for the motion of trajectories in this system. The value of the constant is determined by the initial conditions. Rewriting system (2) as

$$\dot{B} = \alpha B(\Phi_*^p - \Phi^p), \qquad \dot{\Phi} = \gamma(-B_* + B),$$

we can see that the direction vector field in the strictly positive quadrant is as shown in the figure below. Since $p > 1$, $\lim W(B, \Phi) = \infty$ as $B \to 0$, or $\Phi \to 0$, or $B \to \infty$, or $\Phi \to \infty$,

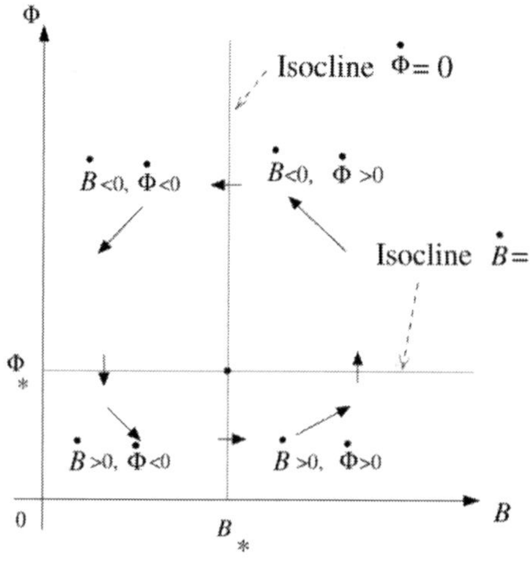

Figure 2. The main isoclines and direction field for system (2)

the trajectories could not explode to infinity. Finally, since the restriction $W(B) = W(B, \Phi_*)$ is strictly monotone for $0 < B < B_*$, we conclude that all trajectories of (2) are periodic, and they orbit anti-clockwise around the equilibrium (B_*, Φ_*).

We can see now that a modified Volterra principle holds for system (2). For a classic Lotka-Volterra system (1), this principle states that the time averages of the population densities will be constant and equal to the equilibrium values, i.e.

$$\frac{1}{T}\int_0^T x(t)\,dt = \bar{x}, \qquad \frac{1}{T}\int_0^T y(t)\,dt = \bar{y},$$

where $x(t)$ and $y(t)$ are the two populations, and T is the period of the solution [3, p.45-46].

For our system, we will have a slightly modified version of Volterra principle:

$$\frac{1}{T}\int_0^T B(t)\,dt = B_*, \qquad \frac{1}{T}\int_0^T \Phi(t)^p\,dt = \Phi_*^p. \tag{6}$$

The first relationship is the same as in standard case and it may be proved as usual. Next, we evaluate

$$\frac{d}{dt}(\log B(t)) = \frac{dB}{dt}\frac{1}{B(t)} = \alpha - \beta\Phi^p,$$

which implies

$$\int_0^T \frac{d}{dt}(\log B(t))\, dt = \int_0^T (\alpha - \beta\Phi(t)^p)\, dt,$$

or

$$\log(B(T)) - \log(B(0)) = \alpha T - \beta \int_0^T \Phi(t)^p\, dt.$$

Taking into account $B(T) = B(0)$, we can write

$$\frac{1}{T}\int_0^T \Phi(t)^p\, dt = \frac{\alpha}{\beta} = (\Phi_*)^p\,,$$

by the second of equations (4).

We now consider the case where we further modify the generalized Lotka-Volterra equations (2) to take into account competition within the species. We assume that the competition between members of the same species for food reduces population growth, but that this effect is relatively small compared to the natural growth rate. We assume therefore that the additional parameters μ and ν are small, and that they affect the square of the population, as in the original Lotka-Volterra equations. We note that Hoffmann says that effectively there is no food competition "food is not a limiting resource for microbes in the marine ecosystem". The analysis of the system with competition is interesting from the mathematical point of view. We will see that any, infinitesimally small competition would change drastically the pattern of longtime behaviour of system.

As we already used modified Lotka-Volterra equations with the powers p for the predators, we therefore include an exponent $2p$ for the predator competition term. This is consistent with our earlier hypothesis that the effective size of the phage population is measured as Φ^p. For instance, suppose that the interactions between two phages and one bacteria are important. In order for there to be a negative impact from competition, we consider the case where two teams of two phages encounter one bacteria. Each pair of phages will effectively be competing with the other for the one prey. Therefore, the competition between the phages will have a negative impact on their growth rate. Also, if pairs of phages are needed to consume each prey, then situations where there are 3 phages and one bacteria will have no role other than the usual interaction of two phages and one bacteria. Hence, we have no term with a cubic power in the corresponding competition model.

Our modified equations are therefore:

$$\dot{B} = B\,(\alpha - \beta\Phi^p - \mu B)\,, \qquad \dot{\Phi} = \Phi^p\,(-\gamma + \delta B - \nu\Phi^p)\,. \tag{7}$$

Following the same procedure as before, we seek a nontrivial equilibrium (B_{**}, Φ_{**}) from the simultaneous equations

$$\alpha - \beta\Phi^p - \mu B = 0, \qquad -\gamma + \delta B - \nu\Phi^p = 0. \tag{8}$$

Therefore,

$$B_{**} = \frac{\beta\gamma + \nu\alpha}{\delta\beta + \nu\mu}, \qquad \Phi_{**} = \left(\frac{\alpha\delta - \mu\gamma}{\delta\beta + \nu\mu}\right)^{1/p},$$

which clearly belongs to the positive quadrant for

$$\alpha\delta > \mu\gamma. \tag{9}$$

Our next step is to test the stability of this equilibrium (B_{**}, Φ_{**}) subject to the condition (9). To do this, we need to construct a Lyapunov function for the system. Recall that the smooth function $V = V(B, \Phi)$ is a Lyapunov function in the positive quadrant for the system

$$\dot{B} = f(B, \Phi), \quad \dot{\Phi} = g(B, \Phi),$$

which has a unique positive equilibrium (B_{**}, Φ_{**}), if the following two conditions hold:

(i) $\dot{V}(B, \Phi) = \dfrac{\partial V}{\partial B} f + \dfrac{\partial V}{\partial \Phi} g < 0$ for $B, \Phi > 0$, and $(B, \Phi) \neq (B_{**}, \Phi_{**})$;

(ii) $\lim V(B, \Phi) = \infty$ as $B \to 0$, or $\Phi \to 0$, or $B \to \infty$, or $\Phi \to \infty$.

The existence of a Lyapunov function guarantees that the equilibrium (B_{**}, Φ_{**}) is globally asymptotically stable in the positive quadrant.

We try to find such a Lyapunov function $V(B, \Phi)$ for the system (7). There is no direct method for finding a Lyapunov function, so, following [3] we refer back to our simpler (non-competing) model. Recall that in equation (5) we used the function

$$W(B, \Phi) = (-\gamma \log B + \delta B(t)) + \left(\alpha \Phi^{1-p}/(p-1) + \beta \Phi\right),$$

which was a constant of motion for all trajectories. By (4) this function may be rewritten as

$$W(B, \Phi) = \delta(-B_* \log B + B) + \beta \left(\Phi_*^p \Phi^{1-p}/(p-1) + \Phi\right).$$

We therefore construct a trial Lyapunov function for system (7) in the same form as above, i.e. we set

$$V(B, \Phi) = \delta(-B_{**} \log B + B) + \beta \left(\Phi_{**}^p \Phi^{1-p}/(p-1) + \Phi\right), \tag{10}$$

which implies immediately that the function V satisfies condition (ii).

It remains to see if it satisfies condition (i). First, we calculate the partial derivatives needed:

$$\frac{\partial V}{\partial B} = -\frac{\delta B_{**}}{B} + \delta, \qquad \frac{\partial V}{\partial \Phi} = -\frac{\beta \Phi_{**}^p}{\Phi^p} + \beta.$$

Now, in our system we have

$$f(B, \Phi) = B\left(\alpha - \beta \Phi^p - \mu B\right), \qquad g(B, \Phi) = \Phi^p\left(-\gamma + \delta B - \nu \Phi^p\right).$$

So for $B, \Phi > 0$ we have

$$\begin{aligned} \frac{\partial V}{\partial B} f(B, \Phi) + \frac{\partial V}{\partial \Phi} g(B, \Phi) &= \delta(-B_{**} + B)(\alpha - \beta \Phi^p - \mu B) \\ &\quad + \beta(\Phi_{**}^p - \Phi^p)\left(-\gamma + \delta B - \nu \Phi^p\right). \end{aligned} \tag{11}$$

On the other hand, from (8) we can write:

$$\alpha = \beta \Phi_{**}^p + \mu B_{**}, \qquad -\gamma = -\delta B_{**} + \nu \Phi_{**}^p. \tag{12}$$

We substitute these expressions for α and $-\gamma$ into (12):

$$\begin{aligned} \frac{\partial V}{\partial B} F(B, \Phi) + \frac{\partial V}{\partial \Phi} G(B, \Phi) &= \delta(-B_{**} + B)\left(\beta(\Phi_{**}^p - \Phi^p) + \mu(B_{**} - B)\right) \\ &\quad + \beta(\Phi_{**}^p - \Phi^p)\left(-\delta(B_{**} - B) + \nu(\Phi_{**}^p - \Phi^p)\right). \end{aligned}$$

This can be rewritten as

$$\frac{\partial V}{\partial B}F(B,\Phi)+\frac{\partial V}{\partial \Phi}G(B,\Phi)=-\delta\mu(B-B_{**})^2-\beta\nu(\Phi^p-\Phi^p_{**})^2,$$

which implies condition (i) immediatelly. We can therefore conclude that we have found the Lyapunov function for this system, and that the equilibrium at (B_{**},Φ_{**}) is globally asymptotically stable in the positive quadrant.

We can generalize equations (2) and (7) further to the form

$$\dot{B}=F(B)\,(\alpha-\beta G(\Phi))\,,\qquad \dot{\Phi}=G(\Phi)\,(-\gamma+\delta F(B))\,. \tag{13}$$

and to the form

$$\dot{B}=F(B)\,(\alpha-\beta G(\Phi)-\mu F(B))\,,\qquad \dot{\Phi}=G(\Phi)\,(-\gamma+\delta F(B)-\nu G(\Phi))\,, \tag{14}$$

where all constants are strictly positive. Here the quantities $F(B)$ and $G(\Phi)$ describe effective sizes of the prey and predator populations. We assume that these functions are strictly monotonic and increase from 0 to ∞. Then there exists a unique positive equilibrium (B_*,Φ_*), to be defined from the equations

$$F\,(B_*)=\gamma/\delta,\qquad G\,(\Phi_*)=\alpha/\beta.$$

For equation (13) there exists a constant of motion given by

$$W(B,\Phi)=\delta(-B_*P(B)+B)+\beta\,(\Phi_*Q(\Phi)+\Phi)$$

with

$$P(B)=\int_{B_*}^{B}1/F(s)\,ds,\qquad Q(\Phi)=\int_{\Phi_*}^{\Phi}1/G(s)\,ds,$$

From now on we assume additionally that

$$F'(0),G'(0)=0,\qquad \text{and}\quad F'(\infty),G'(\infty)=\infty.$$

Then $\lim W(B,\Phi)=-\infty$ as $B\to 0$, or $\Phi\to 0$, or $B\to\infty$, or $\Phi\to\infty$. We conclude again that all trajectories of (13) are periodic, and they orbit anti-clockwise around the equilibrium (B_*,Φ_*). The Volterra principle holds for equations (13) in the form

$$\frac{1}{T}\int_0^T F(B(t))\,dt=F(B_*),\qquad \frac{1}{T}\int_0^T G(\Phi(t))\,dt=G(\Phi_*),$$

where T is period of a solution $(B(t),\Phi(t))$. Equation (14) is asymptotically globally stable in the positive quadrant with the Lyapunov function

$$V(B,\Phi)=\delta(-B_{**}P(B)+B)+\beta\,(\Phi_{**}Q(\Phi)+\Phi)$$

provided that $\alpha\delta>\mu\gamma$. Here the positive equilibrium (B_{**},Φ_{**}), to be defined from the equations

$$F\,(B_{**})=\frac{\beta\gamma+\nu\alpha}{\delta\beta+\nu\mu},\qquad G\,(\Phi_{**})=\frac{\alpha\delta-\mu\gamma}{\delta\beta+\nu\mu}.$$

The proof is as above with a few obvious modifications.

2. Canards and the effects of apparent disappearance in modified L-V equations
Some special solutions of singularly perturbed ordinary differential equations are called *canards or duck–trajectories.* These terms have been introduced by French mathematicians [7, 8, 9]. We first the reader remind in a convenient form of the corresponding definitions.

Let us consider a two-dimensional autonomous system:

$$\dot{x} = f(x, y, \alpha), \qquad \varepsilon\dot{y} = g(x, y, \alpha), \tag{15}$$

where x, y are scalar functions of time, ε is a small positive parameter, and f, g are sufficiently smooth scalar functions. The set of points

$$S_\alpha = \{(x, y) : g(x, y, \alpha) = 0\}$$

of the phase plane is called a *slow curve* of the system (15).

We will make the following assumptions:

1) The curve S_α consists of regular points, i.e. at every point $(x, y) \in S_\alpha$

$$[g_x(x, y, \alpha)]^2 + [g_y(x, y, \alpha)]^2 > 0.$$

holds.

2) Singular points, i.e. points at which $g_y(x, y, \alpha) = 0$, are isolated on S_α.
3) At any singular point the following inequality holds: $g_{yy} \neq 0$.

Definition 1. *A singular point $\mathcal{A}$ of the slow curve S_α is called a* jump point *[13] if*

$$\text{sgn}\,[g_y(\mathcal{A})g_x(\mathcal{A})f(\mathcal{A})] = 1.$$

Parts of S_α which contain only regular points are called regular. *A regular part of S_α, all points of which satisfy the inequality*

$$g_y(x, y, \alpha) < 0 \qquad (g_y(x, y, \alpha) > 0)\,,$$

is called attractive (repulsive).

The attractive and repulsive parts of the slow curve are zeroth order approximations of the corresponding attractive and repulsive integral manifolds. The integral manifolds are located in an ε–neighborhood of the slow curve, except for the jump points, at which existence theorems do not hold.

Definition 2. *Trajectories which at first pass along the attractive integral manifold and then continue for a while along the repulsive integral manifold are called* canards or duck-trajectories.

The existence of canards for the system (15) has been proved originally by techniques from so called *non-standard analysis* [7, 8, 9, 14]. A standard interpretation of the main results can be found in [6, 10, 11, 12]:

The canards and the corresponding values of the parameter α allow asymptotic expansions in powers of the small parameter ε. Near the slow curve the canards are exponentially close, and have the same asymptotic expansion in powers of ε. An analogous assertion is true for corresponding parameter values α. Namely, any two values of the parameter α for which canards exist, have the same asymptotic expansions, and the difference between them is given by $\exp(-1/c\varepsilon)$ where c is some positive number.

As a simple example we mention the equation

$$\dot{x} = 1, \qquad \varepsilon\dot{y} = xy + \alpha. \tag{16}$$

It is clear that for $\alpha = 0$, the trajectory $y = 0$ is a canard. Moreover, for $\alpha = 0$, any trajectory of this system, namely

$$y = y_0 e^{\frac{x^2 - x_0^2}{2\varepsilon}}, \qquad y(x_0) = y_0,$$

is a canard if $x_0 < 0$, and x_0, y_0 are values of the order $O(1)$ as $\varepsilon \to 0$. We can say that the trajectory $y = 0$ is the *longest* canard. Note, that this trajectory can be considered as a slow integral manifold. Note that in this case line $y = 0$ plays the role of a slow curve and of a slow integral manifold, simultaneously. It is illustrated in Fig. 3 that this line is attractive when $0 > x$ (repulsive when $0 < x$).

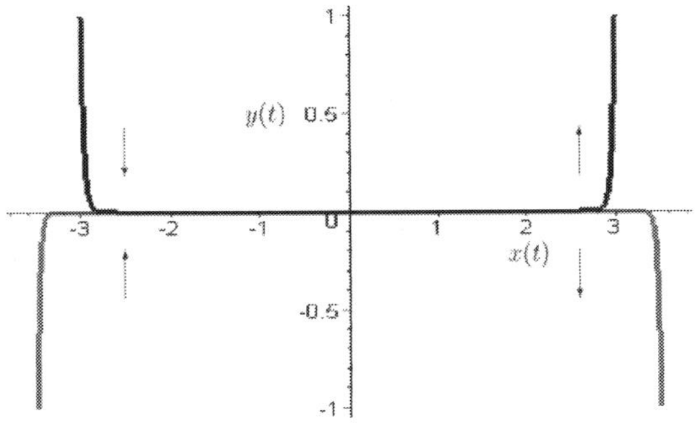

Figure 3. Canard.

To present an example of a periodic canard, we use polar coordinates and the system

$$\dot{\varphi} = 1, \qquad \varepsilon\dot{r} = r(r-1)\sin\varphi.$$

Its solution, satisfying an initial condition $r(\varphi_0) = r_0$, is given by

$$r = 1 + \frac{r_0 - 1}{1 - r_0 + r_0 e^{\frac{\cos\varphi_0 - \cos\varphi}{\varepsilon}}},$$

Therefore, $r = 1$ is a canard.

Now we introduce new types of canards, which are important in the context of this article.

Definition 3. *Trajectories which at first pass along a repulsive integral manifold and then continue for a while along an attractive integral manifold are called* false canards.

To illustrate this definition, consider the following system

$$\dot{x} = 1, \qquad \varepsilon\dot{y} = -xy$$

Any trajectory an with initial condition $y(x_0) = y_0$ can be represented in the form

$$y = y_0 e^{\frac{-(x^2 - x_0^2)}{2\varepsilon}},$$

and the trajectory $y = 0$ is a false canard with the phase diagram represented by Fig. 4. It contains repulsive(attractive) part for $0 > x$ $(0 < x)$.

In the above cases properties of attraction or repulsion can be viewed as exponential attraction and exponential repulsion by analogy with exponential stability. Using an analogy with the asymptotic stability, we introduce the following notion.

Definition 4. *Trajectories which are canards but attraction and repulsion are not exponential are called* feeble canards.

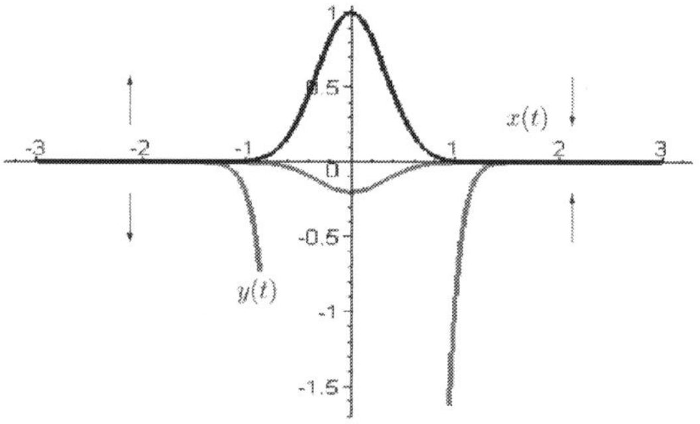

Figure 4. False canard.

As an example, let us consider the following system

$$\dot{x} = 1, \qquad \varepsilon\dot{y} = xy^3$$

which trajectory with an initial condition $y(x_0) = y_0$ is

$$y = \frac{y_0}{[1 + \frac{x^2 - x_0^2}{2\varepsilon} y_0^2]^{1/2}}.$$

In this case the canard $y = 0$ is a feeble canard, which is attractive when $0 > x$ (repulsive when $0 < x$).

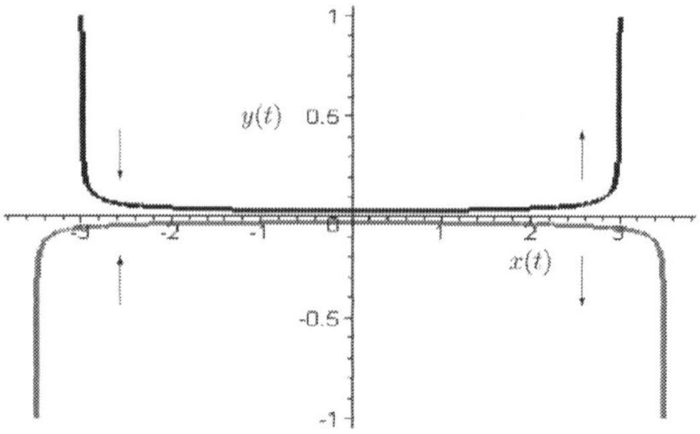

Figure 5. Feeble canard.

Definition 5. *If on one side of the trajectory it looks like a canard, but on the other side it looks like a false canard, we'll say that this trajectory is a* two–faced canard.

For example, the system

$$\dot{x} = 1, \qquad \varepsilon\dot{y} = xy^2$$

whose trajectories are

$$y = \frac{y_0}{1 + \frac{x^2 - x_0^2}{2\varepsilon} y_0},$$

possesses the two–faced feeble canard $y = 0$.

It is easy to give more general examples of feeble canards. The ystem

$$\dot{x} = 1, \qquad \varepsilon\dot{y} = xy^{2k+1}, \qquad k \geq 1,$$

with the trajectories

$$y = \frac{y_0}{\left[1 + \frac{k(x^2 - x_0^2)}{\varepsilon} y_0^{2k}\right]^{1/2k}}, \qquad y(x_0) = y_0,$$

has the feeble canard $y = 0$, and the system

$$\dot{x} = 1, \qquad \varepsilon \dot{y} = xy^{2k}, \qquad k \geq 1,$$

with trajectories

$$y = \frac{y_0}{[1 + \frac{(2k-1)(x^2 - x_0^2)}{2\varepsilon} y_0^{2k-1}]^{1/2k-1}}, \qquad y(x_0) = y_0,$$

possesses the two–faced feeble canard $y = 0$,which is attractive for $0 > x, 0 < y$ and $0 < x, 0 > y$, and repulsive for $0 > x, 0 > y$.

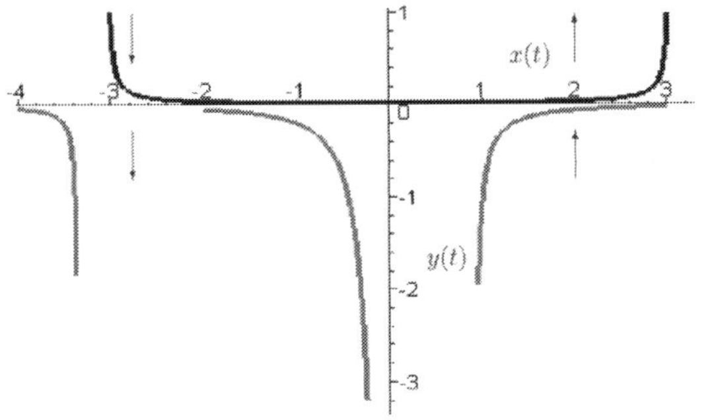

Figure 6. Two-faced feeble canard.

Turning back to the modified Lotka-Volterra we consider two special cases, when one variable (B or Φ) is faster than the other. We can use singularly perturbed differential systems for the modelling such phenomena. The biologically relevant case is "Fast Phages – Slow Bacteria", and the corresponding system is

$$\dot{B} = B(\alpha - \beta\Phi^p), \qquad \varepsilon\dot{\Phi} = \Phi^p(-\gamma + \delta B),$$

with $p > 1$.

Proposition 1. *The trajectory $\Phi = 0$ is a feeble canard with the attractive part $B < \gamma/\delta = B_*$ and the repulsive part $B > \gamma/\delta = B_*$ for positive B and Φ.*

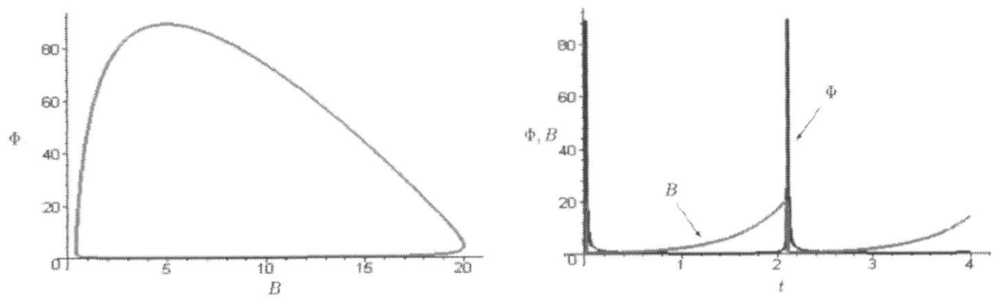

Figure 7. Fast phages and slow microbes.

We can observe here the effect of *apparent disappearance* of phages: they almost disappear for most of the time; however the phages population "explodes" for some short time intervals. A point that should be mentioned is that this feeble canard is two-faced in the cases $p = 2, 4, 6, \ldots$, but that is irrelevant because we are interested in the behavior of our system for positive values of Φ, only.

We also discuss the relationship between minimal $\Phi_{\min}$ and maximal $\Phi_{\max}$ values of the variable Φ in the case of fast phages and slow bacteria. Moreover, by technical reasons, we consider only the case $p = 2$ (the general case $p > 1$ will be discussed in another article). Then

$$\frac{\Phi_{\max}}{\Phi_{\min}} = (1+z)\left(1+z+\sqrt{2z+z^2}\right),$$

where

$$z = \frac{\delta}{\varepsilon\beta}\Psi\left(\frac{\alpha}{\beta}, \frac{\gamma}{\delta}, B_{\max}\right),$$

and

$$\Psi(u,v,w) = \frac{1}{2\sqrt{u}}\left[w - v + v\left(\ln v - \ln w\right)\right].$$

Therefore we arrive the following statement:

Proposition 2. *In the case of fast phages and slow bacteria the relationship*

$$\frac{\Phi_{\min}}{\Phi_{\max}} = O(\varepsilon^2), \quad \text{as} \quad \varepsilon \to 0$$

holds.

We consider, finally, the "Fast Bacteria – Slow Phages" case with the corresponding equations

$$\varepsilon\dot{B} = B(\alpha - \beta\Phi^p), \qquad \dot{\Phi} = \Phi^p(-\gamma + \delta B).$$

We note that this case is probably not interesting from a biological point of view.

Proposition 3. *The trajectory $B = 0$ is a canard with attractive part $\Phi > (\alpha/\beta)^{1/p} = \Phi_*$ and repulsive part $\Phi < (\alpha/\beta)^{1/p} = \Phi_*$.*

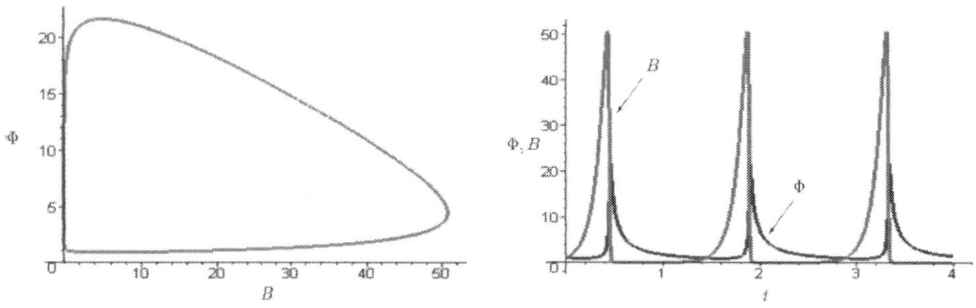

Figure 8. Fast microbes and slow phages.

In this case we observe an effect of *apparent disappearance* of bacteria.

3. Further numerical experiments

We consider modifying our system further to include two phage-like predators and two prey. We assume that both predators will eat either of the two prey. We acknowledge that the usual assumption for phages is that each predator is fairly prey-specific and Hoffmann is clearly talking about multiple independent cycles of predator and prey. Our numerical experiments suggest that this specificity could evolve naturally from a less specific cross-feeding system, as more branched systems are swept away in a long run by prey extinction.

The equations of our system are given by

$$\begin{aligned}\dot{W}(t) &= W(t)(a - bY(t)^2 - cZ(t)^2),\\ \dot{X}(t) &= X(t)(d - eY(t)^2 - fZ(t)^2),\\ \dot{Y}(t) &= Y(t)^2(-g + hW(t) + iX(t)),\\ \dot{Z}(t) &= Z(t)^2(-j + kW(t) + lX(t)),\end{aligned}$$

where we have $W(t)$ and $X(t)$ as prey, and $Y(t)$ and $Z(t)$ as predators, and the parameters a to l are positive constants.

To see the qualitative behavior of the system, we can plot the four populations as functions of time. A typical plot is displayed below.

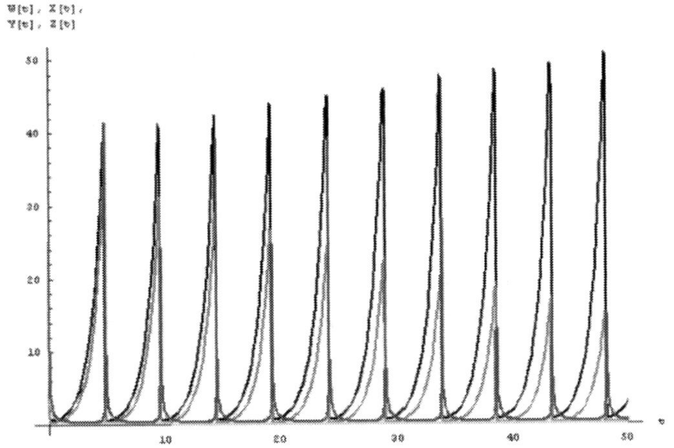

Figure 9. Plot of populations W, X, Y, and Z versus time. The first species $W(t)$ is represented by the black graph, the second species $X(t)$ by the red graph, the third species $Y(t)$ by the green graph, and the fourth species $Z(t)$ by the blue graph.

In this graph we have chosen our parameters a to l to be as follows:

$$\begin{aligned}&a = 1, \quad b = 0.1, \quad c = 0.08, \quad d = 1.1, \quad e = 0.2, \quad f = 0.09,\\ &g = 2, \quad h = 0.05, \quad i = 0.09, \quad j = 1.5, \quad k = 0.11, \quad l = 0.07,\end{aligned}$$

and our initial points and total time τ as follows:

$$W(0) = 10, \quad X(0) = 20, \quad Y(0) = 15, \quad Z(0) = 25, \quad \tau = 100.$$

It is interesting to note that one the populations, namely $Y(t)$, very rapidly dies off to zero. This is a typical behaviour. In fact, if we were to randomly select the parameters a to l and plot the results, we find that in the vast majority of cases at least one of the populations tends to die off very rapidly. This tendency to die off leads us to wonder if it is possible to have an equilibrium with a positive population for all four species. We can test if it is possible by solving the equation

$$\dot{W}(t) = \dot{X}(t) = \dot{Y}(t) = \dot{Z}(t) = 0.$$

Thus a possible positive equilibrium does indeed exist at the point:

$$W_* = \frac{ij - gl}{ik - hl}, \quad X_* = \frac{hj - gk}{-ik + hl}, \quad Y_* = \sqrt{\frac{cd - af}{ce - bf}}, \quad Z_* = \sqrt{\frac{bd - ae}{-ce + bf}}.$$

Hence, if we were to choose our parameters a to l such that the all the coordinates above were positive, we would have a positive equilibrium for all four species.

We now look to include the effects of competition in our system. We modify equations (17) to (17) by adding competition terms. We must take into account two types of competition - competition between two of the same species, and competition between one species and another of the same type, i.e. between the two species of predators, or between the two species of prey. We do this by adding in the terms as shown below:

$$\begin{aligned}
\dot{W}(t) &= W(t)(a - bY(t)^2 - cZ(t)^2 - mX(t) - nW(t)), \\
\dot{X}(t) &= X(t)(d - eY(t)^2 - fZ(t)^2 - oW(t) - pX(t)), \\
\dot{Y}(t) &= Y(t)^2(-g + hW(t) + iX(t) - qZ(t)^2 - rY(t)^2), \\
\dot{Z}(t) &= Z(t)^2(-j + kW(t) + lX(t) - sY(t)^2 - uZ(t)^2).
\end{aligned}$$

We can now look at a numerical example of this system's behavior. We choose the constants a to l to be the same values as given in (3), and the initial values and time to be the same as in (3). We define the remaining constants to be:

$$m = 0.001, \ n = 0.002, \ o = 0.003, p = 0.001, \ q = 0.0022 \ r = 0.0009, \ s = 0.0015, \ u = 0.002.$$

We can now find a numerical solution in the same manner as before. We again use Mathematica's NDSolve command to find interpolating functions, which we can plot as functions of time. This plot is shown below.

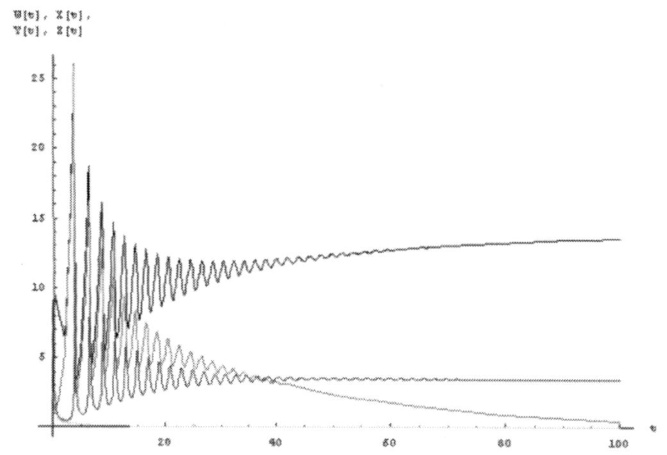

Figure 10. Plot of populations W, X, Y, and Z versus time

We have again kept the same colour convention as before, with $W(t)$ coloured black, $X(t)$ red, $Y(t)$ green, and $Z(t)$ blue. Notice how one species, namely $Y(t)$, dies off very rapidly, while a second, $X(t)$, is also dying out. This appears to mirror the results from the system without competition, that at least one species will typically die out. There is, however, a difference in the system with competition. We can see that the system does seem to be tending towards a stable equilibrium value for the populations, much like the system with only one predator and one prey. We can hypothesize that the tending towards equilibrium is a characteristic of a system with competition. Again, we can use random values of the parameters to generate plots for multiple values of the parameters. We notice that at least one species tends to die off, and that the surviving species tend towards an equilibrium. However, we must see if it is in fact possible to have a system where all four species can coexist together. We hence attempt to find a positive solution of the equations. We can do this analytically easily using Mathematica's Solve function. By examining the results, we do indeed find that there exists a possible equilibrium where all 4 populations are non-zero, although it is a rather cumbersome formula.

Acknowledgments

The authors are grateful to Mark Emmerson, Michael P. Mortell and Robert E. O'Malley for numerous discussions. This research was partially supported by the Boole Centre for Research in Informatics (BCRI), and by the Russian Foundation for Basic Research Grant 06-01-72552.

References

[1] Hoffmann K H, Rodriguez-Brito B, Breitbart M, Bangor D, Angly F, Felts B, Nulton J, Rohwer F and Salamon P 2006 *The Structure of Marine Phage Populations*

[2] Wikipedia contributors 2006 *Wikipedia, The Free Encyclopedia* http://en.wikipedia.org/wiki/Phage

[3] Hofbauer J and Sigmund K 1992 *The Theory of Evolution and Dynamical Systems* (Cambridge: Cambridge University Press)

[4] Dancso A, Farkas H, Farkas M and Szabo G 1990 Investigation on a class of generalyzed two-dimensional Lotka–Volterra schemes *Acta Applicandae Mathematica* **23** 103

[5] Hoffmann K H et al 2005 The structure of marine phage population *Proceedings of ECOS 2005, eds Signe Kjelstrup and Johan Einar Hustad*

[6] Arnold V I, Afraimovich V S, Il'yashenko Yu S and Shil'nikov L P 1994 *Theory of Bifurcations* (Dynamical Systems vol 5 Encyclopedia of Mathematical Sciences) ed V Arnold (New York: Springer-Verlag)

[7] Benoit E, Callot J L, Diener F and Diener M 1981–1982 Chasse au canard *Collect. Math.* **31–32** 37

[8] Benoit E 1983 Systèmes lents-rapides dans R^3 et leurs canards *Société Mathématique de France. Astérisque* **109–110** 159

[9] Diener M 1979 *Nessie et Les Canards* (Strasbourg: Publication IRMA)

[10] Eckhaus W 1983 *Relaxation Oscillations Including a Standard Chase on French Ducks* (Lecture Notes Math. vol 985) ed F Verhulst (Berlin: Springer) p 449

[11] Gorelov G N and Sobolev V A 1992 Duck-trajectories in a thermal explosion problem *Appl. Math. Lett.* **5** 3

[12] Mishchenko E F, Kolesov Yu S, Kolesov A Yu and Rozov N Kh 1995 *Asymptotic Methods in Singularly Perturbed Systems* (New York: Plenum Press)

[13] Mishchenko E F and Rozov N Kh 1980 *Differential Equations with Small Parameters and Relaxation Oscillations* (New York: Plenum Press)

[14] Zvonkin A K and Shubin M A 1984 Non-standard analysis and singular perturbations of ordinary differential equations *Russian Math. Surveys* **39** 69

Institute of Physics Publishing
doi:10.1088/1742-6596/55/1/009
Journal of Physics: Conference Series **55** (2006) 94–104
International Workshop on Multi-Rate Processes and Hysteresis

Maximal combustion temperature estimation

E Golodova and E Shchepakina

Department of Differential Equations and Control Theory, Samara State University, Akademika Pavlova Street, 1, 443011, Russia

E-mail: shchepakina@yahoo.com

Abstract. This work is concerned with the phenomenon of delayed loss of stability and the estimation of the maximal temperature of safe combustion. Using the qualitative theory of singular perturbations and canard techniques we determine the maximal temperature on the trajectories located in the transition region between the slow combustion regime and the explosive one. This approach is used to estimate the maximal temperature of safe combustion in multi–phase combustion models.

1. Introduction and physical background

We consider a model of combustion of a rarefied gas mixture in an inert, porous or dusty medium. We assume that the temperature distribution and phase–to–phase heat exchange are uniform. The chemical conversion kinetics are represented by a one–stage, irreversible reaction. The dimensionless model in the case of an autocatalytic reaction has the form [14]

$$\gamma\frac{d\theta}{d\tau} = \eta(1-\eta)\exp\left(\theta/\left(1+\beta\theta\right)\right) - \alpha(\theta-\theta_c), \tag{1}$$

$$\gamma_c\frac{d\theta_c}{d\tau} = \alpha(\theta-\theta_c), \tag{2}$$

$$\frac{d\eta}{d\tau} = \eta(1-\eta)\exp\left(\theta/\left(1+\beta\theta\right)\right), \tag{3}$$

with the initial conditions

$$\eta(0) = \tilde{\eta}_0/\left(1+\tilde{\eta}_0\right) = \eta_0, \quad \theta(0) = \theta_c(0) = 0.$$

Here, θ and θ_c are the dimensionless temperatures of the reactant and inert phases, respectively; η is the depth of conversion; $\tilde{\eta}_0$ is the criterion for autocatalyticity; the dimensionless parameters β and γ characterize the physical properties of a gas mixture. For real combustible gas mixtures typical values of the parameter γ and β are small [9, 17]. The term $-\alpha(\theta-\theta_c)$ reflects the phase-to-phase heat exchange. The parameter γ_c characterizes the physical features of the inert phase. Depending on the relation between values of the parameters, the chemical reaction either changes to a slow regime with a decay of the reaction, or into a regime of self–acceleration that leads to an explosion. So, if we change the value of one parameter with fixed values of the other parameters we can change the type of chemical reaction. Thus, it is possible to consider this problem as a special control problem.

Note, that the system (1)–(3) possesses a first integral

$$\gamma\theta + \gamma_c\theta_c - \eta = \eta_0,$$

and therefore we obtain on eliminating θ_c

$$\gamma\frac{d\theta}{d\tau} = \eta(1-\eta)\exp\left(\frac{\theta}{1+\beta\theta}\right) - \alpha\left(1+\frac{\gamma}{\gamma_c}\right)\theta + \frac{\alpha}{\gamma_c}(\eta-\eta_0), \tag{4}$$

$$\frac{d\eta}{d\tau} = \eta(1-\eta)\exp\left(\frac{\theta}{1+\beta\theta}\right), \tag{5}$$

with initial conditions

$$\eta(0) = \eta_0, \quad \theta(0) = 0. \tag{6}$$

The chemically relevant phase space Δ of the system is defined by $\Delta := \{(\theta,\eta) \in R^2 : \theta \geq 0, 0 \leq \eta \leq 1\}$. As γ is small (for highly exothermic chemical systems typical values of γ lie in the interval $(0.01, 0.1)$, see, for example, [4]), (4) and (5) represent a singularly perturbed system of autonomous differential equations [15, 16]. A qualitative investigation of this system can be found in [3, 10, 13, 14]. In what follows we recall the main results of these studies.

Consider initial value problem (4)–(6) in Δ for $\alpha \geq 0, \gamma > 0$. There are two equilibria: the point $P = ((1-\eta_0)/(\gamma+\gamma_c), 1)$ is a stable node, the point $Q = (-\eta_0/(\gamma+\gamma_c), 0)$ is a saddle, but only P belongs to Δ.

If we put $\gamma = 0$ in (4) we get the degenerate equation

$$0 = \eta(1-\eta)\exp\left(\frac{\theta}{1+\beta\theta}\right) - \alpha\left(\theta - \frac{\eta-\eta_0}{\gamma_c}\right) = F(\eta,\theta,\alpha), \tag{7}$$

which describes *the slow curve* S_α of (4), (5) (see, for example, [10, 11]).

The subset S_α^s (S_α^u) of S_α with

$$\frac{\partial}{\partial\theta}F(\eta,\theta,\alpha) < 0 \ \ (> 0)$$

is called *the attractive (repulsive) subset of* S_α. A point A on S_α in which $\partial F/\partial\theta = 0$ is called *the breakdown point* [5, 10, 12]. For γ sufficiently small, the system (4), (5) has an attractive slow invariant manifold $S_{\alpha,\gamma}^s$ near S_α^s and a repulsive slow invariant manifold $S_{\alpha,\gamma}^u$ near S_α^u, respectively (see, for example, [12]).

We take α as control parameter with fixed γ_c. The point $\theta = \theta_s$, $\eta = \eta_s$ is the self-intersection point of the slow curve at $\alpha = \alpha_0$. Here, $\alpha = \alpha_0$, $\theta = \theta_s$, $\eta = \eta_s$ satisfy the system

$$F(\eta,\theta,\alpha) = F_\eta(\eta,\theta,\alpha) = F_\theta(\eta,\theta,\alpha) = 0. \tag{8}$$

The dependence of the slow curve S_α on the relation between parameter values gives different forms, see Figure 1.

In the case $\alpha > \alpha_0$ each set S_α^s and S_α^u of S_α consists of a single connected curve, see Figure 1(a). Hence the system has an attractive invariant manifold $S_{\alpha,\gamma}^s$ and a repulsive invariant manifold $S_{\alpha,\gamma}^u$ near S_α^s and S_α^u, respectively.

Since the initial point $(0,\eta_0)$ belongs to the basin of attraction of the set S_α^s, after a short time the trajectory follows the attractive slow invariant manifold $S_{\alpha,\gamma}^s$ and tends to the equilibrium P as τ tends to ∞. This behavior corresponds to *the slow combustion regime* with low temperatures, see Figure 2.

In the case $\alpha < \alpha_0$ each set S_α^s and S_α^u consists of two different components separated by breakdown points A_1 and A_2 with $\theta = \theta_A$ (see Figure 1(b)) and for γ sufficiently small the system has local attractive invariant manifolds $S_{\alpha,\gamma}^{s,i}$ (repulsive invariant manifolds $S_{\alpha,\gamma}^{u,i}$) near

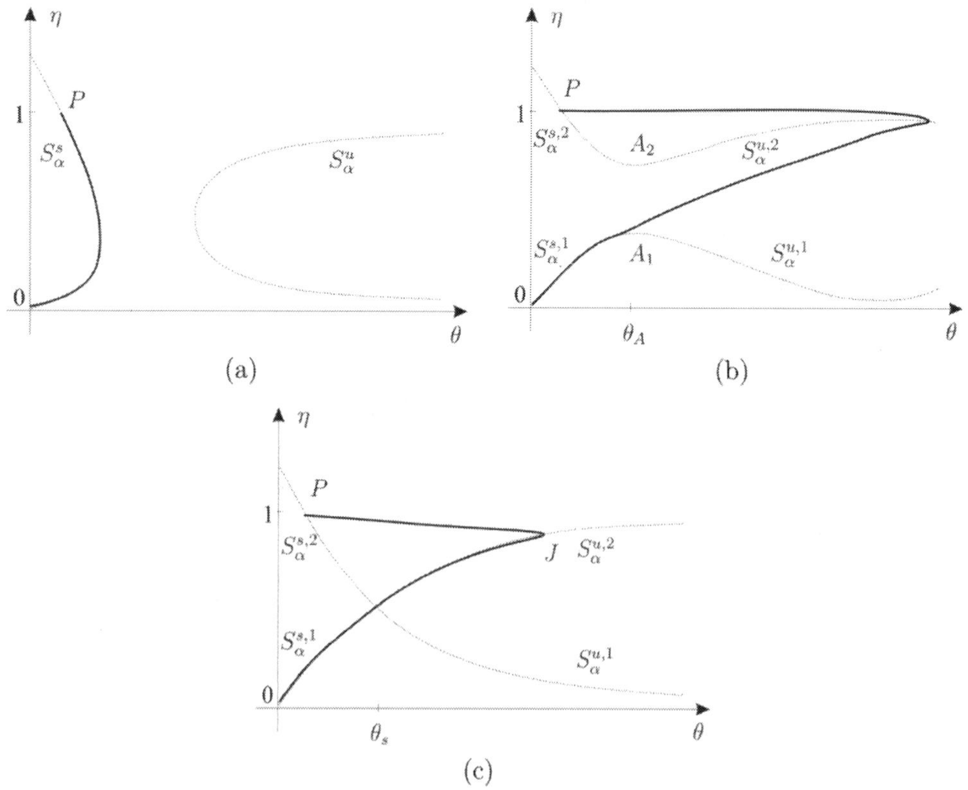

Figure 1. The trajectories (the solid line) of the system (4), (5) and the slow curve (the dashed line) for (a) $\alpha > \alpha_0$, (b) $\alpha < \alpha_0$, (c) $\alpha = \alpha_0$.

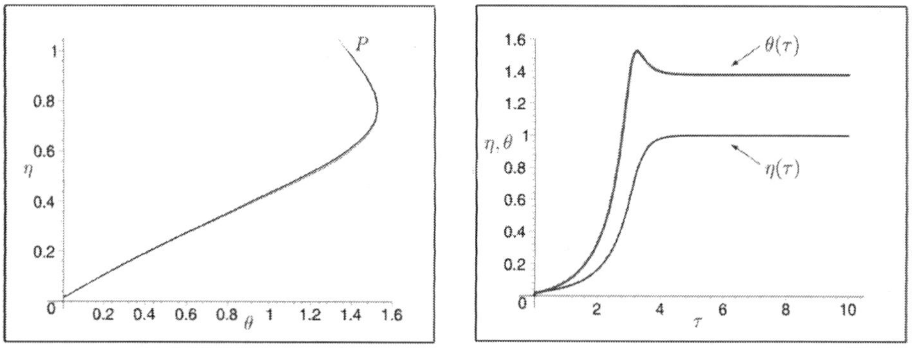

Figure 2. The trajectory of the system (4), (5) and the η– and θ–components of the solution in the case of the slow combustion regime for $\alpha = 1.4$, $\beta = 0.1$, $\gamma = 0.01$, $\gamma_c = 0.7$, $\eta_0 = 0.02$.

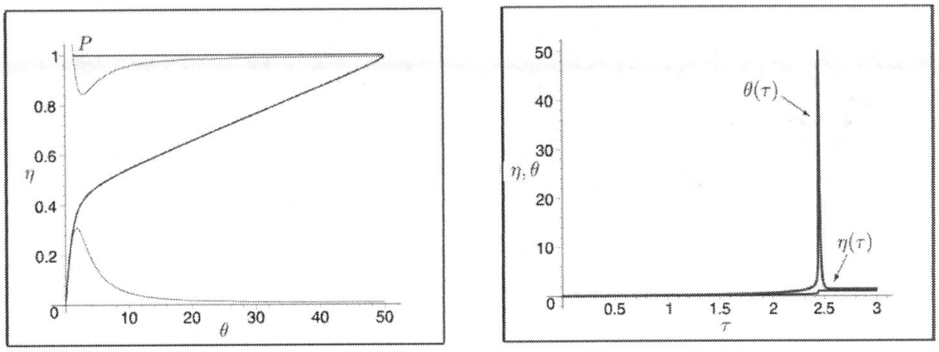

Figure 3. The trajectory of the system (4), (5) and the η– and θ–components of the solution in the case of thermal explosion for $\alpha = 0.7$, $\beta = 0.1$, $\gamma = 0.01$, $\gamma_c = 1.1$, $\eta_0 = 0.02$.

each component of $S_\alpha^{s,i}$ ($S_\alpha^{u,i}$), $i = 1, 2$. In that case, we have the following behavior of the solution of the initial value problem (4)–(6): after a short time, the solution will follow the component of $S_{\alpha,\gamma}^{s,1}$ to breakdown point A_1. After this time, $\theta(\tau)$ will increase rapidly, but the solution remains inside Δ and ultimately tends to the stable equilibrium P. In contrast to the case $\alpha > \alpha_0$ where $\theta(\tau)$ behaves much like $\eta(\tau)$ (see Figure 2), now the temperature rises steeply under insignificant changes of $\eta(\tau)$ (see Figure 3). A drastic growth of the temperature under negligible changes of the concentration (or depth of conversion) of the gas mixture is the characteristic property of a dangerous phenomenon called *thermal explosion* in combustion theory [9].

The transition region from the slow combustion regime to the explosive one exists due to the continuous dependence of our system on the parameters α and γ_c ($\gamma_c > 0$). For $\alpha = \alpha_0$ the slow curve has an intersection point (θ_s, η_s), separating four branches $S_{\alpha_0}^{u,i}$, $S_{\alpha_0}^{s,i}$, $i = 1, 2$, see Figure 1(c). According to (7) and (8), the branch $S_{\alpha_0}^{s,1}$ has the representation

$$S_{\alpha_0}^{s,1} := \{(\theta, \eta) \in \Delta : \eta = \zeta_1^s(\theta),\ 0 \le \theta \le \theta_s\}, \tag{9}$$

where

$$\zeta_1^s(\theta) = \frac{1}{2}\left[1 + \frac{\alpha_0}{\gamma_c}\exp\left(\frac{-\theta}{1+\beta\theta}\right) - \sqrt{\left(1 + \frac{\alpha_0}{\gamma_c}\exp\left(\frac{-\theta}{1+\beta\theta}\right)\right)^2 - 4\alpha_0\exp\left(\frac{-\theta}{1+\beta\theta}\right)\left(\theta + \frac{\eta_0}{\gamma_c}\right)}\right].$$

If we denote by $\psi_1^s(\eta)$ the inverse function of $\zeta_1^s(\theta)$, then $S_{\alpha_0}^{s,1}$ can also be represented in the form

$$S_{\alpha_0}^{s,1} := \{(\theta, \eta) \in \Delta : \theta = \psi_1^s(\eta),\ 0 \le \eta \le \eta_s\}. \tag{10}$$

Similarly, the branch $S_{\alpha_0}^{u,2}$ can be represented as

$$\begin{aligned} S_{\alpha_0}^{u,2} &:= \{(\theta, \eta) \in \Delta : \eta = \zeta_2^u(\theta),\ \theta \ge \theta_s\} = \\ &= \{(\theta, \eta) \in \Delta : \theta = \psi_2^u(\eta),\ \eta_s \le \eta < 1\}, \end{aligned} \tag{11}$$

where

$$\zeta_2^u(\theta) = \frac{1}{2}\left[1 + \frac{\alpha_0}{\gamma_c}\exp\left(\frac{-\theta}{1+\beta\theta}\right) + \sqrt{\left(1 + \frac{\alpha_0}{\gamma_c}\exp\left(\frac{-\theta}{1+\beta\theta}\right)\right)^2 - 4\alpha_0\exp\left(\frac{-\theta}{1+\beta\theta}\right)\left(\theta + \frac{\eta_0}{\gamma_c}\right)}\right],$$

and $\psi_2^u(\eta)$ is the inverse function of $\zeta_2^u(\theta)$.

For $\alpha = \alpha_0$ and γ sufficiently small, the system (4), (5) possesses the local attractive invariant manifolds $S^{s,1}_{\alpha_0,\gamma}$ and $S^{s,2}_{\alpha_0,\gamma}$ near $S^{s,1}_{\alpha_0}$ and $S^{s,2}_{\alpha_0}$, respectively, and the repulsive local invariant manifolds $S^{u,1}_{\alpha_0,\gamma}$ and $S^{u,2}_{\alpha_0,\gamma}$ near $S^{u,1}_{\alpha_0}$ and $S^{u,2}_{\alpha_0}$, respectively.

Using the method of integral manifolds and the canard techniques in [10], it is possible to find the critical value of the parameter α which corresponds to a trajectory modelling *the critical regime* in the chemical system in the form of the asymptotic representation

$$\alpha = \alpha^*(\gamma) = \alpha_0 + \alpha_1\gamma + \alpha_2\gamma^2 + O(\gamma^3),$$

where

$$\alpha_0 = \gamma_c(2\eta_s - 1)\exp\left(\frac{\theta_s}{1+\beta\theta_s}\right).$$

Also (recall that β is a small parameter)

$$\theta_s = \theta_{s0} + \beta\theta_{s1} + O(\beta^2),$$

where

$$\theta_{s0} = \frac{1}{2\gamma_c}\left(1 - 2\eta_0 + \sqrt{1+4\gamma_c^2}\right),$$

$$\theta_{s1} = \frac{\left(1 - 2\eta_0 + \sqrt{1+4\gamma_c^2}\right)^3 - 2\gamma_c(\eta_0^2 - \eta_0)}{2\gamma_c^2\sqrt{1+4\gamma_c^2}};$$

and

$$\eta_s = \eta_{s0} + \beta\eta_{s1} + O(\beta^2),$$

where

$$\eta_{s0} = \gamma_c\theta_{s0} - \gamma_c + \eta_0 = \frac{1}{2}\left(1 + \sqrt{1+4\gamma_c^2}\right) - \gamma_c,$$

$$\eta_{s1} = \gamma_c(\theta_{s1} - 2\theta_{s0}) = \frac{\left(1 - 2\eta_0 + \sqrt{1+4\gamma_c^2}\right)^3 - 2\gamma_c(\eta_0^2 - \eta_0)}{2\gamma_c\sqrt{1+4\gamma_c^2}} - \left(1 - 2\eta_0 + \sqrt{1+4\gamma_c^2}\right).$$

This trajectory after a short time follows the attractive component $S^{s,1}_{\alpha_0}$ until it reaches the value $\theta = \theta_s$. After this moment, it stays near the repulsive component $S^{u,2}_{\alpha_0}$, up to a point J from which the solution "jumps" towards the attractive component $S^{s,2}_{\alpha_0}$ and tends to P as $\tau \to \infty$, see Figure 1(c).

This trajectory contains the attractive/repulsive part of slow motions and thus, is *a canard* [2, 10]. Recall that a canard is a trajectory of a singularly perturbed system of differential equations if it, at first, follows an attractive invariant manifold, and then a repulsive one. In both cases the distances travelled are $O(1)$ as $\gamma \to 0$.

The canard corresponds to a critical chemical reaction separating the domain of self–acceleration reactions ($\alpha < \alpha^*$) and the domain of non–explosive reactions ($\alpha > \alpha^*$). Note, that this regime is not a slow combustion regime since the temperature attains a high value, and is not explosive as the temperature increases at the tempo of the slow variable. Figure 4 shows the results of the numerical investigation of (4)–(6) for the case $\alpha = \alpha^*$.

The sufficient conditions for the existence and the algorithm for the construction of the canard solution of system (4), (5) were given in [14].

Thus, this fact is very important: during the critical regime the temperature attains a high value but without explosion, and this may be the aim of a technological process.

The problem of the location of critical regimes thought of as regimes separating the regions of explosive and nonexplosive chemical reactions is the significant mathematical problem of thermal

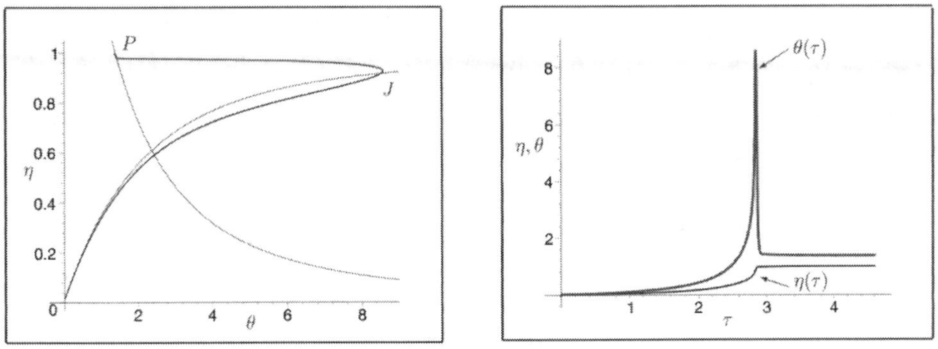

Figure 4. The trajectory of the system (4), (5) and the η– and θ–components of the solution in the case of critical regime for $\alpha = \alpha^* = 0.949$, $\beta = 0.1$, $\gamma = 0.01$, $\gamma_c = 0.7$, $\eta_0 = 0.02$.

explosion theory. The interest in critical phenomena is occasioned by not only for reasons of safety, but in many cases the critical regime is the most effective in technological processes.

So, we are interested in the determination of the value of the maximal temperature during the critical regime as a maximal temperature of safe combustion. In other words, we are interested in the determination of the jump-off point J (the point in which a trajectory modelling the critical regime jumps from the repulsive invariant manifold, see Figure 1(c)), and therefore the maximal temperature on a canard trajectory. To solve this problem we use an estimate of the delay of loss of stability in scalar singularly perturbed equations.

2. Delay of loss of stability in scalar non-autonomous differential equations

We consider the scalar singularly perturbed differential equation

$$\varepsilon \frac{du}{d\eta} = g(u, \eta, \varepsilon) \tag{12}$$

and study the initial value problem

$$u(\eta_0, \varepsilon) = u_0, \quad \eta \in I_\eta := \{\eta \in R : \eta_0 < \eta < \eta_1\} \tag{13}$$

for sufficiently small ε.

If we set $\varepsilon = 0$ in (12) we get the degenerate equation

$$g(u, \eta, 0) = 0.$$

If this equation has a simple isolated root $u = \psi(\eta)$ which is a stable equilibrium of the associated equation

$$\frac{du}{d\tau} = g(u, \eta, 0), \tag{14}$$

and if u_0 belongs to region of attraction of $\psi(\eta_0)$, then the asymptotic behavior of the solution of the initial value problem (12), (13) is uniquely determined by the standard theory of singularly perturbed systems (see, e.g., [16]). In what follows we consider this initial value problem in the case of change of stabilities, that is, we assume that the degenerate equation has two intersecting solutions. Here, we have to distinguish two different situations: immediate loss of stability and delayed loss of stability (see, for example [1, 6]). The latter case is related to the existence of a canard trajectory. We now recall a result concerning the delayed exchange of stabilities in the case of transversal bifurcation.

Let U be an open bounded interval containing the origin, I_{ε_0} is the open interval defined by $I_{\varepsilon_0} := \{\varepsilon \in R : 0 < \varepsilon < \varepsilon_0, 0 < \varepsilon_0 \ll 1\}$.

We consider the initial value problem (12), (13) under the following assumptions:

(A_1). $g : U \times I_\eta \times I_{\varepsilon_0} \to R$ is continuous and twice continuously differentiable with respect to u and ε.

(A_2). $g(0, \eta, \varepsilon) \equiv 0$ for $(\eta, \varepsilon) \in \overline{I}_\eta \times \overline{I}_{\varepsilon_0}$ ($\overline{I}$ means the closure of I).

From (A_1) and (A_2) it follows that a solution of (12) starting at $u = u_0$ remains positive (negative) if u_0 is positive (negative). In the sequel we restrict ourselves to the case $u_0 < 0$. We denote by U^- the set defined by $U^- := \{u \in U : u \leq 0\}$.

(A_3). The solution set of the degenerate equation $g(u, \eta, 0) = 0$ in $\overline{U^- \times I_\eta}$ consists of the two curves $u \equiv 0$ and $u = \psi_-(\eta)$ where ψ_- belongs to the class $C^1([\eta_s, \eta_1], R^-)$ and satisfies $\psi_-(\eta_s) = 0, \psi_-(\eta) < 0$ for $\eta \in (\eta_s, \eta_1]$ (see Figure 5).

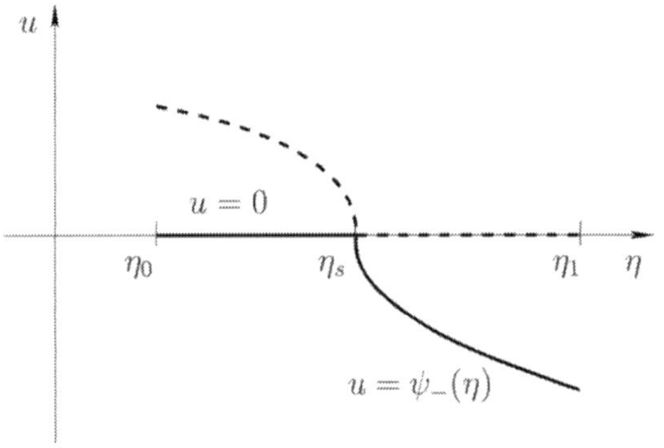

Figure 5. Solution set of $g(u, \eta, 0) = 0$.

(A_4).

$$g_u(0, \eta, 0) \begin{cases} < 0 & \text{for } \eta \in [\eta_0, \eta_s), \\ > 0 & \text{for } \eta \in (\eta_s, \eta_1]. \end{cases}$$

Assumption (A_4) implies that $u = 0$ is an equilibrium point of the associated equation (14) which is exponentially stable for $\eta \in [\eta_0, \eta_s)$ and unstable for $\eta \in (\eta_s, \eta_1]$.

Let

$$G(\eta, \eta_0, \varepsilon) := \int_{\eta_0}^{\eta} g_u(0, \xi, \varepsilon) d\xi.$$

From assumption (A_4) we get that $G(\eta, \eta_0, 0) = 0$ has at most one root $\eta = \eta^*$ in (η_0, η_1). Therefore, we assume

(A_5). $G(\eta, \eta_0, 0) = 0$ has a root η^* in (η_0, η_1).

It is easy to see that η^* is such that

$$\eta^* > \eta_s, \quad G'(\eta^*, \eta_0, 0) > 0$$

holds.

The following assumption on the function g is satisfied if the second derivative of g with respect to u at $u = 0$ is positive for all (η, ε) under consideration.

(A_6). There are sufficiently small positive numbers c_0 and ε_0, such that $[-c_0, c_0] \in U$ and

$$g(u,\eta,\varepsilon) \geq g_u(0,\eta,\varepsilon)u \quad \text{for} \quad \eta_0 \leq \eta \leq \eta^*,\ \varepsilon \in \overline{I}_{\varepsilon_0},\ -c_0 \leq u \leq 0.$$

The following result about the delayed exchange of stabilities has been proved in [7].

Theorem. *Assume the hypotheses* (A_1) - (A_6) *to be valid. Then for sufficiently small* ε *and* $u_0 < 0$ *there exists a unique solution of* (12), (13) *satisfying*

$$\lim_{\varepsilon\to 0} u(\eta,\varepsilon) = \begin{cases} 0 & \text{for} \quad \eta \in [\eta_0, \eta^*), \\ \psi_-(\eta) & \text{for} \quad \eta \in (\eta^*, \eta_1]. \end{cases}$$

After a short time the trajectory of this solution, starting from the initial point, first follows the η-axis until $\eta = \eta^*$. After this the trajectory jumps to the stable part of the curve $u = \psi_-(\eta)$, see Figure 6.

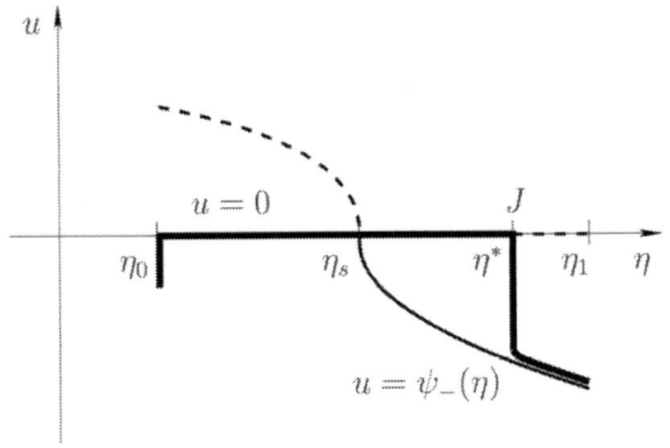

Figure 6. Solution $u = u(\eta,\varepsilon)$ of (12), (13).

3. Maximal temperature of safe combustion

We return to the combustion model.

Our goal is for $\alpha = \alpha^*(\gamma)$ and sufficiently small γ to estimate the maximal temperature θ_γ^{max} on the canard trajectory of system (4), (5), which starts at the given initial point ($\eta = \eta_0, \theta = 0$).

We apply the Theorem above to estimate the maximal temperature on the canard of the equation

$$\gamma\frac{d\theta}{d\eta} = \frac{\eta(1-\eta)\exp\left(\frac{\theta}{1+\beta\theta}\right) - \alpha^*(\gamma)\left(1+\frac{\gamma}{\gamma_c}\right)\theta + \frac{\alpha^*(\gamma)}{\gamma_c}(\eta-\eta_0),}{\eta(1-\eta)\exp\left(\frac{\theta}{1+\beta\theta}\right)}, \tag{15}$$

which follows from (4), (5) with $\alpha = \alpha^*(\gamma)$.

We use the coordinate transformation

$$\theta = \varphi(\eta,\eta_0,\varepsilon) + u, \tag{16}$$

where the function $\varphi(\eta,\eta_0,\varepsilon)$ describes the attractive/repulsive part of slow motions on the canard trajectory (near the attractive component $S_{\alpha_0}^{s,1}$ and the repulsive component $S_{\alpha_0}^{u,2}$, respectively), and

$$\varphi(\eta,\eta_0,\gamma) = \varphi_0(\eta,\eta_0) + \gamma\varphi_1(\eta,\eta_0) + \gamma^2\varphi_2(\eta,\eta_0) + O(\gamma^3),$$

with

$$\varphi_0(\eta,\eta_0) = \begin{cases} \psi_1^s(\eta) & \text{for} \quad 0 < \eta \leq \eta_s, \\ \psi_2^u(\eta) & \text{for} \quad \eta_s \leq \eta < \eta^*. \end{cases} \tag{17}$$

Here, $\eta = \eta^*$ is a coordinate of the point J at which the canard jumps from the slow invariant manifold $S^{u,2}_{\alpha^*,\gamma}$. Recall that this point characterizes the maximal temperature of safe combustion.

The coordinate transformation (16) permits us to rectify the attractive/repulsive part of canard with slow motions. In the new variables the canard trajectory modelling the critical regime converts to the form which is shown on Figure 6.

Our aim is to calculate the value η^* using the Theorem.

From (15) applying the coordinate transformation (16) we get

$$\gamma\frac{du}{d\eta} = \frac{\eta(1-\eta)\exp\left(\frac{\varphi+u}{1+\beta\varphi+\beta u}\right) - \alpha^*(1+\frac{\gamma}{\gamma_c})(\varphi+u) + \frac{\alpha^*}{\gamma_c}(\eta-\eta_0)}{\eta(1-\eta)\exp\left(\frac{\varphi+u}{1+\beta\varphi+\beta u}\right)} - \gamma\frac{d\varphi}{d\eta}.$$

The function $\varphi = \varphi(\eta, \eta_0, \varepsilon)$ is a solution of equation (15). Hence, the last equation can be represented in the form

$$\gamma\frac{du}{d\eta} = \frac{\alpha^*}{\eta(1-\eta)}\left(\left[\left(1+\frac{\gamma}{\gamma_c}\right)\varphi - \frac{\eta-\eta_0}{\gamma_c}\right]\exp\left(\frac{-\varphi}{1+\beta\varphi}\right)\right.$$

$$\left.-\left[\left(1+\frac{\gamma}{\gamma_c}\right)(\varphi+u) - \frac{\eta-\eta_0}{\gamma_c}\right]\exp\left(\frac{-\varphi-u}{1+\beta\varphi+\beta u}\right)\right) \equiv g(u,\eta,\gamma). \tag{18}$$

It is easy to verify that $g(u,\eta,\gamma)$ satisfies all the assumptions of the Theorem for an appropriate choice of η_0 and γ sufficiently small.

According to (18) and the assumption (A_5) of the Theorem, η^* is a root of the equation

$$G(\eta,\eta_0,0) = \int_{\eta_0}^{\eta} g_u(0,\xi,0)d\xi = 0,$$

where

$$g_u(0,\eta,0) = \frac{\alpha_0}{\eta(1-\eta)}\left(\left[\varphi_0 - \frac{\eta-\eta_0}{\gamma_c}\right]\frac{1}{(1+\beta\varphi_0)^2} - 1\right)\exp\left(\frac{-\varphi_0}{1+\beta\varphi_0}\right).$$

From this, (7)–(11), and (17) we get

$$g_u(0,\eta,0) = \frac{1}{(1+\beta\varphi_0)^2} - \frac{1}{\varphi_0 - \frac{\eta-\eta_0}{\gamma_c}}$$

$$\equiv \begin{cases} \frac{1}{(1+\beta\psi_1^s(\eta))^2} - \frac{1}{\psi_1^s(\eta)-\frac{\eta-\eta_0}{\gamma_c}}, & 0<\eta<\eta_s, \\ \frac{1}{(1+\beta\psi_2^u(\eta))^2} - \frac{1}{\psi_2^u(\eta)-\frac{\eta-\eta_0}{\gamma_c}}, & \eta_s<\eta<\eta^*. \end{cases}$$

Hence,

$$G(\eta,\eta_0,0) = \int_{\eta_0}^{\eta_s}\left(\frac{1}{(1+\beta\psi_1^s(\xi))^2} - \frac{1}{\psi_1^s(\xi)-\frac{\xi-\eta_0}{\gamma_c}}\right)d\xi$$

$$+\int_{\eta_s}^{\eta}\left(\frac{1}{(1+\beta\psi_2^u(\xi))^2} - \frac{1}{\psi_2^u(\xi)-\frac{\xi-\eta_0}{\gamma_c}}\right)d\xi.$$

We have by (9)–(11)

$$G(\eta,\eta_0,0) = \int_{\theta_0}^{\theta_s}\left(\frac{1}{(1+\beta\sigma)^2} - \frac{1}{\sigma-\frac{\zeta_1^s(\sigma)-\eta_0}{\gamma_c}}\right)(\zeta_1^s)'(\sigma)d\sigma$$

$$+\int_{\theta_s}^{\theta}\left(\frac{1}{(1+\beta\sigma)^2}-\frac{1}{\sigma-\frac{\zeta_2^u(\sigma)-\eta_0}{\gamma_c}}\right)(\zeta_2^u)'(\sigma)d\sigma \equiv \tilde{G}(\theta,\theta_0,0). \tag{19}$$

Here, θ_0 is the projection of the initial point of the canard on the zeroth order approximations ($\gamma = 0$) of the slow invariant manifold, i.e. on $S^{s,1}_{\alpha_0}$, and according to (9), (10) we have $\eta_0 = \zeta_1^s(\theta_0)$. The equation

$$\tilde{G}(\theta,\theta_0,0) = 0$$

gives a value $\theta = \theta^*$ which can be calculated numerically. As follows from (19), the value θ^* depends on θ_0. The fact of this dependence is shown in Figure 7. It should be noted that if we take the initial point close to the origin, we can obtain a high value of the maximal temperature of safe combustion. The closer to the origin, the higher value of the maximal temperature.

Thus, the maximal temperature of safe combustion in the case of an autocatalytic reaction is the maximal temperature θ^{max}_γ on the canard trajectory starting at the point $(\eta = \eta_0, \theta = 0)$ with $0 < \eta_0 < \eta_s$ and

$$\lim_{\gamma\to 0}\theta^{max}_\gamma = \theta^*(\theta_0).$$

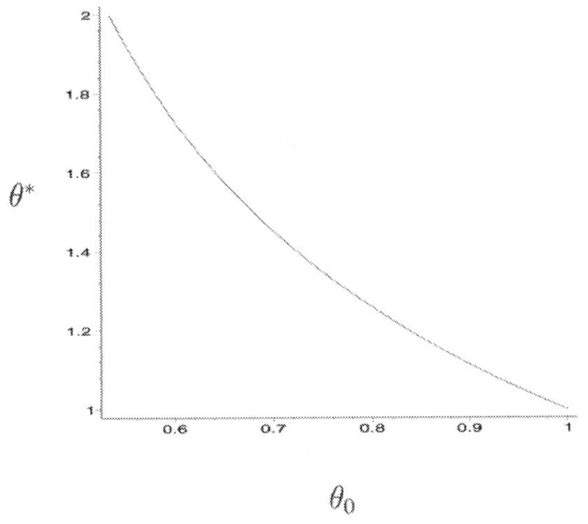

Figure 7. Dependence of maximal temperature on θ_0.

4. Conclusion

The model of combustion of a rarefied gas mixture in an inert medium in the case of an autocatalytic reaction is considered. The connection between the phenomenon of delayed loss of stability in a singularly perturbed differential equation and the estimation of the maximal temperature of safe combustion is stated. It is shown that the maximal temperature of safe combustion is attained during the critical regime separating the domain of explosive chemical reactions and the domain of non–explosive chemical reactions with low temperatures. The critical regime is modelled by the mathematical object called a canard in the modern mathematical literature. Such trajectories pass from the attractive slow invariant manifold to the repulsive one.

The maximal temperature of safe combustion corresponds to the point at which the canard jumps from the repulsive slow invariant manifold. The dependence of maximal temperature on the initial data is obtained.

It should be noted that such an approach was used in [8] to determine the maximal combustion temperature for classic model of autocatalytic combustion.

Acknowledgments

The authors are grateful to K. R. Schneider, V. A. Sobolev, and M. P. Mortell for helpful discussions. This work was supported in part by the Boole Centre for Research in Informatics, University College Cork, and by the Russian Foundation for Basic Research under grant no. 04–01–96515.

References

[1] Arnold V I, Afraimovich V S, Il'yashenko Yu S and Shil'nikov L P 1994 *Theory of Bifurcations* (Dynamical Systems vol 5 Encyclopedia of Mathematical Sciences) ed V Arnold (New York: Springer-Verlag)

[2] Benoit E, Callot J L, Diener F and Diener M 1981–1982 Chasse au canard *Collect. Math.* **31–32** 37

[3] Gorelov G N, Sobolev V A and Shchepakina E A 2006 Canards and critical behaviour in autocatalytic combustion models *J. of Engineering Math.* (in press)

[4] Gray B F 1973 Critical behaviour in chemical reacting systems: 2. An exactly soluble model *Combust. Flame* **21** 317

[5] Mishchenko E F and Rozov N Kh 1980 *Differential Equations with Small Parameters and Relaxation Oscillations* (New York: Plenum Press)

[6] Mishchenko E F, Kolesov Yu S, Kolesov A Yu and Rozov N Kh 1995 *Asymptotic Methods in Singularly Perturbed Systems* (New York: Plenum Press)

[7] Nefedov N N and Schneider K R 2003 On immediate-delayed exchange of stabilities and periodic forced canards *Preprint of Weierstraß–Institut für Angewandte Analysis und Stochastik, Berlin* 872

[8] Schneider K R and Shchepakina E A 2003 Maximal temperature of safe combustion in case of an autocatalytic reaction *Preprint of Weierstraß–Institut für Angewandte Analysis und Stochastik, Berlin* 890

[9] Semenov N N 1928 Zur theorie des verbrennungsprozesses *Z. Physik. Chem.* **48** 571

[10] Shchepakina E and Sobolev V 2005 *Black Swans and Canards in Laser and Combustion Models* (Singular Perturbation and Hysteresis) ed M P Mortell *et al* (Philadelphia: SIAM) pp 207–255

[11] Shchepakina E and Sobolev V 2001 Integral manifolds, canards and black swans *Nonlinear Analysis* A **44** 897

[12] Sobolev V 2005 *Geometry of Singular Perturbations: Critical Cases* (Singular Perturbation and Hysteresis) ed M P Mortell *et al* (Philadelphia: SIAM) pp 153–206

[13] Sobolev V A and Shchepakina E A 1996 Duck trajectories in a problem of combustion theory *Differential Equations* **32** 1177

[14] Sobolev V A and Shchepakina E A 1993 Self–ignition of laden medium *J. Combustion, Explosion and Shock Waves* **29** 378

[15] O'Malley R E 1991 *Singular Perturbation Methods for Ordinary Differential Equations* (Appl. Math. Sci. vol 89) (New-York: Springer–Verlag)

[16] Vasilieva A B, Butuzov V F and Kalachev L V 1995 *The Boundary Function Method for Singular Perturbation Problems* (Series in Applied Mathematics vol 14) (Philadelphia: SIAM)

[17] Zeldovich Ya B, Barenblatt G I, Librovich V B and Makhviladze G M 1985 *The Mathematical Theory of Combustion and Explosions* (New York: Consultants Bureau)

Institute of Physics Publishing
doi:10.1088/1742-6596/55/1/010

Journal of Physics: Conference Series **55** (2006) 105–113
International Workshop on Multi-Rate Processes and Hysteresis

Rate-independent linear operators on various function spaces

F Holland
Mathematics Department, University College, Cork, Ireland
E-mail: f.holland@ucc.ie

Abstract. It is shown that the only rate-independent continuous linear operators acting on the Banach space L^1 of functions defined on the real line, are scalar multiples of the identity operator; and that the only such operators, acting on the corresponding Hilbert space L^2, are linear combinations of the Hilbert transform and the identity.
Keywords: rate-independent operators, hysteresis, function spaces, affine invariance, Hilbert transform

1. Introduction

An input-output relationship, K, which pairs functions defined on the real line, $\mathbb{R}$, is said to be *rate-independent* if, whenever $g = Kf$, the function $t \to g(at+b)$ is the output corresponding to the input $t \to f(at+b)$ for every positive number a and every real number b. The importance of this concept in systems with hysteresis was recognised a long time ago by M. A. Krasnosel'skii and his co-workers. An account of its role in their development of a mathematical theory of hysteresis operators between suitable function spaces can be found in [4].

It's an easy matter to provide examples of *non-linear* rate-independent operators. For instance, if p is any polynomial with real or complex coefficients, then the composition map $f \to p \circ f$ takes the space of all functions f defined on $\mathbb{R}$ to itself, and is rate-independent to boot. However, unless p is a multiple of the identity, this map is plainly non-linear. On the other hand, interesting examples of linear operators which are not of this kind, and possess the property, are hard to come by.

Several years ago, I asked myself the question: What reasonably behaved non-zero linear integral-operators are rate-independent? I discovered to my surprise that, essentially, only multiples of the Hilbert transform fit the bill, and outlined a proof of this to Alexei Pokrovskii, who ever since has attached the footnote[1] to several of his publications [5, 6]! This statement can also be found on the website http://euclid.ucc.ie/hysteresis (which is worth visiting, as it contains a good introduction to the theory of hysteresis).

The purpose of this article is to provide a proper context within which to substantiate this footnote. Now, along with elephants, integral operators are easily recognised on sight, but they are not so easy to describe! For this reason, we have chosen instead to consider bounded linear transformations that act on appropriate normed vector spaces of functions, and to single

[1] We note, in passing, a result of F. Holland(a personal communication): the only linear integral operator satisfying the rate independent property is the Hilbert transform.

out those that are rate-independent. To emphasise that we are excluding rate-independent *non-linear* transformations from our treatment, we will refer to the former as *affine-invariant operators*, a terminology that will be introduced and developed in the next sections.

2. Groups of affine functions

With each point $z = x + iy$ in the open right half-plane

$$\mathbb{C}_+ = \{z \in \mathbb{C} : x = Re z > 0\},$$

of the complex plane $\mathbb{C}$, we associate the function $z : \mathbb{R} \to \mathbb{R}$ defined by

$$z(t) = xt + y, \ -\infty < t < \infty.$$

(It's for notational convenience that we use the same letter for the function z and the point z. Hopefully, this won't cause any confusion.) This defines a homeomorphism of $\mathbb{R}$ onto $\mathbb{R}$. The collection $\{z : z \in \mathbb{C}_+\}$ of such mappings is a subgroup of the group, under the operation of composition, of all homeomorphisms sending $\mathbb{R}$ to itself. We call the members of this subgroup *affine homeomorphisms.*

Each affine homeomorphism, z, induces a linear mapping τ_z on the vector space of functions defined on the real line as follows: if $f : \mathbb{R} \to \mathbb{C}$, $\tau_z f$ is defined to be the composition $f \circ z$, so that, for $z \in \mathbb{C}_+$,

$$(\tau_z f)(t) = f(z(t)) = f(xt + y), \ -\infty < t < \infty.$$

Clearly, τ_z is linear:

$$\tau_z(g + \lambda g) = \tau_z f + \lambda \tau_z g,$$

for all functions f, g, and all scalars λ. We call τ_z an *affine transformation.* Plainly, the family of $\{\tau_z : z \in \mathbb{C}_+\}$ is a non-commutative group under composition of operators.

Note that, if we restrict z to the line $x = 1$, the resulting mappings generate the familiar commutative group of translation transformations indexed on $\mathbb{R}$:

$$(\tau_{(1,y)} f)(t) = f(t + y), \ -\infty < t < \infty, \ -\infty < y < \infty.$$

3. Affine-invariant vector spaces

Definition 1 *We call a vector space E of complex functions defined on $\mathbb{R}$* affine-invariant *if, for all $z \in \mathbb{C}_+$, $\tau_z f \in E$ whenever $f \in E$.*

Of course, the vector space of complex functions defined on the real line is affine-invariant. But certain metrizable subspaces of this are also affine-invariant. In particular, if $0 < p \leq \infty$, the L^p-spaces are. (Recall that, if $0 < p < \infty$, L^p consists of Lebesgue measurable functions f for which

$$||f||_p = \left(\int_{-\infty}^{\infty} |f(t)|^p \, dt \right)^{1/p} < \infty;$$

while L^∞ is the space of bounded Lebesgue measurable functions f with

$$||f||_\infty = \sup\{|f(t)| : -\infty < t < \infty\}.)$$

Indeed, it's easily verified that, if $0 < p \leq \infty$, then

$$||\tau_z f||_p = x^{-1/p} ||f||_p, \ \forall f \in L^p, \ \forall z \in \mathbb{C}_+.$$

In case $1 \leq p \leq \infty$, L^p is a Banach space with norm $f \longrightarrow ||f||_p$. In what ensues, only the Banach spaces L^1, L^2 and L^∞ will be of major interest to us, and we observe especially that L^2 is a Hilbert space endowed with its usual inner-product

$$< f, g >= \int_{-\infty}^{\infty} f(t)\overline{g(t)}\, dt,$$

so that

$$||f||_2 = \sqrt{< f, f >},\ \forall f \in L^2.$$

Remark. The notion of affine invariance can be extended to any collection of functions defined on $\mathbb{R}$, as follows: A collection X of functions defined on $\mathbb{R}$ is said to be affine-invariant if, for all $z \in \mathbb{C}_+$, $\tau_z f \in X$ whenever $f \in X$.

Definition 2 *Suppose E is an affine-invariant function space on $\mathbb{R}$. We say that a linear transformation A on E to E is an* affine-invariant transformation *if it commutes with all members of the family $\{\tau_z : z \in \mathbb{C}_+\}$, i.e., if*

$$A\tau_z = \tau_z A,\ \forall z \in \mathbb{C}_+.$$

More generally, if X, Y are two affine-invariant collections of functions on $\mathbb{R}$, and $K : X \longrightarrow Y$, we say that K is affine-invariant if it belongs to the commutant of the family $\{\tau_z : z \in \mathbb{C}_+\}$, i.e., commutes with all members of it. Thus, such mappings—in general non-linear—are what we previously described as rate-independent; but from now on we'll adhere to the terminology just introduced.

The identity transformation on any vector space of functions defined on $\mathbb{R}$ is affine-invariant. Note too that an affine-invariant transformation A commutes with translations:

$$A\tau_{(1,y)} = \tau_{(1,y)} A,\ \forall y \in \mathbb{R}.$$

The aim of this note is to characterise those affine-invariant transformations on the spaces L^1 and L^2, introduced above, that are, in addition, *continuous*; we refer to these as *affine-invariant operators*.

4. Affine-invariant operators on L^1.

Theorem 1 *Suppose A is an affine-invariant operator on L^1. Then A is a scalar multiple of the identity.*

Proof. To begin with, since A commutes with all translations, and belongs to $B(L^1)$, the normed vector space of continuous linear transformations on L^1 to itself, with the operator norm, there is a finite measure μ on $\mathbb{R}$ such that

$$Af = f * \mu,\ \forall f \in L^1,$$

where $*$ denotes convolution (see [7], p. 75). We claim that μ is concentrated at the origin. To see this, we apply the Fourier transform to the last identity, noting that, for $f \in L^1$, and all $z \in \mathbb{C}_+$, the Fourier transform of $\tau_z f$, evaluated at any real number ω, is given by

$$\widehat{(\tau_z f)}(\omega) = \int_{-\infty}^{\infty} f(z(t))e^{-i\omega t}\, dt = \int_{-\infty}^{\infty} f(xt + y)e^{-i\omega t}\, dt = \frac{1}{x} e^{\frac{iy\omega}{x}} \hat{f}(\frac{\omega}{x}).$$

Bearing in mind that the Fourier transform of a convolution of an L^1-function and a complex measure is the product of their Fourier transforms, it follows that $\widehat{f * \mu} = \hat{f}\,\hat{\mu}$, for all $f \in L^1$. Hence, for all $x > 0$, and all $f \in L^1$,

$$\begin{aligned}
\frac{1}{x}\hat{f}(\frac{\omega}{x})\,\hat{\mu}(\omega) &= (\widehat{\tau_{(x,0)}f})(\omega)\hat{\mu}(\omega) \\
&= ((\widehat{\tau_{(x,0)}f) * \mu})(\omega) \\
&= \widehat{A(\tau_{(x,0)}f)}(\omega) \\
&= \widehat{\tau_{(x,0)}(Af)}(\omega) \\
&= \widehat{\tau_{(x,0)}(f * \mu)}(\omega) \\
&= \frac{1}{x}\widehat{f * \mu}(\frac{\omega}{x}) \\
&= \frac{1}{x}\hat{f}(\frac{\omega}{x})\,\hat{\mu}(\frac{\omega}{x}).
\end{aligned}$$

Hence, for all $x > 0$ and all real ω,

$$\hat{\mu}(\omega) = \hat{\mu}(\frac{\omega}{x}).$$

Fixing $\omega \neq 0$ and letting $x \to \infty$, we deduce that

$$\hat{\mu}(\omega) = \lim_{x\to\infty}\hat{\mu}(\frac{\omega}{x}) = \hat{\mu}(0),$$

by continuity of $\hat{\mu}$. In other words, $\hat{\mu}$ is a constant λ, say, i.e.,

$$\lambda = \int_{\mathbb{R}} e^{-i\omega t}\,d\mu,\ \forall\omega \in \mathbb{R}.$$

Thus, μ is concentrated at the origin, i.e., $\mu(\mathbb{R}) = \mu(\{0\}) = \lambda$, and hence $Af = \lambda f$, $\forall f \in L^1$. This ends the proof.

5. The Hilbert transform

If $1 \leq p < \infty$, and $f \in L^p$, the limit

$$\lim_{\delta\to 0^+}\frac{1}{\pi}\int_{|s-t|\geq\delta}\frac{f(t)}{s-t}\,dt$$

exists and is finite for almost all real s [8, 3]. The limit function is denoted by $\mathbb{H}f$, and called the Hilbert transform of f. Clearly, the correspondence between f and $\mathbb{H}f$ is linear.

Moreover, arguing formally, we see that, if $x + iy = z \in \mathbb{C}_+$, then

$$\begin{aligned}
(\mathbb{H}\tau_z f)(s) &= \frac{1}{\pi}\int_{-\infty}^{\infty}\frac{f(xt+y)}{s-t}\,dt \\
&= \frac{1}{\pi}\int_{-\infty}^{\infty}\frac{f(xt+y)}{xs+y-(xt+y)}\,d(xt+y) \\
&= \frac{1}{\pi}\int_{-\infty}^{\infty}\frac{f(t)}{xs+y-t}\,dt \\
&= \mathbb{H}f(xs+y) \\
&= (\tau_z\mathbb{H}f)(s).
\end{aligned}$$

In other words, $\mathbb{H}$ is an affine-invariant linear transformation.

(Incidentally, the same heuristic argument may lead one to believe that the "convolution-type integral operator" that sends f to g, where

$$g(y) = \int_{-\infty}^{\infty} \frac{f(x)}{|y-x|}\, dx, \quad -\infty < y < \infty,$$

is also of this kind. However, unlike the Hilbert transform, the latter may be infinite on a set of positive measure for functions f that belong to every L^p-space. For instance, if $0 \le y \le 1$,

$$\int_0^1 \frac{1}{|y-x|}\, dx = \infty.$$

Hence, it is not an operator on such spaces in the accepted sense.)

Note, too, that $\mathbb{H}$ is not a multiple of the identity—for example, the function $s \to -s(1+s^2)^{-1}$ is the Hilbert transform of $t \to (1+t^2)^{-1}$. It therefore follows from the previous section that $\mathbb{H}$ is not continuous on L^1. (Of course, the example just given shows this directly.) We record this fact as

Corollary 1 *The restriction of $\mathbb{H}$ to L^1 is unbounded.*

However, according to a celebrated theorem of Marcel Riesz, $\mathbb{H}$ is a bounded transformation on L^p if $1 < p < \infty$ [8, 3]. Hence, if $1 < p < \infty$, the Hilbert transform is an example of a nontrivial affine-invariant operator on L^p. In particular, it is an example of an interesting affine-invariant operator acting on L^2.

We note, for future reference, a few additional facts about the restriction of $\mathbb{H}$ to L^2. In particular, if $f \in L^2$, then, for almost all $\omega \in \mathbb{R}$,

$$\widehat{\mathbb{H}f}(\omega) = isgn(\omega)\hat{f}(\omega),$$

whence

$$\widehat{\mathbb{H}^2 f}(\omega) = isgn(\omega)\widehat{\mathbb{H}f}(\omega) = (isgn(\omega))^2\hat{f}(\omega) = -\hat{f}(\omega).$$

In other words, applying the inverse Fourier transform, we see that $\mathbb{H}^2 = -I$. Thus, $\mathbb{H}$ is a square-root of the identity operator I.

6. Affine-invariant operators on L^2.

Theorem 2 *Suppose A is an affine-invariant operator on L^2. Then A is a linear combination of the identity I, and the Hilbert transform $\mathbb{H}$.*

Proof. We set the problem in the frequency domain, and, to do so, introduce the following family of mappings $\{\zeta_z : z \in \mathbb{C}_+\}$ on L^2: if $g \in L^2$ and $z = x + iy \in \mathbb{C}_+$, then

$$(\zeta_z g)(\omega) = \frac{1}{x} e^{\frac{iy\omega}{x}} g(\frac{\omega}{x}) \equiv \frac{1}{x} e_y(\frac{\omega}{x}) g(\frac{\omega}{x}),$$

where, for $y \in \mathbb{R}$, the exponential function e_y is defined by

$$e_y(t) = e^{iyt}, \quad -\infty < t < \infty.$$

Define T on L^2 as follows: if $f \in L^2$, set

$$T\hat{f} = \widehat{Af}.$$

Since the correspondence between a function and its Fourier transform is linear, and A is linear, then T is, and, furthermore,

$$||T\hat{f}||_2 = ||Af||_2 \le ||A||\,||f||_2 = ||A||\,||\hat{f}||_2,$$

since the Fourier transform is a unitary operator on L^2, and A is bounded with norm $||A||$. It follows that T is bounded on L^2, and its norm doesn't exceed that of A. Note next that

$$T(\zeta_z \hat{f}) = T\widehat{\tau_z f} = \widehat{A\tau_z f} = \widehat{\tau_z A f} = \zeta_z \widehat{Af} = \zeta_z T\hat{f}.$$

In other words, T commutes with the family of linear operators $\{\zeta_z : z \in \mathbb{C}_+\}$.

In particular, $T\zeta_{(1,y)}g = \zeta_{(1,y)}Tg$, $\forall g \in L^2$, and all real y, i.e.,

$$T(e_y g) = e_y Tg, \ \forall g \in L^2, \ \forall y \in \mathbb{R}.$$

Hence, for every trigonometric polynomial p of the form

$$p(\omega) = \sum c_k e_{y_k}(\omega) = \sum c_k e^{iy_k\omega},$$

we have $T(pg) = pTg$. Choose any non-zero function $g_0 \in L^2$. Then $\{e_y g_0 : y \in \mathbb{R}\}$ is dense in L^2. (Because, if, for some $f \in L^2$, $0 =< e_y g_0, f >$, $\forall y \in \mathbb{R}$, the Fourier transform of $g_0\bar{f}$ is the zero function, whence $f = 0$, since $g_0 \neq 0$ by hypothesis.) In other words, the subspace $\{pg_0 : p \text{ a trigonometric polynomial}\}$ is dense in L^2. Hence, given any $g \in L^2$, there is a sequence of trigonometric polynomials p_n such that $||g - p_n g_0||_2 \to 0$ as $n \to \infty$. From the continuity of T, and the fact that it commutes with polynomials, we deduce that

$$Tg = \lim T(p_n g_0) = \lim p_n T g_0,$$

in the L^2-norm, whence

$$\begin{aligned}
\int_{-\infty}^{\infty} Tg(\omega)\, g_0(\omega)\, d\omega &= \ < Tg, \bar{g_0} > \\
&= \lim < p_n T g_0, \bar{g_0} > \\
&= \lim < p_n g_0, \overline{Tg_0} > \\
&= \ < g, \overline{Tg_0} > \\
&= \int_{-\infty}^{\infty} g(\omega) T g_0(\omega)\, d\omega.
\end{aligned}$$

More generally, in a similar manner, we see that the Fourier transforms of the L^1-functions $g_0 Tg$ and gTg_0 agree on $\mathbb{R}$. Hence, they are equal a.e. In particular, if $h_0 \in L^2$, $h_0 \neq 0$ a.e.,

$$h_0 T g_0 = g_0 T h_0.$$

Accordingly, setting

$$\phi = \frac{Tg_0}{g_0},$$

we see that ϕ is measurable, independent of g_0 and $Tg = \phi g$, $\forall g \in L^2$; and $\phi g \in L^2$, whenever $g \in L^2$. Moreover,

$$\int_{-\infty}^{\infty} |\phi(\omega) g(\omega)|^2\, d\omega = ||\phi g||_2^2 = ||Tg||_2^2 \le ||T||\,||g||_2^2, \ \forall g \in L^2.$$

Iterating this we infer that

$$\int_{-\infty}^{\infty} |\phi(\omega)|^{2n} h(\omega)\, d\omega \leq ||T||^{2n} ||h||_1, \ n = 1, 2, \ldots,$$

for all non-negative $h \in L^1$.

Hence, if $||h||_1 = 1$, and $h \geq 0$, then

$$\sqrt[n]{\int_{-\infty}^{\infty} |\phi(\omega)|^{2n} h(\omega)\, d\omega} \leq ||T||^2, \ n = 1, 2, \ldots$$

It follows that $\phi \in L^\infty$ and $||\phi||_\infty \leq ||T||$. In other words, T is multiplication by ϕ, and $||T|| = ||\phi||_\infty$.

To finish off, the further condition that $T\zeta_{(x,0)} = \zeta_{(x,0)}T$, $\forall x > 0$, implies that, for any pair of functions $f, g \in L^2$,

$$\begin{aligned}
\int_{-\infty}^{\infty} \phi(\omega) g(\omega) \bar{f}(\omega)\, d\omega &= <\phi g, f> \\
&= <Tg, f> \\
&= x <\zeta_{(x,0)} Tg, \zeta_{(x,0)} f> \\
&= x <T\zeta_{(x,0)} g, \zeta_{(x,0)} f> \\
&= x <\phi\zeta_{(x,0)} g, \zeta_{(x,0)} f> \\
&= \int_{-\infty}^{\infty} \phi(x\omega) g(\omega) \bar{f}(\omega)\, d\omega,
\end{aligned}$$

for all $x > 0$. Hence, for all $h \in L^1$,

$$\int_{-\infty}^{\infty} \phi(\omega) h(\omega)\, d\omega = \int_{-\infty}^{\infty} \phi(x\omega) h(\omega)\, d\omega,$$

for all $x > 0$. Choosing h to be the characteristic (or indicator) function of the interval $(0, 1)$, we deduce that

$$\int_0^x \phi(\omega)\, d\omega = x \int_0^1 \phi(\omega)\, d\omega,$$

whence, by Lebesgue's theorem on differentiation [9],

$$\phi(x) = \int_0^1 \phi(\omega)\, d\omega = a,$$

say, for almost all $x > 0$. Similarly,

$$\phi(x) = \int_{-1}^0 \phi(\omega)\, d\omega = b,$$

say, for almost all $x < 0$. Letting

$$\lambda = \frac{a+b}{2}, \ \mu = \frac{a-b}{2i},$$

we can say that

$$\phi(\omega) = \lambda + i\mu sgn(\omega), \ \text{a.e. on } (-\infty, \infty).$$

Hence, reverting to A,

$$\widehat{Af}(\omega) = (T\hat{f})(\omega) = \phi(\omega)\hat{f}(\omega) = \lambda\hat{f}(\omega) + i\mu sgn(\omega)\hat{f}(\omega) = \lambda\hat{f}(\omega) + i\mu\widehat{\mathbb{H}f},$$

whence, by the Fourier Inversion Theorem,

$$Af = \lambda f + \mu\mathbb{H}f, \ \forall f \in L^2.$$

In other words, A is the sum of a multiple of the identity operator and a multiple of the Hilbert transform. This ends the proof.

7. Some additional remarks

It may be of interest to workers in the field of hysteresis to point out that we can now use the Hilbert transform to manufacture *non-linear* affine-invariant transformations that send L^2 to itself, (or, indeed, more generally, that map L^p to L^p, if $1 < p < \infty$). For instance, the following mappings are examples of this kind:

$$f \to |\mathbb{H}f|, \ \ f \to \mathbb{H}|f|, \ \ f \to \mathbb{H}f + |\mathbb{H}|f||.$$

It may be wort remarking as well that $\mathbb{H}$ preserves the Lip_α classes[8]. More precisely, if $f \in L^p$, for $p > 1$, and satisfies the Lipschitz condition

$$||\tau_{(1,y)}f - f||_\infty = O(|y|^\alpha) \ (0 < \alpha < 1),$$

then g, the Hilbert transform of f, is defined on all of $\mathbb{R}$, belongs to L^p, and satisfies the same Lipschitz condition:

$$||\tau_{(1,y)}g - f||_\infty = O(|y|^\alpha).$$

Hence, since $|f|$ belongs to Lip_α whenever f does, it's clear, from what we've just said, that each of the displayed non-linear operators leave the intersection of L^p and Lip_α invariant if $p > 1$.

8. Open problems

As we've previously remarked, the Hilbert transform is an example of an affine-invariant operator that maps L^p to L^p, for every $p \in (1, \infty)$. This raises the question: If $1 < p < \infty$, but $p \neq 2$, what are the affine-invariant operators that map L^p to L^p? It's to be expected that the analogue of Theorem 2 holds in this case. However, the methods used to establish this theorem are heavily reliant on the properties of the Fourier transform—which do not extend to all the L^p spaces; for instance, it's not even definable in the ordinary sense if $p > 2$—and a different approach is needed to tackle the question. The first hurdle that must be overcome, is a description of the translation-invariant operators acting on an arbitrary L^p space, and a solution to this problem is furnished by Schwarz's theory of distributions [2]. Indeed, this problem was explicitly solved by Schwarz himself [1]. I suspect that one can build on his result, and confirm my belief that the only affine-invariant operators on L^p are linear combinations of the identity and the Hilbert transform, but I've not carried out the analysis.

What about L^1? As we know, the Hilbert transform is defined on this space, and is affine-invariant as a linear transformation from this space to the space of measurable functions whose distributions functions are small at infinity. (More precisely, there is an absolute constant $c > 0$, such that, if $f \in L^1$, and $\lambda > 0$, then,

$$m\left(\{x \in \mathbb{R} : |\mathbb{H}f(x)| \geq \lambda\}\right) \leq \frac{c||f||_1}{\lambda},$$

where m denotes Lebesgue measure on $\mathbb{R}$, a result due to Kolmogorov [3].) The difficulty, however, is that $\mathbb{H}$ is not continuous on this space. But it does leave the subspace

$$M = \{f \in L^1 : ||\mathbb{H}f||_1 < \infty\}$$

invariant. If we give M the norm $f \rightarrow ||f||_1 + ||\mathbb{H}f||_1$, are linear combinations of the identity and the Hilbert transform the only affine-invariant operators on it?

Finally, we ask: what are the affine-invariant operators between different L^p spaces?

References

[1] Donoughue W F Jr 1969 *Distributions and Fourier Transforms* (Academic Press)
[2] Katznelson Y 1968 *An Introduction to Harmonic Analysis* (John Wiley & Sons, Inc.)
[3] Koosis P 1980 *Introduction to* H_p *Spaces* (Cambridge University Press)
[4] Krasnosel'skii M A and Pokrovskii A V 1989 *Systems with Hysteresis* (Berlin: Springer-Verlag)
[5] Pokrovskii A and Sobolev V 2005 A naive view of time relaxation and hysteresis *Singular Perturbations and Hysteresis* (Philadelphia: SIAM)
[6] Brokate M, Pokrovskii A, Rachinskii D and Rasskazov O 2006 Differential equations with hysteresis via a canonical example *The Science of Hysteresis* vol 1 (Elsevier)
[7] Rudin W 1962 *Fourier Analysis on Groups* (Interscience)
[8] Titchmarsh E C 1948 *Introduction to the Theory of Fourier Integrals* Second edition (Oxford University Press) .
[9] Titchmarsh E C 1939 *The Theory of Functions* second edition (Oxford University Press)

Institute of Physics Publishing
doi:10.1088/1742-6596/55/1/011

Journal of Physics: Conference Series **55** (2006) 114–129
International Workshop on Multi-Rate Processes and Hysteresis

Reduced model of neurotransmitter transport in the presence of generic receptors and transporters

L V Kalachev

Department of Mathematical Sciences, University of Montana, Missoula, MT 59812, USA

E-mail: `kalachev@mso.umt.edu`

Abstract. We present a description of a reduction procedure for a model of glutamate transport in a synaptic cleft in the presence of a finite number of receptors and transporters with different kinetics properties. Under certain conditions a system of equations for glutamate, receptors and transporters concentrations can be substituted by a single equation for glutamate with effective diffusion and modified kinetics term. The results of model reduction are illustrated with an example showing good agreement between the behavior of the original model and the reduced model.

Keywords: model reduction, neurotransmitters, glutamate transport, synaptic cleft, effective diffusion, receptors, transporters

1. Introduction

We consider a general reduction procedure for a model of neurotransmitter transport in the presence of receptors and transporters. While the approach that we describe is applicable to a wide class of models of molecular transport in the presence of fast forward and reverse binding reactions, here we are mostly interested in a particular case related to glutamate transport.

Glutamate molecules are neurotransmitters that carry chemical signals between neurons through synaptic clefts, thin regions ($\approx$ 20 nm) separating consecutive neurons. A certain number of glutamate molecules ($\approx$ 3000) is released from a synaptic vesicle at the pre-synaptic boundary. These molecules diffuse through the cleft, activate receptors located at the postsynaptic neuron boundary, leading to the appearance of ionic currents through the postsynaptic neuron membrane. Also, glutamate is being removed from the cleft via transporters located at certain regions of the postsynaptic boundary, as well as via diffusion through the side boundaries of the synapse.

We consider a procedure that under certain conditions reduces the spatially 3-dimensional model of glutamate transport to a spatially 2-dimensional one. The ultimate goal of the analysis is to produce an adequate model for fast computation of glutamate concentration distributions (as a function of time and spatial variables) for various initial concentrations of glutamate released from the vesicles in the interior of the cleft as well as for various spatial distributions of receptors and transmitters at the postsynaptic boundary of the cleft. The resulting model will allow us to include in the computations any finite number of receptors and transmitters of different types.

Another goal in obtaining a reduced model is to produce an explicit expression for such a characteristic of glutamate transport as the *effective diffusion* that must change depending on the local concentration of receptors and transporters. Analysis of various limiting cases using this expression will allow one to elucidate the effects of changing receptors and transporters concentration and positions on glutamate transport.

In this presentation we are only interested in characterizing the glutamate dynamics (and effects of receptor and transporter presence on this dynamics), and we are not going to address, e.g., the analysis of ionic currents produced by activated receptors, etc. Thus, here we are going to discuss the case of some *generic* (effective) receptors and transmitters with variable kinetics characteristics that affect glutamate transport properties. The derivation of more precise reduced models for some particular types of receptors and transmitters (for which the kinetic schemes are considerably more complex; see, e.g., Wadiche and Kavanaugh [11], Franks et al. [2], Larsson et al. [5]) will be performed in the near future and published elsewhere.

We note that recent numerical modeling of neurotransmitters transport (including modeling of glutamate in a synaptic cleft, with analysis of possible spill over of glutamate in the extra-synaptic regions) has attracted considerable attention (see, e.g., Rusakov and Kullman [7], Nielsen and DiGregorio [6]; for Monte Carlo simulation approaches see, e.g., Franks et al. [2], and Cogan et al. [1]). However, the reduction procedures for the models used in such numerical simulations are not addressed in the literature. Reduced models can help one to understand which parameters and parameter combinations in the system can be estimated from experiments, and which ones, in principle, cannot be estimated for given sensitivities of measuring devices.

Mathematical justifications of the procedures discussed in this paper can be performed using the methods of classical perturbation analysis (see, e.g., Vasil'eva et al. [10]), and they are not included in the current presentation.

2. Statement of the problem: geometry of the synaptic cleft region and assumptions

We initially represent the synaptic cleft as a cylindrical domain (e.g., in a Cartesian coordinate system): $(x, y) \in \Omega$, $0 \leq z \leq h$. The height h of this domain is assumed to be much smaller compared to the characteristic length L of its base Ω: $0 < h \ll L$. In what follows we will use this inequality to perform the original model simplification.

Let us formulate a set of assumptions:

(a) *In the absence of receptors and transporters the diffusion coefficient for glutamate is the same (i.e., it is a constant) throughout the cleft region.*

(b) *When the glutamate molecules arrive at the side boundaries of the domain, $(x, y) \in \partial\Omega$, $0 \leq z \leq h$, they are immediately washed out: this means that glutamate concentration at these boundaries is equal to zero (no glutamate molecules re-enter the region from outside).*

(c) *The receptors and transporters affect the glutamate molecules in a thin region near the postsynaptic boundary of the cleft. We denote the effective height of this near-boundary region by δ. Evidently, $0 < \delta \ll h$.*

(d) *Presynaptic boundary is not permeable for glutamate molecules.*

Certain additional assumptions related to glutamate binding and transport dynamics due to the presence of receptors and transporters will be specified later.

3. Statement of the problem: equations, initial and boundary conditions

The following model will correspond to the assumptions stated above.

In the interior of the cleft the glutamate concentration $u(x, y, z, t)$ is described by the diffusion equation (D is diffusion coefficient):

$$\frac{\partial u}{\partial t} = D\left(\frac{\partial^2 u}{\partial x^2} + \frac{\partial^2 u}{\partial y^2} + \frac{\partial^2 u}{\partial z^2}\right). \qquad (1)$$

The boundary conditions for u are formulated as follows:

$$u(x,y,z,t)=0 \quad \text{for} \quad (x,y)\in\partial\Omega, \quad 0\le z\le h, \tag{2}$$

(washout at the side boundaries);

$$\frac{\partial u}{\partial z}(x,y,h,t)=0 \quad \text{(impermeable presynaptic boundary)}; \tag{3}$$

$$-D\frac{\partial u}{\partial z}(x,y,0,t)=\delta\frac{\partial u}{\partial t}(x,y,0,t). \tag{4}$$

(receptors/transporters related dynamics at the post-synaptic boundary).

Here

$$\frac{\partial u}{\partial t}(x,y,0,t)=\sum_{i=1}^{n} f_i(u(x,y,0,t),v_i,x,y)+\sum_{j=1}^{m} g_j(u(x,y,0,t),w_j,x,y). \tag{5}$$

For different receptor concentrations v_i ($i = 1,\ldots,n$), and different transporter concentrations w_j ($j=1,\ldots,m$), we have the equations

$$\frac{\partial v_i}{\partial t}(x,y,t)=f_i(u(x,y,0,t),v_i,x,y), \quad (i=1,\ldots,n), \tag{6}$$

$$\frac{\partial w_j}{\partial t}(x,y,t)=G_j(u(x,y,0,t),w_j,x,y), \quad (j=1,\ldots,m). \tag{7}$$

Equation (5) describes glutamate dynamics (forward/reverse binding, loss, etc.) in a thin layer (of effective depth δ) near the postsynaptic boundary of the cleft due to the presence of receptors and transporters (f_i with $i=1,\ldots,n$ describe the kinetics associated with n different types of receptors, and g_j with $j=1,\ldots,m$ describe the kinetics associated with m different types of transporters distributed over the postsynaptic boundary surface).

A particular form of the right hand side of the boundary condition (4) at $z=0$, which is proportional to the right hand side of (5), as well as the right hand sides of the equations (6), (7) for surface receptor concentration v_i and surface transporter concentration w_j, respectively, will be specified later after additional assumptions on receptor and transporter kinetics are introduced. We note that in our model $G_j\neq g_j$ (see discussion in Section 5 below).

The initial conditions have the form:

$$u(x,y,z,0)=U^*(x,y,z) \quad \text{(initial glutamate distribution in the cleft)}; \tag{8}$$

$$v_i(x,y,0)=V_i^*(x,y) \quad (i=1,\ldots,n) \tag{9}$$

(initial spatial distribution of receptors);

$$w_j(x,y,0)=W_j^*(x,y) \quad (j=1,\ldots,m) \tag{10}$$

(initial spatial distribution of transporters).

4. Step 1 of model reduction: derivation of spatially 2-dimensional equation for glutamate concentration

Let us average the equation for glutamate concentration u over the height of the synaptic cleft. To do so we integrate both sides of the equation (1) with respect to z from 0 to h, and divide the result by h:

$$\begin{aligned}\frac{\partial \bar{u}}{\partial t} &= D\left(\frac{\partial^2 \bar{u}}{\partial x^2} + \frac{\partial^2 \bar{u}}{\partial y^2} + \frac{1}{h}\int_0^h \frac{\partial^2 u}{\partial z^2} dz\right) \\ &= D\left(\frac{\partial^2 \bar{u}}{\partial x^2} + \frac{\partial^2 \bar{u}}{\partial y^2} + \frac{1}{h}\frac{\partial u}{\partial z}(x,y,h,t) - \frac{1}{h}\frac{\partial u}{\partial z}(x,y,0,t)\right).\end{aligned} \tag{11}$$

Here we use notation

$$\bar{u}(x,y,t) = \frac{1}{h}\int_0^h u(x,y,z,t)dz \tag{12}$$

for the values of u averaged over z.

Substituting conditions (3) and (4) in the right hand side of (11), we obtain:

$$\begin{aligned}\frac{\partial \bar{u}}{\partial t} &= D\left(\frac{\partial^2 \bar{u}}{\partial x^2} + \frac{\partial^2 \bar{u}}{\partial y^2}\right) + \frac{\delta}{h}\frac{\partial u}{\partial t}(x,y,0,t) \\ &= D\left(\frac{\partial^2 \bar{u}}{\partial x^2} + \frac{\partial^2 \bar{u}}{\partial y^2}\right) \\ &+ \frac{\delta}{h}\left(\sum_{i=1}^{n} f_i(u(x,y,0,t), v_i, x, y) + \sum_{j=1}^{m} g_j(u(x,y,0,t), w_j, x, y)\right),\end{aligned} \tag{13}$$

where we also used relation (5).

From (12) it follows that

$$\begin{aligned}\bar{u}(x,y,t) &= \frac{1}{h}\int_0^h \left(u(x,y,0,t) + \frac{\partial u}{\partial z}(x,y,0,y)z + \ldots\right)dz \\ &= u(x,y,0,t) + \frac{\partial u}{\partial z}(x,y,0,t)\frac{h}{2} + \ldots.\end{aligned} \tag{14}$$

Assuming that h is sufficiently small (compared to L), $\partial u/\partial z(x,y,0,t)$ is moderate , δ/h is finite, f_i, g_j are sufficiently smooth, and introducing the new notation

$$\tilde{u}(x,y,t) = u(x,y,0,t), \tag{15}$$

we arrive at the approximate equation

$$\frac{\partial \tilde{u}}{\partial t} = D\left(\frac{\partial^2 \tilde{u}}{\partial x^2} + \frac{\partial^2 \tilde{u}}{\partial y^2}\right) + \frac{\delta}{h}\left(\sum_{i=1}^{n} f_i(\tilde{u}, v_i, x, y) + \sum_{j=1}^{m} g_j(\tilde{u}, w_j, x, y)\right). \tag{16}$$

It follows from (2) and (8) that we must solve (16), together with the system (6) and (7) for v_i $(i = 1, \ldots, n)$ and w_j $(j = 1, \ldots, m)$, with the following boundary and initial conditions for $\tilde{u}$:

$$\tilde{u}(x, y, t) = 0 \quad \text{for} \quad (x, y) \in \partial\Omega, \tag{17}$$

$$\tilde{u}(x, y, 0) = \frac{1}{h}\int_0^h U^*(x, y, z)dz = \tilde{U}^*(x, y), \tag{18}$$

and initial conditions (9), (10) for v_i, w_j, respectively.

5. Receptor and transporter kinetics

Let us use the notation U for a molecule of glutamate. We consider the following generic reaction scheme for i-th receptor:

$$U + V_i \overset{k_i^+}{\to} C_i, \quad C_i \overset{k_i^-}{\to} U + V_i. \tag{19}$$

Here V_i denotes the free i-th receptor, and C_i denotes the corresponding occupied receptor (in what follows we use the notation v_i and c_i for concentrations of V_i and C_i, respectively); $k_i^{\pm}$ are the rate constants of the forward and reverse receptor binding reactions.

Also, we assume the following generic reaction kinetics for the j-th transporter:

$$U + W_j \overset{K_j^+}{\to} E_j \overset{\lambda_j}{\to} W_j + U_{removed}, \quad E_j \overset{K_j^-}{\to} U + W_j. \tag{20}$$

Here W_j denotes the free j-th transporter, and E_j denotes the corresponding occupied transporter (in what follows we use the notation w_j and e_j for concentrations of W_j and E_j, respectively); $U_{removed}$ represents glutamate molecules that are removed from the cleft via transporters; $K_j^{\pm}$ are the rate constants of the forward and reverse transporter binding reactions; λ_j are the rate constants of glutamate removal reactions.

Using the Law of Mass Action, we write the equations (at the postsynaptic boundary) for $v_i(x, y, t)$, $c_i(x, y, t)$, $w_j(x, y, t)$, and $e_j(x, y, t)$:

$$\frac{\partial v_i}{\partial t} = -k^+\tilde{u}v_i + k^- c_i; \tag{21}$$

$$\frac{\partial c_i}{\partial t} = +k^+\tilde{u}v_i - k^- c_i; \tag{22}$$

$$\frac{\partial w_j}{\partial t} = -K_j^+\tilde{u}w_j + K^- e_j + \lambda_j e_j; \tag{23}$$

$$\frac{\partial e_j}{\partial t} = +K_j^+\tilde{u}w_j - K_j^- e_j - \lambda_j e_j. \tag{24}$$

Corresponding initial conditions describe the initial spatial distribution of v_i, c_i, w_j, and e_j at the postsynaptic boundary:

$$v_i(x, y, 0) = V_i^*(x, y), \quad c_i(x, y, 0) = 0, \tag{25}$$

$$w_j(x, y, 0) = W_j^*(x, y), \quad e_j(x, y, 0) = 0. \tag{26}$$

Equations (21), (22) with corresponding conditions from (25) can be used to eliminate c_i:

$$\frac{\partial v_i}{\partial t} = -k_i^+\tilde{u}v_i + k_i^-(V_i^*(x, y) - v_i) := f_i(\tilde{u}, v_i, x, y). \tag{27}$$

Similarly, equations (23), (24) with corresponding conditions from (26) can be used to eliminate e_j:

$$\begin{aligned}\frac{\partial w_j}{\partial t} &= -K_j^+ \tilde{u} w_j + K_j^-(W_j^*(x,y) - w_j) + \lambda_j(W_j^*(x,y) - w_j) \\ &= G_j(\tilde{u}, w_j, x, y) := g_j(\tilde{u}, w_j, x, y) + \lambda_j(W_j^*(x,y) - w_j).\end{aligned} \tag{28}$$

We note that the functions f_i and g_j defined in (27) and (28), respectively, are exactly the ones that enter (5) (and thus, equation (16) for $\tilde{u}$) and (6), the functions G_j are the ones that enter (7).

So, now our model consists of a quasi-linear parabolic equation (16), with conditions (17), (18), and equations (27), (28), with conditions (9), (10).

6. Step 2 of model reduction: making use of fast receptor and transporter kinetics

Let us make additional assumptions related to receptor and transporter kinetics that will allow us to further simplify the model. First, we assume that

(e) *distributions of various types of receptors and transporters at the postsynaptic boundary are uniform* (i.e., $V_i^*(x,y) = V_i^*$ =const, and $W_i^*(x,y) = W_i^*$ =const). This assumption is only needed to clarify the relations between various characteristic time scales in the model. After the derivation, the reduced model may be used for non-uniform spatial receptor and transporter distributions as well. However, we will then need to take into account the fact that in the regions without receptors and/or transporters corresponding values of V_i and/or W_j will be zero, and local transport in these regions will only be associated with the original glutamate diffusion.

Let us introduce several time scales characteristic of our problem:

- characteristic diffusion time $\tau_D = L^2/D$;
- characteristic times of forward and reverse binding reactions for receptors, $\tau_{k_i^+} = 1/(k_i^+ V_i^*)$ and $\tau_{k_i^-} = 1/(k_i^-)$;
- characteristic times of forward and reverse binding reactions for transporters, $\tau_{K_j^+} = 1/(K_j^+ W_j^*)$ and $\tau_{K_j^-} = 1/(K_j^-)$;
- characteristic times of glutamate loss via transporters, $\tau_{\lambda_j} = 1/\lambda_j$.

Next, we formulate the assumptions that will allow us to reduce the system of equations for concentrations of glutamate, receptors and transporters to one equation for glutamate with an *effective* diffusion coefficient that depends nonlinearly on the concentration of free glutamate. If only a portion of conditions formulated below is satisfied for a certain number of receptors and transporters, then only this number of equations can be eliminated, and the reduced system will have fewer equations compared to the original one. If none of the conditions formulated below is satisfied, then further reduction of the system (16), (27), (28) is not possible.

(f) *We assume that the forward and reverse receptor binding reactions are fast compared to diffusion, i.e.,* $\tau_{k_i^\pm} \ll \tau_D$. *Also, we assume that the ratios* $\tau_{k_i^-}/\tau_{k_i^+}$ *are moderate.*

(g) *We assume that the forward and reverse transporter binding reactions are fast compared to diffusion, i.e.,* $\tau_{K_j^\pm} \ll \tau_D$. *We also assume that the ratios* $\tau_{K_i^-}/\tau_{K_i^+}$ *are moderate.*

(h) *In addition, we consider the case where* τ_{λ_j}/τ_D *are moderate.*

In the Appendix we present the non-dimensionalization procedure (needed for the rigorous construction of the asymptotic approximation of the solution), as well as the details of mathematical model reduction algorithm. Here we only mention characteristic model reduction steps written in terms of the original variables and parameters, and present the final reduced model formulation.

Substituting expressions (27) and (28) for f_i and g_i,

$$f_i(\tilde{u}, v_i, x, y) = \frac{\partial v_i}{\partial t}, \tag{29}$$

$$g_j(\tilde{u}, w_j, x, y) = \frac{\partial w_j}{\partial t} - \lambda_j(W_j^* - w_j), \tag{30}$$

into (16), we obtain:

$$\begin{aligned} \frac{\partial \tilde{u}}{\partial t} &= D\left(\frac{\partial^2 \tilde{u}}{\partial x^2} + \frac{\partial^2 \tilde{u}}{\partial y^2}\right) \\ &+ \frac{\delta}{h}\left(\sum_{i=1}^{n} \frac{\partial v_i}{\partial t} + \sum_{j=1}^{m}\left[\frac{\partial w_j}{\partial t} - \lambda_j(W_j^* - w_j)\right]\right). \end{aligned} \tag{31}$$

In the quasi-steady state (or, in the leading order approximation following from the asymptotic algorithm; see Appendix), from (27), (28), we have

$$0 = -k_i^+ \tilde{u} v_i + k_i^-(V_i^* - v_i). \tag{32}$$

Thus, from (32), we can express v_i in terms of $\tilde{u}$ as follows:

$$v_i = \frac{k_i^- V_i^*}{k_i^- + k_i^+ \tilde{u}}. \tag{33}$$

Similarly, in the quasi-steady state (or, in the leading order approximation), from (28), we get

$$0 = -K_j^+ \tilde{u} w_j + K_j^-(W_j^* - w_j). \tag{34}$$

Thus, from (34), we can express w_j in terms of $\tilde{u}$ as follows:

$$w_j = \frac{K_j^- W_j^*}{K_j^- + K_j^+ \tilde{u}}. \tag{35}$$

Differentiating (33) and (35) with respect to t, we obtain:

$$\frac{\partial v_i}{\partial t} = -\frac{k_i^- k_i^+ V_i^*}{(k_i^- + k_i^+ \tilde{u})^2} \frac{\partial \tilde{u}}{\partial t}, \tag{36}$$

$$\frac{\partial w_j}{\partial t} = -\frac{K_j^- K_j^+ W_j^*}{(K_j^- + K_j^+ \tilde{u})^2} \frac{\partial \tilde{u}}{\partial t}. \tag{37}$$

Substituting (35), (36), (37) into (31), and re-arranging terms, we write:

$$\begin{aligned} &\frac{\partial \tilde{u}}{\partial t}\left\{1 + \frac{\delta}{h}\left(\sum_{i=1}^{n} \frac{k_i^- k_i^+ V_i^*}{(k_i^- + k_i^+ \tilde{u})^2} + \sum_{j=1}^{m} \frac{K_j^- K_j^+ W_j^*}{(K_j^- + K_j^+ \tilde{u})^2}\right)\right\} \\ &= D\left(\frac{\partial^2 \tilde{u}}{\partial x^2} + \frac{\partial^2 \tilde{u}}{\partial y^2}\right) - \frac{\delta}{h}\sum_{j=1}^{m} \lambda_j\left(W_j^* - \frac{K_j^- W_j^*}{K_j^- + K_j^+ \tilde{u}}\right). \end{aligned} \tag{38}$$

Finally, from (38), after some more re-arrangement of terms, we obtain the reduced equation for $\tilde{u}(x, y, t)$:

$$
\frac{\partial \tilde{u}}{\partial t} = \frac{D\left(\dfrac{\partial^2 \tilde{u}}{\partial x^2} + \dfrac{\partial^2 \tilde{u}}{\partial y^2}\right)}{\left\{1 + \dfrac{\delta}{h}\left(\displaystyle\sum_{i=1}^{n} \frac{(k_i^+/k_i^-)V_i^*}{(1+(k_i^+/k_i^-)\tilde{u})^2} + \sum_{j=1}^{m} \frac{(K_j^+/K_j^-)W_j^*}{(1+(K_j^+/K_j^-)\tilde{u})^2}\right)\right\}}
- \frac{\dfrac{\delta}{h}\displaystyle\sum_{j=1}^{m} \lambda_j W_j^* \left(\frac{(K_j^+/K_j^-)\tilde{u}}{1+(K_j^+/K_j^-)\tilde{u}}\right)}{\left\{1 + \dfrac{\delta}{h}\left(\displaystyle\sum_{i=1}^{n} \frac{(k_i^+/k_i^-)V_i^*}{(1+(k_i^+/k_i^-)\tilde{u})^2} + \sum_{j=1}^{m} \frac{(K_j^+/K_j^-)W_j^*}{(1+(K_j^+/K_j^-)\tilde{u})^2}\right)\right\}}. \tag{39}
$$

This equation must be solved with boundary conditions (17) and initial condition (18); see the detailed discussion about the choice of asymptotically correct initial conditions in the Appendix. When $u(x, y, t)$ is known, $v_i(x, y, t)$ $(i = 0, \ldots, n)$ and $w_j(x, y, t)$ $(j = 0, \ldots, m)$ are found from (33) and (35), respectively.

We note that the actual number of parameters (or parameter combinations) that enter (39) is smaller compared to the system (16), (27), (28): e.g., $k_i^{\pm}$ $(i = 0, \ldots, n)$ appear only in combinations (k_i^+/k_i^-), $K_j^{\pm}$ $(j = 0, \ldots, m)$ appear as the combinations (K_j^+/K_j^-). If the conditions for which the reduction is possible are satisfied, and if the time sensitivity of measuring devices is of the order of the (slow) characteristic diffusion time, τ_D, then $k_i^{\pm}$, $K_j^{\pm}$ cannot be independently identified from an experiment, and only combinations (k_i^+/k_i^-) and (K_j^+/K_j^-) can, in principle, be determined.

Another important observation: it follows from (39) that glutamate transport in the presence of receptors and transporters is characterized by the new *effective* diffusion coefficient that is now not constant (it depends on $\tilde{u}(x, y, t)$ and the original distribution of free receptors V_i^* and free transporters W_j^*):

$$
D_{effective} = \frac{D}{\left\{1 + \dfrac{\delta}{h}\left(\displaystyle\sum_{i=1}^{n} \frac{(k_i^+/k_i^-)V_i^*}{(1+(k_i^+/k_i^-)\tilde{u})^2} + \sum_{j=1}^{m} \frac{(K_j^+/K_j^-)W_j^*}{(1+(K_j^+/K_j^-)\tilde{u})^2}\right)\right\}}. \tag{40}
$$

It can be easily seen from (40) that for any value of $\tilde{u}$ we have $D_{effective} \leq D$. Also, as $\tilde{u} \to \infty$, the effective diffusion coefficient $D_{effective} \to D$. Finally, the minimal value of $D^{min}_{effective}$ is observed when $\tilde{u} \to 0$:

$$
D^{min}_{effective} = \frac{D}{\left\{1 + \dfrac{\delta}{h}\left(\displaystyle\sum_{i=1}^{n} (k_i^+/k_i^-)V_i^* + \sum_{j=1}^{m} (K_j^+/K_j^-)W_j^*\right)\right\}}. \tag{41}
$$

Remark. The asymptotic procedure described in the Appendix is based on the *Boundary Function Method* algorithm (see Vasil'eva et al. [10]). Our results are also related to those presented in Haario and Kalachev [3], Keener and Sneyd [4], Smith et al. [8], and Sneyd et al. [9].

Next, we consider an illustrative example where for some particular choice of parameter values we compare a solution of a model system consisting of a radially symmetric (and thus, spatially 1-dimensional) equation for free glutamate concentration coupled with a transporter equation and a solution of the corresponding reduced model consisting of one equation for free glutamate.

7. Illustrative example

Let us consider the case where one transporter is uniformly distributed throughout the postsynaptic boundary of the cleft, region Ω has the shape of a disk with radius $R = 2$ μm, and the problem is radially symmetric (the glutamate vesicle opens in the center of the disk, and the initial shape of the released glutamate concentration is radially symmetric). For the radially symmetric case the Laplace operator $(\partial^2\tilde{u}/\partial x^2 + \partial^2\tilde{u}/\partial y^2)$ will have the form $(\partial^2\tilde{u}/\partial r^2 + (1/r)\partial\tilde{u}/\partial r)$, where r is the radial distance from the center of the disk.

We will call the system for concentrations $\tilde{u}$ and w_1 averaged over the height of the cleft ($h = 20$ nm) and the corresponding additional conditions the *original model*. It consists of the following equations:

$$\frac{\partial \tilde{u}}{\partial t} = D\left(\frac{\partial^2 \tilde{u}}{\partial r^2} + \frac{1}{r}\frac{\partial \tilde{u}}{\partial r}\right) + \frac{\delta}{h}[-K_1^+\tilde{u}w_1 + K_1^-(W_1^* - w_1)]; \tag{42}$$

$$\frac{\partial w_1}{\partial t} = -K_1^+\tilde{u}w_1 + K_1^-(W_1^* - w_1) + \lambda_1(W_1^* - w_1). \tag{43}$$

The boundary and initial conditions are written as follows:

$$\tilde{u}(R,t) = 0, \tag{44}$$

$$\tilde{u}(r,0) = \tilde{U}^*(r) = 0.0004 \times \exp(-5r^2), \tag{45}$$

$$w_1(r,0) = W_1^* = 0.0001. \tag{46}$$

All concentrations are measured in moles per liter (M), and time is measured in msec. Other sample parameter values are chosen as follows: $D = 0.2$ μm^2/msec; $K_1^+ = 1.8 \times 10^4$ M/msec; $K_1^- = 1.8$ 1/msec; $\lambda_1 = 0.18$ 1/msec; $\delta = 2$ nm.

For the above choice of parameter values the characteristic times defined in the previous section are: $\tau_D = 20 \gg \tau_{K_1^+} = \tau_{K_1^-} \approx 0.55$, $\tau_{\lambda_1} \approx 5.55$. The conditions for which the reduction of the model is possible are (approximately) satisfied.

The reduced model will consist of the equation

$$\frac{\partial \tilde{u}}{\partial t} = \frac{(1 + (K_1^+/K_1^-)\tilde{u})^2 D}{(1 + (K_1^+/K_1^-)\tilde{u})^2 + (\delta/h)(K_1^+/K_1^-)W_1^*}\left(\frac{\partial^2 \tilde{u}}{\partial r^2} + \frac{1}{r}\frac{\partial \tilde{u}}{\partial r}\right)$$
$$- \frac{(\delta/h)\lambda_1(K_1^+/K_1^-)W_1^*(1 + (K_1^+/K_1^-)\tilde{u})\,\tilde{u}}{(1 + (K_1^+/K_1^-)\tilde{u})^2 + (\delta/h)(K_1^+/K_1^-)W_1^*}, \tag{47}$$

and additional conditions (44), (45).

When $\tilde{u}$ is known, we also know (see (35)):

$$w_1 = \frac{K_1^- W_1^*}{K_1^- + K_1^+\tilde{u}}. \tag{48}$$

Solutions for the *original* model and for the *reduced* model were obtained numerically using the method of lines (programs were written in MATLAB). The radially symmetric concentration profiles obtained by numerical computations for $t = 0$, 0.5, 1.0 msec are shown in Figures 1 (a), (b), (c), respectively.

At $t = 0$ the initial conditions for both models are the same. We note that for $t = 0.5$ and $t = 1.0$ the curves representing concentrations of free glutamate obtained for the original and the reduced models are almost undistinguishable.

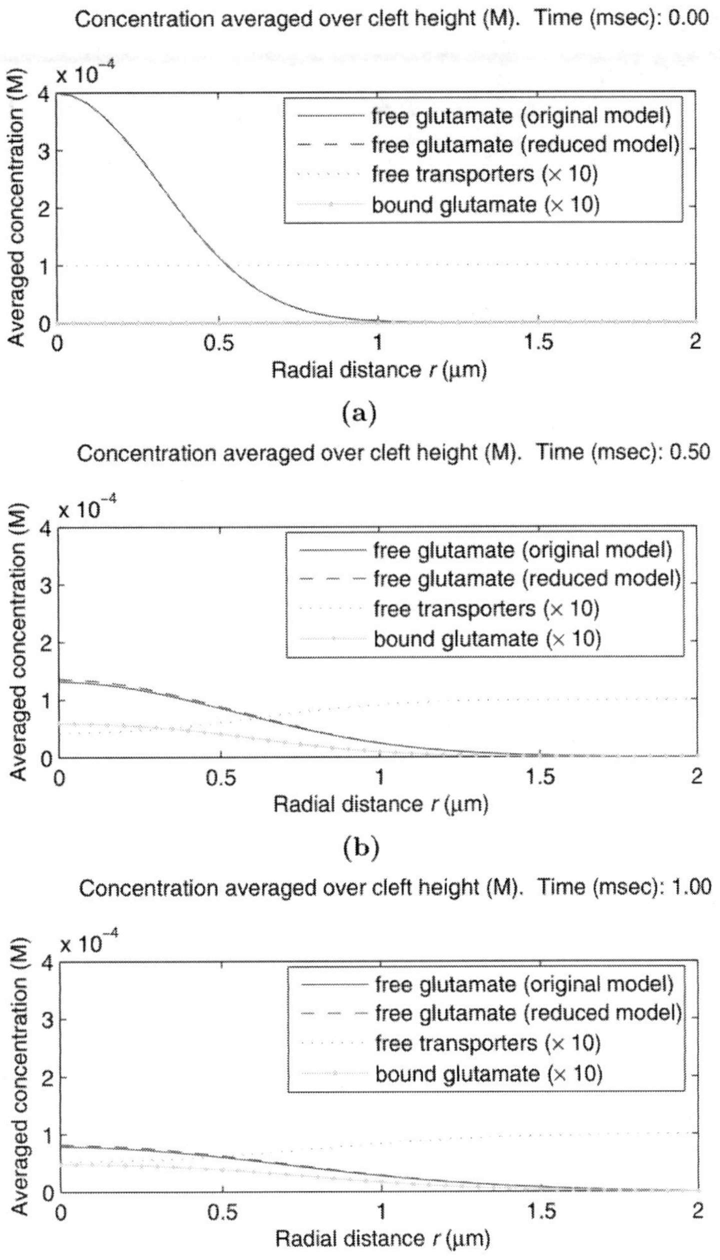

(c)

Figure 1. Example: comparison of (radially dependent) concentration profiles of free glutamate for the original and the reduced models. The concentration of free transporters and bound glutamate for the original model are also shown; (a) initial instant of time $t = 0$ msec (initial conditions are the same for the original and for the reduced model); (b) instant of time $t = 0.5$ msec; (c) instant of time $t = 1.0$ msec.

Remark. The numerical results shown in Figures 1 (a), (b) and (c) qualitatively stay the same for the set of parameter values found in the literature (see, e.g., Rusakov and Kullman [7]): $K_1^+ = 6.0 \times 10^3$ M/msec; $K_1^- = 0.1$ 1/msec; $\lambda_1 = 0.02$ 1/msec.

8. Conclusion

We presented a general model reduction procedure for neurotransmitter transport in the presence of a finite number of generic receptors and transporters. It is important to emphasize that here we did not perform numerical experiments for various distributions of receptors and transporters to elucidate the details of glutamate transport in a synaptic cleft. Such numerical experiments as well as experimental data fitting using a set of reduced models of the type presented above will be done in the near future and published elsewhere. Our main goal was to illustrate the reduction procedure and specify conditions under which various steps of the reduction can be performed.

Comparison of the solutions of the *original* model and the *reduced* model for an example with a particular sample choice of parameter values that satisfied the conditions imposed shows good agreement between the two models.

Appendix

Here we present the asymptotic reduction algorithm for a spatially 2-dimensional model in the case where forward/reverse binding reactions involving receptors and transmitters are assumed to be fast in comparison with diffusion and trans-membrane neurotransmitter transport. In what follows to simplify the explanation of the reduction procedure, we assume that receptors and transporters are uniformly distributed, i.e., V_i^*, W_j^* =const ($i = 1, \dots, n$; $j = 1, \dots, m$).

Earlier, we have introduced the following characteristic time scales: characteristic diffusion time $\tau_D = L^2/D$; characteristic times of forward and reverse binding reactions for receptors, $\tau_{k_i^+} = 1/(k_i^+ V_i^*)$ and $\tau_{k_i^-} = 1/(k_i^-)$; characteristic times of forward and reverse binding reactions for transporters, $\tau_{K_j^+} = 1/(K_j^+ W_j^*)$ and $\tau_{K_j^-} = 1/(K_j^-)$; characteristic times of glutamate loss via transporters (i.e., trans-membrane transport), $\tau_{\lambda_j} = 1/\lambda_j$. Let us introduce the following re-scaled quantities and auxiliary notations: we define non-dimensional time $\hat{t} = t/t_D$, non-dimensional spatial variables $\hat{x} = x/L$, $\hat{y} = y/L$, where L is characteristic lateral dimension of the synaptic cleft region. We define re-scaled concentrations ($i = 1, \dots, n$; $j = 1, \dots, m$):

$$\hat{u} = \frac{\tilde{u}}{\max_{(x,y)\in\Omega} \tilde{U}^*(x,y)}; \quad \hat{U}^* = \frac{\tilde{U}^*(x,y)}{\max_{(x,y)\in\Omega} \tilde{U}^*(x,y)} \leq 1;$$

$$\hat{v}_i = \frac{\tilde{v}_i}{V_i^*} \leq 1; \quad \hat{w}_j = \frac{\tilde{w}_j}{W_j^*}; \leq 1; \quad \hat{\lambda}_j = \lambda_j \cdot \tau_D.$$

Let us re-formulate assumptions (f), (g), (h) introduced in Section 6 using a small parameter $0 < \varepsilon \ll 1$. Without loss of generality, we define

$$\varepsilon = \tau_{k_1^+}/\tau_D. \tag{49}$$

Then assumptions (f), (g), (h) are equivalent to the following set of asymptotic relations:

$$\frac{\tau_D}{\tau_{k_i^+}} \frac{\max_{(x,y)\in\Omega} \tilde{U}^*(x,y)}{V_i^*} = \frac{a_i}{\varepsilon} = O\left(\frac{1}{\varepsilon}\right); \quad \frac{\tau_D}{\tau_{k_i^-}} = \frac{b_i}{\varepsilon} = O\left(\frac{1}{\varepsilon}\right);$$

$$\frac{\tau_D}{\tau_{K_j^+}} \frac{\max_{(x,y)\in\Omega} \tilde{U}^*(x,y)}{W_j^*} = \frac{A_j}{\varepsilon} = O\left(\frac{1}{\varepsilon}\right); \quad \frac{\tau_D}{\tau_{K_j^-}} = \frac{B_j}{\varepsilon} = O\left(\frac{1}{\varepsilon}\right); \quad \hat{\lambda}_j = O(1).$$

Here a_i, b_i, A_j, $B_j = O(1)$ $(i = 1, \ldots, n;\ j = 1, \ldots, m)$.

Using the new notation introduced above, we re-write the problem for (16), (27), (28) with corresponding conditions in the form:

$$\varepsilon\frac{\partial \hat{u}}{\partial \hat{t}} = \varepsilon\left(\frac{\partial^2 \hat{u}}{\partial \hat{x}^2} + \frac{\partial^2 \hat{u}}{\partial \hat{y}^2}\right) + \frac{\delta}{h}\left(\sum_{i=1}^{n}[-a_i\hat{u}\hat{v}_i + b_i(1-\hat{v}_i)] + \sum_{j=1}^{m}[-A_j\hat{u}\hat{w}_j + B_j(1-\hat{w}_j)]\right). \tag{50}$$

$$\varepsilon\frac{\partial \hat{v}_i}{\partial \hat{t}} = -a_i\hat{u}\hat{v}_i + b_i(1-\hat{v}_i); \tag{51}$$

$$\varepsilon\frac{\partial \hat{w}_j}{\partial \hat{t}} = -A_j\hat{u}\hat{w}_j + B_j(1-\hat{w}_j) + \varepsilon\hat{\lambda}_j(1-\hat{w}_j); \tag{52}$$

$$\hat{u}(\hat{x},\hat{y},0) = \hat{U}^*(\hat{x},\hat{y}) \quad \text{for} \quad (\hat{x},\hat{y}) \in \hat{\Omega}, \tag{53}$$

where $\hat{\Omega}$ is re-scaled Ω; $\hat{U}^*(\hat{x},\hat{y}) = 0$ for $(\hat{x},\hat{y}) \in \partial\hat{\Omega}$;

$$\hat{v}_i(\hat{x},\hat{y},0) = 1, \quad \hat{w}_j(\hat{x},\hat{y},0) = 1 \quad \text{for} \quad (\hat{x},\hat{y}) \in \hat{\Omega}; \tag{54}$$

$$\hat{u}(\hat{x},\hat{y},t) = 0 \quad \text{for} \quad (\hat{x},\hat{y}) \in \partial\hat{\Omega}. \tag{55}$$

To simplify the notation, let us omit hats ($\hat{}$) in (50) – (55).

According to the *Boundary Function Method* algorithm (see Vasil'eva et al. [10]), we seek approximate solution to (50) – (55) in the form of an asymptotic expansion

$$u(x,y,t) = \bar{u}_0(x,t) + \varepsilon\bar{u}_1(x,t) + \ldots + \Pi_0 u(x,\tau) + \varepsilon\Pi_1 u(x,\tau) + \ldots, \tag{56}$$

with similar expressions for v_i and w_j. Here $\bar{u}_0$, $\bar{u}_1$ are the functions that describe the, so-called, regular part of the asymptotic expansion and are important for $t > 0$ away from the initial layer; $\Pi_0 u$, $\Pi_1 u$ are the boundary layer functions important in the initial layer near $t = 0$ (for $\tau \geq 0$, where $\tau = t/\varepsilon$ is the, so-called, stretched variable).

Substituting (56) and similar expressions for v_i and w_j into (50) – (55), and equating at every order of ε terms of the asymptotic expansions separately for regular and boundary functions, we obtain the problems for $\bar{u}_0$, $\bar{u}_1$, etc.

In the leading order approximation, for $\bar{u}_0$, $\bar{v}_{i0}$, $\bar{w}_{j0}$, we have

$$0 = -a_i\bar{u}_0\bar{v}_{i0} + b_i(1-\bar{v}_{i0}); \tag{57}$$

$$0 = -A_j\bar{u}_0\bar{w}_{j0} + B_j(1-\bar{w}_{j0}). \tag{58}$$

From (57) and (58) we write expressions for $\bar{v}_{i0}$, $\bar{w}_{j0}$ in terms of $\bar{u}_0$:

$$\bar{v}_{i0} = \frac{b_i}{b_i + a_i\bar{u}_0}; \tag{59}$$

$$\bar{w}_{j0} = \frac{B_j}{B_j + A_j\bar{u}_0}. \tag{60}$$

In the first order approximation we have the following equations for the regular functions $\bar{u}_1$, $\bar{v}_{i1}$, $\bar{w}_{j1}$:

$$\frac{\partial \bar{u}_0}{\partial t} - \left(\frac{\partial^2 \bar{u}_0}{\partial x^2} + \frac{\partial^2 \bar{u}_0}{\partial y^2}\right) = \frac{\delta}{h}\left(\sum_{i=1}^{n}[-a_i\bar{u}_1\bar{v}_{i0} - a_i\bar{u}_0\bar{v}_{i1} - b_i\bar{v}_{i1}]\right.$$

$$\left. + \sum_{j=1}^{m}[-A_j\bar{u}_1\bar{w}_{j0} - A_j\bar{u}_0\bar{w}_{j1} - B_j\bar{w}_{j1}]\right); \tag{61}$$

$$-a_i\bar{u}_1\bar{v}_{i0} - a_i\bar{u}_0\bar{v}_{i1} - b_i\bar{v}_{i1} = \frac{\partial \bar{v}_{i0}}{\partial t}; \tag{62}$$

$$-A_j\bar{u}_1\bar{w}_{j0} - A_j\bar{u}_0\bar{w}_{j1} - B_j\bar{w}_{j1} = \frac{\partial \bar{w}_{j0}}{\partial t} - \lambda_j(1 - \bar{w}_{j0}). \tag{63}$$

The solvability condition for the linear algebraic system (61) – (63) is written in the form:

$$\frac{\partial \bar{u}_0}{\partial t} - \left(\frac{\partial^2 \bar{u}_0}{\partial x^2} + \frac{\partial^2 \bar{u}_0}{\partial y^2}\right) = \frac{\delta}{h}\left(\sum_{i=1}^{n}\frac{\partial \bar{v}_{i0}}{\partial t}\right.$$

$$\left. + \sum_{j=1}^{m}\left[\frac{\partial \bar{w}_{j0}}{\partial t} - \lambda_j(1 - \bar{w}_{j0})\right]\right). \tag{64}$$

From (59), (60), for the derivatives of $\bar{v}_{i0}$, $\bar{w}_{j0}$, we have

$$\frac{\partial \bar{v}_{i0}}{\partial t} = -\frac{a_i b_i}{(b_i + a_i\bar{u}_0)^2}\frac{\partial \bar{u}_0}{\partial t}; \tag{65}$$

$$\frac{\partial \bar{w}_{j0}}{\partial t} = -\frac{A_j B_j}{(B_j + A_j\bar{u}_0)^2}\frac{\partial \bar{u}_0}{\partial t}. \tag{66}$$

Substituting (60), (65), and (66) into (64), we obtain the equation for $\bar{u}_0$:

$$\frac{\partial \bar{u}_0}{\partial t} - \left(\frac{\partial^2 \bar{u}_0}{\partial x^2} + \frac{\partial^2 \bar{u}_0}{\partial y^2}\right) = \frac{\delta}{h}\left(\sum_{i=1}^{n}\left[-\frac{a_i b_i}{(b_i + a_i\bar{u}_0)^2}\frac{\partial \bar{u}_0}{\partial t}\right]\right.$$

$$\left. + \sum_{j=1}^{m}\left[-\frac{A_j B_j}{(B_j + A_j\bar{u}_0)^2}\frac{\partial \bar{u}_0}{\partial t} - \lambda_j\left(1 - \frac{B_j}{B_j + A_j\bar{u}_0}\right)\right]\right). \tag{67}$$

From the above equation, after moving all the terms containing derivatives of $\bar{u}_0$ to the left hand side of (67), we obtain:

$$\left[1 + \frac{\delta}{h}\sum_{i=1}^{n}\frac{a_i b_i}{(b_i + a_i\bar{u}_0)^2} + \frac{\delta}{h}\sum_{i=1}^{m}\frac{A_j B_j}{(B_j + A_j\bar{u}_0)^2}\right]\frac{\partial \bar{u}_0}{\partial t}$$

$$-\left(\frac{\partial^2 \bar{u}_0}{\partial x^2} + \frac{\partial^2 \bar{u}_0}{\partial y^2}\right) = -\frac{\delta}{h}\sum_{j=1}^{m}\lambda_j\left(1 - \frac{B_j}{B_j + A_j\bar{u}_0}\right). \tag{68}$$

Dividing both sides of equation (68) by the term in the square brackets, we arrive at the partial differential equation for $\bar{u}_0$ with effective diffusion and effective kinetics term:

$$\frac{\partial \bar{u}_0}{\partial t} - \frac{1}{\left[1 + \frac{\delta}{h}\sum_{i=1}^{n}\frac{a_i b_i}{(b_i + a_i\bar{u}_0)^2} + \frac{\delta}{h}\sum_{i=1}^{m}\frac{A_j B_j}{(B_j + A_j\bar{u}_0)^2}\right]}\left(\frac{\partial^2 \bar{u}_0}{\partial x^2} + \frac{\partial^2 \bar{u}_0}{\partial y^2}\right)$$

$$= -\frac{\delta}{h}\frac{\sum_{j=1}^{m}\lambda_j\left(1 - \frac{B_j}{B_j + A_j\bar{u}_0}\right)}{\left[1 + \frac{\delta}{h}\sum_{i=1}^{n}\frac{a_i b_i}{(b_i + a_i\bar{u}_0)^2} + \frac{\delta}{h}\sum_{i=1}^{m}\frac{A_j B_j}{(B_j + A_j\bar{u}_0)^2}\right]}. \quad (69)$$

Returning to the original variables in (69), we immediately reproduce (39).
The boundary condition for (69) follows from (55):

$$\bar{u}_0(x, y, t) = 0 \quad \text{for} \quad (x, y) \in \partial\hat{\Omega}. \quad (70)$$

The initial conditions for (69) are found in the process of the construction of the boundary functions. For $\Pi_0 u$, $\Pi_{i0} v$, $\Pi_{j0} w$ we have the following system of equations:

$$\frac{\partial \Pi_0 u}{\partial \tau} = \frac{\delta}{h}\left(\sum_{i=1}^{n}[-a_i\bar{u}_0(x, y, 0)\Pi_{i0}v - a_i\bar{v}_{i0}(x, y, 0)\Pi_0 u - a_i\Pi_0 u\Pi_{i0}v - b_i\Pi_{i0}v_i]\right.$$
$$\left. + \sum_{j=1}^{m}[-A_j\bar{u}_0(x, y, 0)\Pi_{j0}w - A_j\bar{w}_0(x, y, 0)\Pi_0 u - A_j\Pi_0 u\Pi_{j0}w - B_j\Pi_{j0}w]\right). \quad (71)$$

$$\frac{\partial \Pi_{i0} v}{\partial \tau} = -a_i\bar{u}_0(x, y, 0)\Pi_{i0}v - a_i\bar{v}_{i0}(x, y, 0)\Pi_0 u$$
$$-a_i\Pi_0 u\Pi_{i0}v - b_i\Pi_{i0}v_i; \quad (72)$$

$$\frac{\partial \Pi_{j0} w}{\partial \tau} = -A_j\bar{u}_0(x, y, 0)\Pi_{j0}w - A_j\bar{w}_0(x, y, 0)\Pi_0 u$$
$$-A_j\Pi_0 u\Pi_{j0}w - B_j\Pi_{j0}w. \quad (73)$$

Together with the leading order regular functions the boundary functions $\Pi_0 u$, $\Pi_{i0} v$, $\Pi_{j0} w$ must satisfy the initial conditions (see (53), (54)):

$$\bar{u}_0(x, y, 0) + \Pi_0 u(x, y, 0) = U^*,$$
$$\bar{v}_{i0}(x, y, 0) + \Pi_{i0} v(x, y, 0) = 1,$$
$$\bar{w}_{j0}(x, y, 0) + \Pi_{j0} w(x, y, 0) = 1, \quad (74)$$
$$\text{for} \quad (x, y) \in \Omega.$$

Also, the boundary functions at any order must decay to zero as $\tau \to \infty$:

$$\Pi_0 u(x, y, \infty) = 0, \quad \Pi_{i0} v(x, y, \infty) = 0, \quad \Pi_{j0} w(x, y, \infty) = 0. \quad (75)$$

It follows from (71), (72), (73) that

$$\frac{\partial \Pi_0 u}{\partial \tau} = \frac{\delta}{h}\left(\sum_{i=1}^{n} \frac{\partial \Pi_{i0} v}{\partial \tau} + \sum_{j=1}^{m} \frac{\partial \Pi_{j0} w}{\partial \tau}\right). \tag{76}$$

Integrating (76)) with conditions (75), we arrive at the relation:

$$\Pi_0 u = \frac{\delta}{h}\left(\sum_{i=1}^{n} \Pi_{i0} v + \sum_{j=1}^{m} \Pi_{j0} w\right). \tag{77}$$

Substituting (59), (60), and (77) into the system describing initial conditions in the leading order approximation (74), we obtain:

$$\begin{aligned} &\bar{u}_0(x,y,0) + \frac{\delta}{h}\left(\sum_{i=1}^{n} \Pi_{i0} v(x,y,0) + \sum_{j=1}^{m} \Pi_{j0} w(x,y,0)\right) = U^*, \\ &\frac{b_i}{b_i + a_i \bar{u}_0(x,y,0)} + \Pi_{i0} v(x,y,0) = 1, \\ &\frac{B_j}{B_j + A_j \bar{u}_0(x,y,0)} + \Pi_{j0} w(x,y,0) = 1. \end{aligned} \tag{78}$$

This is a system of $(n+m+1)$ equations for $(n+m+1)$ unknowns, the initial conditions for the regular function $\bar{u}_0$ and the boundary functions $\Pi_{i0} v$, $\Pi_{j0} w$.

From (78) we express

$$\begin{aligned} \Pi_{i0} v(x,y,0) &= 1 - \frac{b_i}{b_i + a_i \bar{u}_0(x,y,0)} = \frac{a_i \bar{u}_0(x,y,0)}{b_i + a_i \bar{u}_0(x,y,0)}, \\ \Pi_{j0} w(x,y,0) &= 1 - \frac{B_j}{B_j + A_j \bar{u}_0(x,y,0)} = \frac{A_j \bar{u}_0(x,y,0)}{B_j + A_j \bar{u}_0(x,y,0)}. \end{aligned} \tag{79}$$

Substituting (79) into the first equation of (78), we obtain one equation for $\bar{u}_0(x,y,0)$:

$$\bar{u}_0(x,y,0) + \frac{\delta}{h}\left(\sum_{i=1}^{n} \frac{a_i \bar{u}_0(x,y,0)}{b_i + a_i \bar{u}_0(x,y,0)} + \sum_{j=1}^{m} \frac{A_j \bar{u}_0(x,y,0)}{B_j + A_j \bar{u}_0(x,y,0)}\right) = U^*, \tag{80}$$

which we re-write for convenience as

$$F(\bar{u}_0(x,y,0)) = U^*, \tag{81}$$

where the expression for F corresponds to the left hand side of equation (80).

The function $F(u)$ has the following properties: $F(0) = 0$, $\lim_{u\to\infty} F(u) \to \infty$. Also,

$$F'(u) = 1 + \frac{\delta}{h}\left(\sum_{i=1}^{n} \frac{a_i b_i}{(b_i + a_i u)^2} + \sum_{j=1}^{m} \frac{A_j B_j}{(B_j + A_j u)^2}\right) > 0 \tag{82}$$

for all u.

It follows from the above properties of F that equation (81) has a unique solution $\bar{u}_0(x,y,0)$ for any choice of $U^* \geq 0$ (geometrically this means that the increasing function $F(u)$ intersects the horizontal line with the ordinate $U^* \geq 0$ only at one point).

Now the initial condition for $\bar{u}_0$ is known. The problem for $\bar{u}_0$ is completely defined. It consists of the equation (69), boundary condition (70), and the initial condition that solves (81) (which is equivalent to (80)).

When $\bar{u}_0(x, y, 0)$ is known, the initial conditions for the boundary functions are found from (79). It can be shown that the equations (72), (73) for $\Pi_{i0}v$, $\Pi_{j0}w$, where the expression (77) is substituted for $\Pi_0 u$, with corresponding initial conditions have solutions exponentially decaying to zero as $\tau \to \infty$.

Remark. If $\delta/h \ll 1$, then equation (80) for $\bar{u}_0(x, y, 0)$ may be replaced by its approximation

$$\bar{u}_0(x, y, 0) \approx U^*. \tag{83}$$

So, in this case the initial condition for $\bar{u}_0$ is approximated by the original initial condition value U^*. Such an approximation was used in the illustrative example discussed in this paper.

References

[1] Coggan J S, Bartol T M, Esquenazi E, Stiles J L, Lamont S, Martone M E, Berg D K, Ellisman M H and Sejnowski T J 2005 Evidence for ectopic neurotransmission at a neuronal syapse *Science* **309** 446

[2] Franks K M, Bartol T M and Sejnowski T J 2002 A Monte Carlo model reveals indepedent signaling at central glutamatergic synapses *Biophys. J.* **83** 2333

[3] Haario H and Kalachev L 1999 Model reductions and identifications for multi-phase phenomena *Math. Engrng Ind.* **7(4)** 457

[4] Keener J and Sneyd J 1998 *Mathematical Physiology* (Berlin, Heidelberg, New York: Springer-Verlag)

[5] Larsson H P, Tzingounis A V, Koch S P and Kavanaugh M P 2004 Fluorometric measurements of conformational changes in glutamate transporters *PNAS* **101(11)** 3951

[6] Nielsen T A and DiGregorio D A 2004 Modulation of glutamate mobility reveals the mechanism undelying slow-rising AMPAR EPSCs and the diffusion coefficient in the synaptic cleft, *Neuron* **42** 757

[7] Rusakov D A and Kullman D M 1998 Extrasynaptic glutamate diffusion in the hippocampus: ultrastructural constraints, uptake, and receptor activation *J. Neurosci.* **18(9)** 5158

[8] Smith G D, Dai L, Miura R M and Sherman A 2001 Asymptotic analysis of buffered calcium diffusion near a point source *SIAM J. Appl. Math.* **61(5)** 1816

[9] Sneyd J, Dale P D and Duffy A 1998 Traveling waves in buffered systems: application to calcium waves *SIAM J. Appl. Math.* **58(4)** 1178

[10] Vasil'eva A B, Butuzov V F and Kalachev L V 1995 *The Boundary Function Method for Singular Perturbation Problems* (Philadelphia: SIAM)

[11] Wadiche J I and Kavanaugh M P 1998 Macroscopic and microscopic properties of a cloned glutamate transporter/chloride channel *J. Neurosci.* **18(19)** 7650

Institute of Physics Publishing
doi:10.1088/1742-6596/55/1/012

Journal of Physics: Conference Series **55** (2006) 130–134
International Workshop on Multi-Rate Processes and Hysteresis

Hysteresis in biological models

J Kopfová

Mathematical Institute of the Silesian University in Opava

E-mail: jana.kopfova@math.slu.cz

Abstract. The paper gives an overview of results for partial differential equations with hysteresis whose motivation comes from biology.

1. Introduction

In the biological literature, there are examples of processes whose state variables change due to a change of parameters in such a way that when the parameters go back to the old values the system does not follow its steps in reverse and thus a hysteresis loop is formed. Also many biological problems involve a fold catastrophe regime which can be replaced by a model involving hysteresis [8].

Although during the last decades there has been a steady growth in the mathematical study of various hysteresis operators and applications, the mathematical treatment of biological problems with hysteresis has been considered so far in few papers.

It is the purpose of the present paper to give an overview of such biological problems studied from a mathematical point of view.

2. Model for bacterial growth

A model for bacterial growth patterns was the first biological model involving hysteresis which was studied mathematically [5], [6], [7].

Similar to the classical experiment for Liesegang phenomena in chemical precipitation, concentrical growth rings were observed in response to a diffusing front of histidine auxotrophic salmonella typhimurium spreading from the center of a Petri dish to its boundary. Here the bacteria are immobile, they are fixed on an agar gel containing all chemicals necessary for growth except the missing amino acid. Therefore the spatial interaction is caused only by diffusion of the nutrients and the buffer neutralized by acids produced as byproducts of the cell growth.

Let us note that in an attempt to better understand the mechanism for the formation of concentric rings of growth, additional insight into the process was sought using the technique of mathematical modelling, see e.g. [13]. To explain the periodic structure a mathematical model with hysteresis was suggested in [5].

With the assumption that there exist thresholds for growth expressed in terms of the concentration of chemicals, i.e., there is no growth until a certain threshold is reached and the growth continues even as the system is falling below this threshold until a second threshold is reached where the growth stops, the model was described mathematically in [6] by a system of partial differential equations of the following type:

$$\frac{\partial B}{\partial t} = \alpha VB, \qquad \text{in } \Omega \times (0,T) \tag{1}$$
$$\frac{\partial H}{\partial t} = D_H \triangle H - \beta VB, \qquad \text{in } \Omega \times (0,T) \tag{2}$$
$$\frac{\partial G}{\partial t} = D_G \triangle G - \gamma VB, \qquad \text{in } \Omega \times (0,T), \tag{3}$$

where H describes the histidine concentration, G the concentration of the growth mediums buffer and B the size of the bacterial population. V is a function describing the metabolic activity of bacteria. It is assumed that the cell growth continues until the combination of H and G reaches a threshold value at which cell growth stops, growth does not begin again until a higher threshold is reached. This is modelled with a hysteresis operator of relay type.

Here $\alpha, \beta, \gamma, D_H, D_G$ are given positive constants, large compared to the positive diffusivity constants D_H, D_G, and Ω is a bounded region in R^n with smooth boundary $\partial\Omega$.

The system (1)-(3) was coupled with the following boundary and initial conditions:

$$\frac{\partial H}{\partial \nu} = \frac{\partial G}{\partial \nu} = 0 \text{ on } \partial\Omega \times (0,T), \tag{4}$$

$$B(x,0) = B_0(x), H(x,0) = H_0(x), G(x,0) = G_0(x) \text{ on } \Omega, \tag{5}$$

where ν is the outward unit normal vector to the boundary.

In [6], [5] numerical results for this model were presented giving a very good correspondence with biological experiments. The mathematical problem, however, was never solved. An attempt has been made by A. Visintin, [11], where a simplified model

$$\frac{\partial B}{\partial t} = \alpha V \tag{6}$$
$$\frac{\partial H}{\partial t} = D_H \triangle H - \beta V \tag{7}$$
$$\frac{\partial G}{\partial t} = D_G \triangle G - \gamma V \tag{8}$$

was shown to have a solution by approximation, a priori estimates and a limit procedure. An existence result for the system (1)-(3) was not proved until recently, [1]. Note that uniqueness, because of the presence of the hysteresis operator of the type of completed relay in the source term, is not expected, see [12].

The hysteresis relation can be described mathematically as follows: Let ψ_{on}, ψ_{off} be functions on R^2 satisfying $\psi_{off} \geq \psi_{on}$ on R^2, ψ_{off} will denote the threshold at which cell growth stops and ψ_{on} denotes that at which cells begin to grow again. We will use the following notation

$$\Gamma_{on} = \{(H,G); \psi_{on}(H,G) = 0\}, \tag{9}$$
$$\Gamma_{off} = \{(H,G); \psi_{off}(H,G) = 0\}, \tag{10}$$
$$M_{on} = \{(H,G); \psi_{on}(H,G) > 0\}, \tag{11}$$
$$M_{off} = \{(H,G); \psi_{off}(H,G) < 0\}, \tag{12}$$
$$M_{on-off} = \{(H,G); 0 \leq \psi_{off}(H,G), \psi_{on}(H,G) \leq 0\}, \tag{13}$$
$$M^*_{on} = M_{on} \cup \Gamma_{on}, \tag{14}$$
$$M^*_{off} = M_{off} \cup (\Gamma_{off}/\Gamma_{on}), \tag{15}$$
$$M = \{(H,G); 0 < \psi_{off}(H,G), \psi_{on}(H,.G) < 0\}. \tag{16}$$

Definition 2.1. *For any $(H,G) \in [C^0([0,T])]^2$ and any initial state $\xi \in \{0,1\}$ we define the (single valued) function $S = S(H,G,\xi) : [0,T] \to \{0,1\}$ as follows:*

$$S(0) = \begin{cases} 0 & \textit{if } (G(0),H(0)) \in M^*_{off} \\ \xi & \textit{if } (H(0),G(0)) \in M \\ 1 & \textit{if } (H(0),G(0)) \in M^*_{on} \end{cases}$$

for any $t \in (0,T]$, and setting $X_t = \{\tau \in (0,T],\ (H(t),G(t)) \in \Gamma_{on} \text{ or } \Gamma_{off}\}$,

$$S(t) = \begin{cases} S(0) & \textit{if } X_t = \emptyset \\ 0 & \textit{if } X_t \neq \emptyset \textit{ and } (H(maxX_t),G(maxX_t)) \in \Gamma_{off} \\ 1 & \textit{if } X_t \neq \emptyset \textit{ and } (H(maxX_t),G(maxX_t)) \in \Gamma_{on}, \end{cases}$$

see Figure 1.

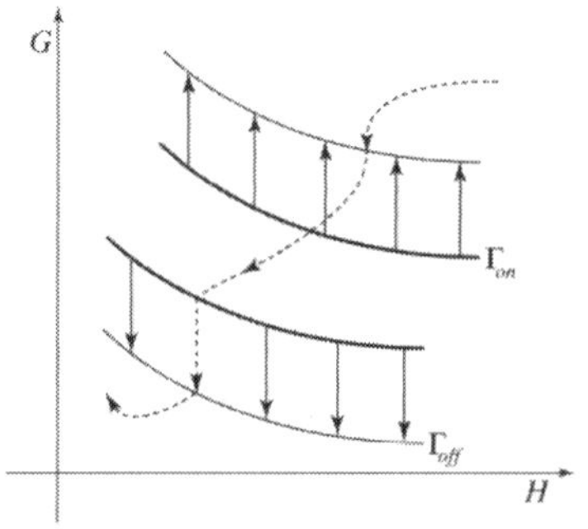

Figure 1. The hysteresis operator

Notice the similarity with the definition of a classical relay operator, see e.g. [12] for more details.

It can be shown that the operator S is not closed with respect to the strong topology of $[C^0([0,T])]^2$ for the input (H,G) and the weak star topology of $BV(0,T)$ for the output S.

Then the operator $\tilde{S} : [C^0([0,T])]^2 \times [0,1] \to \mathcal{P}(BV(0,T))$ is defined as follows:

$$\tilde{S}(0) = \begin{cases} \{0\} & \text{if } (H(0),G(0)) \in M_{off} \\ [0,\xi] & \text{if } (H(0),G(0)) \in \Gamma_{off} \\ \{\xi\} & \text{if } (H(0),G(0)) \in M \\ [\xi,1] & \text{if } (H(0),G(0)) \in \Gamma_{on} \\ \{1\} & \text{if } (H(0),G(0)) \in M_{on} \end{cases}$$

$$\tilde{S}(t) = \begin{cases} \{0\} & \text{if } (H(t),G(t)) \in M_{off} \\ [0,1] & \text{if } (H(t),G(t)) \in M_{on-off} \\ \{1\} & \text{if } (H(t),G(t)) \in M_{on} \end{cases}$$

$$\begin{cases} \text{if } (H(t),G(t)) \neq \Gamma_{on},\Gamma_{off}, & \text{then } \tilde{S} \text{ is constant in a neighbourhood of } t \\ \text{if } (H(t),G(t)) \in \Gamma_{off}, & \text{then } \tilde{S} \text{ is nonincreasing in a neighbourhood of } t \\ \text{if } (H(t),G(t)) \in \Gamma_{on}, & \text{then } \tilde{S} \text{ is nondecreasing in a neighbourhood of } t. \end{cases} \quad (17)$$

The hysteresis relation is assumed to hold pointwise in space $V(x,t) = \tilde{S}(H(x,\cdot),G(x,\cdot))(t)$.

For the existence proof, the relay type hysteresis operator is in [1] approximated by a sequence of operators of play type which can be represented as differential inclusions:

$$\frac{\partial V}{\partial t} + \partial I_\epsilon(G, H; V) \ni 0, \tag{18}$$

where $I_\epsilon(G, H; \cdot)$ is the indicator function of the closed interval $[f_{*\epsilon}, f^*_\epsilon]$, f_* and f^* are functions on R^2 such that $f_* = 1$ on M^*_{on} and $f_* = 0$ otherwise, $f^* = 0$ on M^*_{off} and $f^* = 1$ otherwise, and $\{f_{*\epsilon}\}$ and $\{f^*_\epsilon\}$ are sequences of $C^2(R^2)$ functions approximating it, i.e. $f_{*\epsilon}(x) \to f_*(x)$ and $f^*_\epsilon(x) \to f^*(x)$ as $\epsilon \to 0$ for each $x \in R^2$. It is well known that the differential inclusion (18) is equivalent to the classical play operator.

Theorem 2.1. *If the Assumptions* $(A1)$ - $(A3)$

(A1) $B_0, H_0, G_0 \in L^\infty(\Omega)$

(A2) $B_0 \geq 0, H_0 \geq 0, G_0 \geq 0.$

(A3) $f_*(H_0, G_0) \leq V_0 \leq f^*(H_0, G_0).$

are satisfied, there exists at least one solution of the system (1) – (3) *coupled with initial conditions* (5) *and boundary conditions* (4) *and* V *a solution of (18).*

The theorem is proved using the standard technique of Yosida approximation, deriving uniform bounds and limit procedure.

3. Prey-predator systems with hysteresis

In [2] a system of nonlinear PDEs with diffusive as well as hysteresis effects is suggested to model the evolution of populations. The model originates from a prey-predator model of the following type:

$$\sigma = \lambda(u) \quad \text{in } \Omega \times (0, T) \tag{19}$$

$$\frac{\partial u}{\partial t} - \Delta u = h(\sigma, u, v) \quad \text{in } \Omega \times (0, T) \tag{20}$$

$$\frac{\partial v}{\partial t} - \Delta v = g(\sigma, u, v) \quad \text{in } \Omega \times (0, T). \tag{21}$$

$$\tag{22}$$

Here σ denotes the density of the quality of food for the prey, u and v are densities of the prey and predator, respectively and Ω is a bounded region in R^n with smooth boundary $\partial\Omega$.

The speed of change of the density of food when the density of the prey decreases is different from the speed when the density of the prey increases, i.e. the state of the system depends on the previous evolution of data. Also when a small diffusive effect for the food of the prey is considered a more complicated model is obtained:

$$\frac{\partial \sigma}{\partial t} - \frac{\partial \lambda(u)}{\partial t} - k\Delta\sigma + \partial I_{u,v}(\sigma) \ni F(\sigma, u, v) \quad \text{in } \Omega \times (0, T) \tag{23}$$

$$\frac{\partial u}{\partial t} - \Delta u = h(\sigma, u, v) \quad \text{in } \Omega \times (0, T) \tag{24}$$

$$\frac{\partial v}{\partial t} - \Delta v = g(\sigma, u, v) \quad \text{in } \Omega \times (0, T), \tag{25}$$

where $I_{u,v}(\cdot)$ denotes the indicator function of a closed interval $[f_*(u, v), f^*(u, v)]$, $f_*(u, v)$ and $f^*(u, v)$ are C^2 functions, $0 \leq f_*(u, v) \leq f^*(u, v) \leq 1$, $\partial I_{u,v}(\cdot)$ is the subdifferential of $I_{u,v}(\cdot)$

Let us note that if σ is the solution of the differential inclusion

$$\frac{\partial \sigma}{\partial t} - \frac{\partial u}{\partial t} + \partial I_u(\sigma) \ni 0 \text{ in } \Omega \times (0, T), \tag{26}$$

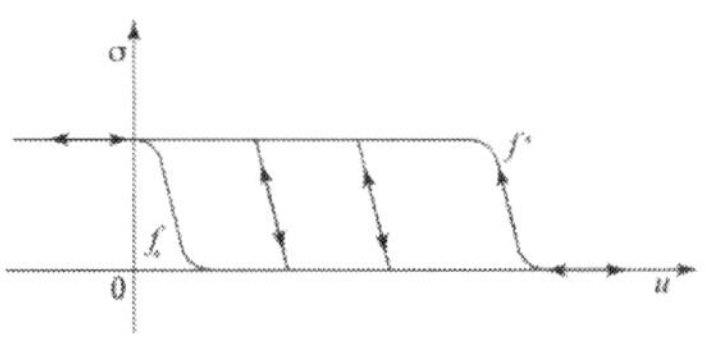

Figure 2. Generalized stop operator

then σ is the output of the hysteresis operator of stop type, see Figure 2.

This fact was pointed in [12] and used in many papers for the analysis of nonlinear phenomena, e.g. in real-time control problems, or solid-liquid phase transition where hysteresis effects are taken into account. The paper [2] is devoted to a detailed analysis of the system (23) -(25). Results for positivity, boundedness, existence and uniqueness of solutions of the prey-predator model are obtained. Under appropriate assumptions, using the method of Yosida approximation combined with the derivation of appropriate uniform bounds, the authors prove that there exists at least one solution of the system. Uniqueness is obtained for $n \leq 3$.

In [10] a more complicated system for population dynamics is analyzed, taking also convective terms into account:

$$\frac{\partial \sigma}{\partial t} - \nabla \cdot (\nabla \sigma + \lambda(\sigma)) + \partial I_U(\sigma) \ni g(\sigma, U) \quad \text{in } \Omega \times (0,T) \tag{27}$$

$$\frac{\partial u_i}{\partial t} - \nabla \cdot (\nabla u_i + \mu_i(u_i)) = h_i(\sigma, U) \quad \text{in } \Omega \times (0,T), \tag{28}$$

where $U = (u_1, \ldots, u_m)$.

Results for non-negativity, boundedness and existence of at least one solution of the population model with hysteresis effect are obtained. The main tools are the method of Yosida approximation, the energy method for parabolic systems and fixed point arguments.

Acknowledgments

This work was supported by the project MSM4781305904 of the Czech Ministry of Education.

References

[1] Aiki T and Kopfová J A hysteresis model for bacterial growth (in preparation)
[2] Aiki T and Minchev E 2005 A prey-predator model with hysteresis effect *SIAM J. Math. Anal.* **36** 2020–32
[3] Brokate M and Sprekels J 1996 *Hysteresis and Phase Transitions* Appl. Math. Sci. Vol. 121 (Springer)
[4] Colli P, Kenmochi N and Kubo M 2001 A phase-field model with temperature dependent constraint *J. Math. Anal. Appl.* **256** 668–85
[5] Hoppensteadt F C and Jäger W 1980 Pattern formation by bacteria *Lecture Notes in Biomathematics* **38** 68–81
[6] Hoppensteadt F C, W.Jäger and Poppe C 1984 A hysteresis model for bacterial growth patterns *Modelling of Patterns in Space and Time, Lecture Notes in Biomath.* **55** (Springer) pp 123–34
[7] Jäger W 1981 A diffusion-reaction system modelling spatial patterns *Proc. Equadiff* **5** (Bratislava) pp 151–8
[8] Krasnosel'skii M A and Pokrovskii A V 1989 *Systems with Hysteresis* (Springer)
[9] Krejčí P 1996 *Hysteresis, Convexity and Dissipation in Hyperbolic Equations* Gakuto Int. Series Math. Sci. Appl., vol 8 (Gakkötosho)
[10] Minchev E 2004 A diffusion-convection prey-predator model with hysteresis *Math. J. Toyama Univ.* **27** 51–69
[11] Visintin A 1986 Evolution problems with hysteresis in the source term *SIAM J. Math. Anal* **17** 1113–38
[12] Visintin 1994 *Differential Models of Hysteresis* (Springer)
[13] Wimpeny J W T, Jaffe S and Coombs J P Periodic growth phenomena in spatially organized microbial systems *Modelling of Patterns in Space and Time, Lecture Notes in Biomath.* **55** (Springer) pp 388–403

Institute of Physics Publishing
doi:10.1088/1742-6596/55/1/013

Journal of Physics: Conference Series **55** (2006) 135–143
International Workshop on Multi-Rate Processes and Hysteresis

Quasilinear hyperbolic equations with hysteresis

P Kordulová

Mathematical Institute, Silesian University in Opava, 746 01 Opava, Czech Republic

E-mail: Petra.Kordulova@math.slu.cz

Abstract. This paper is devoted to the summary of results about a quasilinear hyperbolic partial differential equation of first order with a hysteresis operator $v = \mathcal{F}[u]$:

$$\frac{\partial(u+v)}{\partial t} + \frac{\partial u}{\partial x} = 0.$$

Hysteresis is represented by functional describing adsorption and desorption on the particles of the substance and is a possibly discontinuous generalized play operator. The results can be extended to possibly discontinuous generalized Prandtl-Ishlinskii operators of play type.

1. Introduction

We study the partial differential equation

$$\frac{\partial(u+v)}{\partial t} + \frac{\partial u}{\partial x} = 0, v = \mathcal{F}[u] \text{ in } (0,L) \times [0,T], \tag{1}$$

as a generic model for the transport and adsorption of a chemical of concentration $u(x,t)$ carried in a solution with constant unit velocity in a tube $x \in (0,L)$ for $t > 0$. Here $\mathcal{F}[\cdot]$ is a general functional describing adsorption and desorption of the chemical on the particles of solid filling up the tube. In the classical case $\mathcal{F}[\cdot]$ is given by an isotherm, a real-valued function corresponding to an experimentally determined relation between the concentration $v(x,t)$ of the adsorbed species on the surface of the particles and the concentration $u(x,t)$ in solution. In the general situation considered here, the adsorption-desorption functional $\mathcal{F}[\cdot]$ exhibits hysteresis, i.e., the relations between u and v for the case when u is increasing (adsorption) and decreasing (desorption) follow different curves. The motivation for our study comes from applications in chemical and geological engineering. Hysteresis coupled to transport phenomena occurs also in the modelling of oil-water interaction and of waste treatment in subsurface reservoirs [6]. The phenomenon of hysteresis in adsorption has been observed and studied for many years (see [4, 7]), frequently in parallel with capillary condensation hysteresis (see [5]).

The equation (1) was studied in [8, 9, 18, 20, 21]. Visintin investigated the Cauchy problem for equation (1) with the hysteresis functional represented by a possibly discontinuous generalized play operator by using the semigroup approach and got the existence and uniqueness of the integral solution in L^1. These results can be extended to possibly discontinuous generalized Prandtl-Ishlinskii operators of play type, which includes the case of possibly discontinuous Preisach operators. In his book [20] it was posed as an open problem whether the integral solution of (1) with hysteresis satisfies an entropy condition introduced by Kružkov. Such an entropy condition was derived by Kopfová [8]. Visintin also considered the Cauchy problem for the completed relay operator and its regularization and proved the existence of a weak solution.

Showalter and Peszynska [17] obtained the existence and uniqueness of differentiable solutions by the theory of nonlinear semigroups in a Hilbert space L^2. They assumed that hysteresis is represented by a classical play operator and also by a more general case of a convex adsorption-desorption hysteresis functional. Kordulová [10] resumed the investigation of continuous solutions with a slightly modified classical play operator and found that even a small change of the hysteresis operator can cause a discontinuity in the solution. In [11] an asymptotic stability of the solution of the equation (1) in L^1 is derived.

The results can be extended to the more general quasilinear partial differential hyperbolic equation

$$\frac{\partial(u+v)}{\partial t} + \sum_{j=1}^{N} \frac{\partial}{\partial x_j}(b_j u) + cu \;=\; f, \tag{2}$$

where b_j and c are given smooth functions, [9, 20].

The paper is organized as follows. In Section 2 the definitions of hysteresis operators are recalled. In Section 3 Visintin's results about the existence and uniqueness of an integral solution and the existence of a weak solution are summed up. The entropy condition for equation (1) is established. Section 4 is devoted to the existence and uniqueness of a differentiable solution of (1) with a classical play operator. Finally the asymptotic behaviour of the solutions is characterized in Section 5 and results of [10] are described.

2. Hysteresis operators

In this section the definitions of hysteresis operators are briefly reviewed, for more details see [12, 20].

We consider equation (1) and couple it with the hysteresis relation

$$v(x,t) = [\mathcal{E}(u(x,.), v_0(x)]\,(t) \quad \text{in } [0,T], \text{ a.e. in } \Omega\ . \tag{3}$$

Here $\mathcal{E}$ is a multivalued functional, $v_0(x)$ a given initial value. Its values depend not only on the current value of $u(.,t)$ at $t > 0$, but on the past history $u(.,s), 0 < s < t$. We consider at first $\mathcal{E}$ to be a generalized play operator.

We set $\widetilde{\mathbb{R}} := [-\infty, +\infty]$, $\widetilde{\mathbb{R}}^+ := [0 + \infty]$, $\widetilde{\mathbb{R}}^- := [-\infty, 0]$.

Let $\gamma_l, \gamma_r : \mathbb{R} \to \mathcal{P}(\widetilde{\mathbb{R}})$ be maximal monotone (possibly multivalued) functions with

$$\inf \gamma_r(u) \leq \sup \gamma_l(u), \quad \forall u \in \mathbb{R}. \tag{4}$$

Let Λ denote the hysteresis region, i.e., the subset of $\mathbb{R}^2$ of admissible pairs (u, v) such that $\inf \gamma_r(u) \leq v \leq \sup \gamma_l(u)$.

Now, given $v_0 \in \mathbb{R}$, we construct the hysteresis operator $\mathcal{E}(\cdot, v_0)$ as follows. Let u be any continuous, piecewise linear function on $\mathbb{R}^+$ such that u is linear on $[t_{i-1}, t_i]$ for $i = 1, 2, \ldots$. We then define $v := \mathcal{E}(u, v_0) : \mathbb{R}^+ \to \mathbb{R}$ by

$$v(t) := \begin{cases} \min\{\gamma_l(u(0)), \max\{\gamma_r(u(0)), v_0\}\} & \text{if} \quad t = 0 \\ \min\{\gamma_l(u(t)), \max\{\gamma_r(u(t)), v(t_{i-1})\}\} & \text{if} \quad t \in (t_{i-1}, t_i], i = 1, 2, \ldots. \end{cases}$$

Note that $v(0) = v_0$ only if $\gamma_r(u(0)) \leq v_0 \leq \gamma_l(u(0))$.

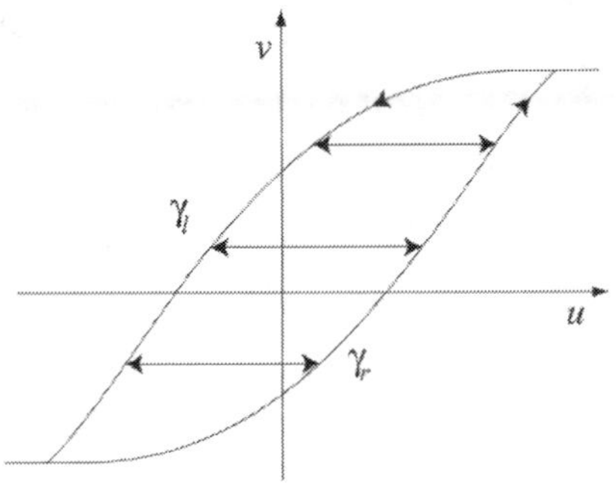

Figure 1. Generalized play operator

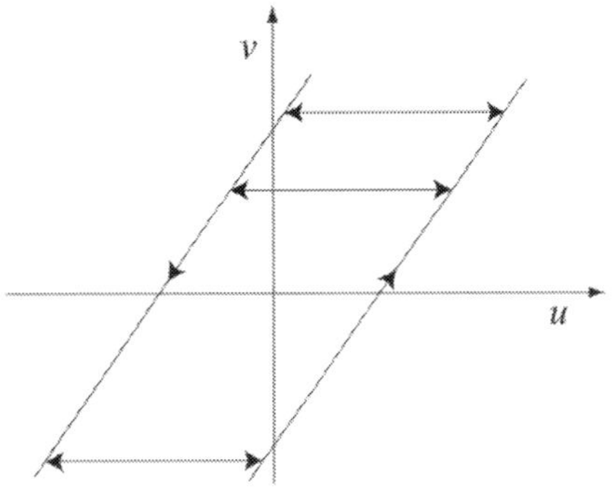

Figure 2. Classical play operator

As proved in Visintin [20], Section III.2, $\mathcal{E}(\cdot, v_0)$ can be extended to $C^0([0,T])$ by continuity. This operator is called a generalized play [12] (see Figure 1).

A special example of a generalized play operator, where $\gamma_r = u - r$, $\gamma_l = u + r$ is called the classical play operator (Figure 2).

Let us fix any pair $\rho := (\rho_1, \rho_2) \in \mathbb{R}^2$, with $\rho_1 < \rho_2$. If the right hysteresis boundary curve is defined as

$$\gamma_r(u) \in \begin{cases} -1 & \text{if} \quad u < \rho_2 \\ [-1,\ 1] & \text{if} \quad u = \rho_2 \\ 1 & \text{if} \quad u > \rho_2, \end{cases} \tag{5}$$

and the left boundary curve is given by

$$\gamma_l(u) \in \begin{cases} -1 & \text{if} \quad u < \rho_1 \\ [-1,\ 1] & \text{if} \quad u = \rho_1 \\ 1 & \text{if} \quad u > \rho_1, \end{cases} \tag{6}$$

then the generalized play operator is the completed relay operator [21] (see Figure 3).

To define the generalized Prandtl-Ishlinskii operator of play type, let us assume that we are given a measure space $(\mathcal{P}, \mathcal{A}, \mu)$, where μ is a finite Borel measure. For almost any $\sigma \in \mathcal{P}$, let $(\gamma_{\sigma l}, \gamma_{\sigma r})$ be a pair of functions $\mathbb{R} \to \mathbb{R}$, satisfying (4), and for each $\sigma \in \mathcal{P}$ let $v_{\sigma 0} \in \mathbb{R}$, be a given initial value. Let $\mathcal{E}_\sigma(\cdot, v_{\sigma 0})$ be the corresponding generalized play operator corresponding

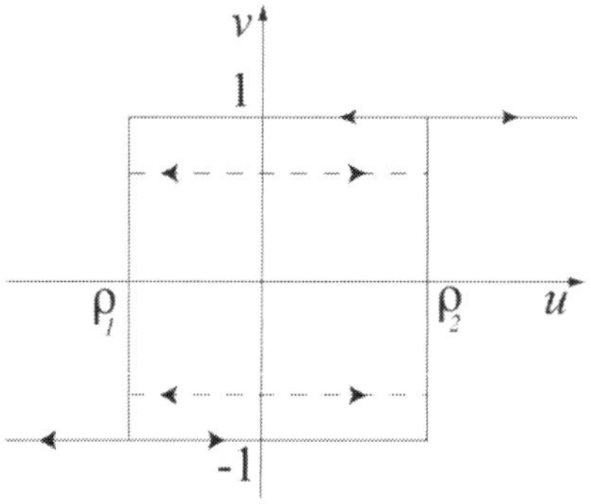

Figure 3. Completed relay operator

to the couple $(\gamma_{\sigma l}, \gamma_{\sigma r})$. Then the operator defined as

$$\tilde{\mathcal{E}}_\mu \left(u, \{v_{\sigma 0}\}_{\sigma\in\mathcal{P}}\right) = \int_{\mathcal{P}} \mathcal{E}_\sigma \left(u, v_{\sigma 0}\right) d\mu(\sigma)$$

is called a generalized Prandtl-Ishlinskii operator of play type.

If boundary curves $\gamma_{\sigma r}$ and $\gamma_{\sigma l}$ are defined as (5) and (6) then the generalized Prandtl-Ishlinskii operator corresponds with the Preisach operator [20].

3. Integral solution

Visintin formulated equation (1) by using a system of differential inclusions containing an accretive operator for a possibly discontinuous generalized play operator. Equation (1) is equivalent to the Cauchy problem:

$$\begin{cases} \frac{\partial U}{\partial t} + \mathcal{A}(U) + \Re(U) & \ni & 0, \quad \text{in } \Omega \times [0,T] \\ U(0) & = & U_0, \end{cases} \tag{7}$$

where

$$\begin{aligned} \Omega &:= (0, L) \\ \mathcal{D}(\mathcal{A}) &:= \{U := (u,v) \in \mathbb{R}^2 : \inf \gamma_r(u) \le v \le \sup \gamma_l(u)\} \\ \mathcal{A}(U) &:= \{(\xi, -\xi) \in \mathbb{R}^2 : \xi \in \phi(U) \cap \mathbb{R}\} \ \forall U \in \mathcal{D}(\mathcal{A})\} && (8) \\ \mathcal{B}(u) &:= \frac{\partial u}{\partial x} \\ \Re(U) &:= (\mathcal{B}(u), 0) && (9) \\ \mathcal{D}(\Re) &:= \{U \in L^1(\Omega, \mathbb{R}_1^2) : \mathcal{B}u \in L^1(\Omega)\} \\ \mathcal{Q}(U) &:= \mathcal{A}(U) + \Re(U) \\ \mathcal{D}(\mathcal{Q}) &:= \{U := (u,v) \in L^1(\Omega, \mathbb{R}_1^2) : U \in \mathcal{D}(\mathcal{A}) \text{ a.e. in } \Omega, \\ & \quad u \in W^{1,1}(\Omega), u(0) = 0\}. \end{aligned}$$

(10)

By setting

$$U := (u, v), \quad U_0 := (u_0, v_0),$$

$$\phi(u,v) = \begin{cases} \{+\infty\} & \text{if} \quad v < \inf \gamma_r(u) \\ \widetilde{\mathbb{R}}^+ & \text{if} \quad v \in \gamma_r(u) \setminus \gamma_l(u) \\ \{0\} & \text{if} \quad \sup \gamma_r(u) < v < \inf \gamma_l(u) \\ \widetilde{\mathbb{R}}^- & \text{if} \quad v \in \gamma_l(u) \setminus \gamma_r(u) \\ \{-\infty\} & \text{if} \quad v > \sup \gamma_l(u) \\ \widetilde{\mathbb{R}} & \text{if} \quad v \in \gamma_l(u) \cap \gamma_r(u), \end{cases} \tag{11}$$

the following result is proved in [20]:

Theorem 3.1 *Assume that $\gamma_l(u)$, $\gamma_r(u)$ fulfil (4) and are affinely bounded, that is, there exist constants C_1, $C_2 > 0$, such that $\forall w \in \mathbb{R}$, $\forall z \in \gamma_h(w)$*

$$|z| \leq C_1|w| + C_2, \quad (h = l, r). \tag{12}$$

Then the operator $\mathcal{A} + \Re$ is m- and T-accretive in $L^1(\Omega, \mathbb{R}_1^2)$.

Classical results of the theory of nonlinear semigroups can be applied yielding the following existence and uniqueness result.

Theorem 3.2 *Let Ω be an open subset of $\mathbb{R}^N (N \geq 1)$ of Lipschitz class. Let $L^1(\Omega, \mathbb{R}^2)$ be endowed with the norm*

$$\| U \|_{L^1(\Omega,\mathbb{R}^2)} := \int_\Omega (| u(x) | + | v(x) |) dx$$

$$\forall U := (u, v) \in L^1(\Omega, \mathbb{R}^2).$$

Assume that (12) holds. Take any $U_0 := (u_0, v_0) \in L^1(\Omega, \mathbb{R}_1^2)$ such that $U_0 \in \mathcal{D}(\phi)$ a.e. in Ω.

Then the Cauchy problem (7) has one and only one integral solution $U : [0, T] \to L^1(\Omega, \mathbb{R}_1^2)$, which depends continuously on data u_0, v_0. Moreover, if $\Re(u_0) \in L^1(\Omega)$, then U is Lipschitz continuous.

The theory of nonlinear semigroups is used to prove the existence and uniqueness of the integral solution of (7). The exact result arises from Theorem 3.2. It was expected, see [20], that the integral solution of (1) fulfills a condition of the type introduced by Kružkov. Visintin formulated it as an open problem in his book. Kopfová [8] showed that an integral solution of (1) defined by the semigroup theory (Theorem 3.2) satisfies an entropy condition introduced by Kružkov if $v = \mathcal{F}[u]$ is represented by a (possibly discontinuous) generalized play or Prandtl-Ishlinskii hysteresis operator of play type. This is proved without the assumption of the symmetry of hysteresis boundary curves, unlike in [8].

Theorem 3.3 *Let the assumptions of Theorem 3.2 hold. Let $\mathcal{A}_0 U = \mathcal{A}(U) + \mathcal{R}(U)$ on $\mathcal{D}(\mathcal{A}_0)$, and let $\mathcal{S}(t) = (u, v)$ be the corresponding semigroup of contractions.*
Let $w \in \overline{\mathcal{D}(\mathcal{A})}$ and $t \geq 0$. Then if $w = (u_0, v_0) \in L^\infty(\Omega) \times L^\infty(\Omega)$,

$$\int_0^T \int_\Omega |u - k| \psi_t(x,t) dx\, dt + \int_0^T \int_\Omega |v - \tilde{k}| \psi_t(x,t) dx\, dt + \int_0^T \int_\Omega |u - k| \psi_x(x,t) dx\, dt \geq 0$$

for every $\psi(x,t) \in C_0^\infty((0,T) \times \Omega)$ such that $\psi \geq 0$ and every $k, \tilde{k} \in \Lambda$ and $T > 0$.

A generalization of this theorem, which also includes the equation (2), is proved in [9].

Visintin dealt with the Cauchy problem for the quasilinear hyperbolic equation of first order that contains an either continuous or discontinuous hysteresis operator $\mathcal{F}[u]$:

$$\begin{cases} \frac{\partial}{\partial t}(u+\mathcal{F}[u]) + \frac{\partial u}{\partial x} = f & \text{in} \quad \mathbb{R}\times(0,T) \\ (u+v)(x,0) = u(0)+v(0) & \text{in} \quad \mathbb{R}, \end{cases} \tag{13}$$

where $\mathcal{F}[u]$ is represented either by a relay operator or by its regularization, see Figure 4.

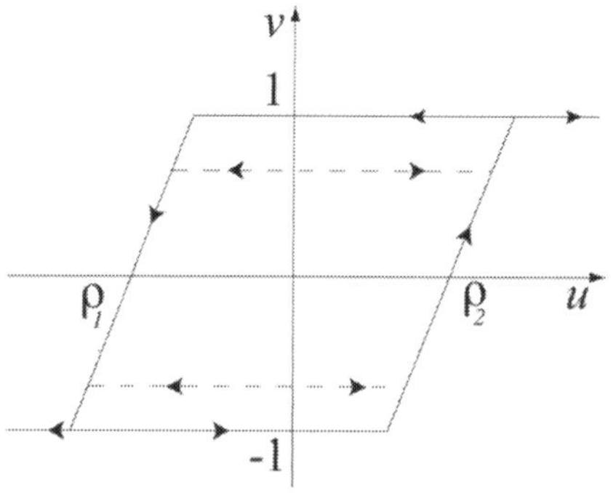

Figure 4. Regularized relay operator

Because of the discontinuity of the relay it is necessary to replace it by its closure with respect to appropriate topologies, i.e., the completed relay. For more details, see [20]. The existence of a weak solution is proved for $\mathcal{F}[u]$ equal either to a completed relay operator or to its regularization. Further, under the assumption that $f \equiv 0$, continuous monotone dependence of the solution on the initial data, and hence uniqueness, is proved. These results might be extended to the larger class of either continuous or discontinuous Preisach operators.

The precise result is based on the equivalent definition of the relay operator (Figure 3)

$$|\, v \,| \leq 1, \quad \begin{cases} (v-1)(u-\rho_2) \geq 0 \\ (v+1)(u-\rho_1) \geq 0, & \text{a.e. in } (0,T), \end{cases} \tag{14}$$

$$\int_0^t u\,\mathrm{d}v \geq \int_0^t \rho_2\,\mathrm{d}v^+ - \int_0^t \rho_1\,\mathrm{d}v^- =: \Psi^0_\rho(v;[0,t]) \quad \forall t \in (0,T)$$

(these are Stieltjes integrals). Notice that $\Psi^0_\rho(v;[0,t])$ is finite whenever $\frac{\partial v}{\partial t} \in C^0\overline{(\mathbb{R}_T)}'$. The reformulated problem looks like:

Problem 3.1 *Find $u \in L^\infty(0,T;L^2(\mathbb{R}))$ and $v \in L^\infty(\mathbb{R}_T)$ such that*

$$|v| \leq 1 a.e. \ in\ \mathbb{R}_T, \quad \frac{\partial v}{\partial t} \in C^0\overline{(\mathbb{R}_T)}', \tag{15}$$

$$\iint\limits_{\mathbb{R}_T} \Big((u+v-u(0)-v(0))\frac{\partial w}{\partial t} + u\frac{\partial w}{\partial x} + fv\Big)\,\mathrm{d}x\mathrm{d}y = 0$$

$$\forall w \in H^1(\mathbb{R}_T)\cap W^{1,1}(\mathbb{R}_T), \quad w(\cdot,T) = 0, \tag{16}$$

(17)

$$\begin{cases} (v-1)(u-\rho_2) \geq 0 \\ (v+1)(u-\rho_1) \geq 0, \quad a.e.\ in\ (\mathbb{R}_T), \end{cases} \tag{18}$$

$$\frac{1}{2}\int_{\mathbb{R}} [u(x,t)^2 - u(x,0)^2]\,\mathrm{d}x + \int_{\mathbb{R}} \Psi^0_\rho(v;[0,t])\,\mathrm{d}x \leq \int\int_{\mathbb{R}_t} fu\,\mathrm{d}x\mathrm{d}\tau$$
$$\textit{for a.e. } t \in (0,T). \tag{19}$$

Visintin proved the existence of a solution of Problem 3.1 via time discretization, and the derivation of a priori estimates.

Theorem 3.4 *Set $\mathbb{R}_t := \mathbb{R} \times (0,t)$ for any $t > 0$, fix any $T > 0$, and assume that*

$$u(0), v(0) \in L^2(\mathbb{R}), \quad |v(0)| \leq 1 \ a.e.\ in\ \mathbb{R}, \quad f \in L^1(\mathbb{R}_T) \cap L^2(\mathbb{R}_T).$$

Then Problem 3.1 has a solution. Moreover, for any $p \in [1, +\infty]$,

$$\begin{aligned} &u(0), v(0) \in BV(\mathbb{R}), \quad f \in L^p(0,T;BV(\mathbb{R})) \Rightarrow \\ &\quad u, v \in L^\infty(0,T;BV(\mathbb{R})) \cap W^{1,p}(0,T;\overline{C^0(\mathbb{R})}'). \end{aligned} \tag{20}$$

4. Differentiable solution

Showalter and Peszynska [18] considered an equation in the form

$$\frac{\partial(u+v)}{\partial t} + \frac{\partial u}{\partial x} = 0, \ v = \mathcal{F}[u], \ x \in (0,L), \ t > 0 \tag{21}$$

with the boundary condition

$$u(0,t) = \varphi(t), \ t > 0,$$

and the initial condition

$$u(x,0) = v(x,0) = 0, \ x \in (0,L).$$

They assumed that $v = \mathcal{F}[u]$ is the elementary case of those hysteresis functionals whose sides are bounded by parallel lines, i.e., the classical play. Firstly, they computed an explicit solution. Then they proved the existence and uniqueness of a differentiable solution by using the theory of nonlinear semigroups in L^2. This is possible because of the special form of the hysteresis operator.

The classical play operator is characterized by the evolution equation

$$v_t + sgn^{-1}(v-u) \ni 0, \ t > 0, \tag{22}$$

$sgn^{-1}(.)$ is a maximal monotone graph in $\mathbb{R}$.

In addition to the usual formulation of the existence-uniqueness results, a second one is suggested by switching the variables x and t to represent the solution by a semigroup of breakthrough curves. The spatial variable x is treated as another time variable, and therefore it is renamed: $x = \tau$. Then the system is given by

$$u_\tau + u_t - sgn^{-1}(v-u) \ni 0, \tag{23}$$
$$v_t + sgn^{-1}(v-u) \ni 0, \tag{24}$$
$$u(0,t) = \varphi(t), \tag{25}$$
$$u(\tau,0) = 0 = v(\tau,0). \tag{26}$$

In order to define $\mathcal{A}$, set $\mathcal{D}(\mathcal{A}) = \{u \in H^1(0,T) : u(0) = 0\}$. Note that for each such $u \in \mathcal{D}(\mathcal{A})$, the Cauchy problem

$$\begin{aligned}(v-u)_t + sgn^{-1}(v-u) &\ni -u_t, \\ (v-u)(0) &= 0\end{aligned}$$

has a unique solution $v \in H^1(0,T)$. Then $\mathcal{A}$ is defined as $\mathcal{A}(u) = u_t + v_t$ with $u \in \mathcal{D}(\mathcal{A})$ and v as given.

It can be proved that $\mathcal{A}$ is m–accretive on $L^2(0,T)$ see [18] and they have the following result:

Theorem 4.1 *If $\varphi \in W^{1,2}(0,T) = H^1(0,T)$ there is a unique solution of (23 - 26) with $u_x \in L^\infty(0,L;L^2(0,T))$, and for all x, $u(x,.) \in H^1(0,T)$, $u(x,0) = 0$.*

5. Discontinuity and stability

Now there is a question whether we can get discontinuous solutions, similar to the quasilinear case without hysteresis, see [19]. We consider the partial differential equation (1) with the given initial conditions

$$u(x,0) = u_0(x), \quad v(x,0) = v_0(x).$$

In order to illustrate the typical features of hysteresis, the cases were studied where the functional is first represented by the play hysteresis operator whose sides are bounded by parallel lines [18]. Secondly a slightly different (nonconvex) hysteresis model is considered. Thirdly we presumed a play operator with a skip. The purpose was to study the corresponding classical initial value problem where the functional $\mathcal{F}[u]$ is represented by a convex play operator and additionally by a nonconvex hysteresis operator. The solution of this problem was investigated coupled with each model, i.e., we computed an explicit solution by the method of characteristics.

It follows from direct computations that in the case of a regularized relay (see Figure 4) the equation exhibits similar properties as a quasilinear hyperbolic equation of first order, i.e., the characteristics cross and the solution has to be discontinuous.

The paper [11] is concerned with the asymptotic behaviour of the solution of (1). We used the theory of nonlinear semigroups in L^1 and an asymptotic result of Wittbold [22]. The precise result is as follows.

Theorem 5.1 *Suppose all conditions of Theorem 3.1 are satisfied, i.e., the operator $\mathcal{A} + \Re$ is m- and T-accretive in $L^1(\Omega, \mathbb{R}^2_1)$. Suppose also that $u_0 \in \overline{\mathcal{D}(\mathcal{A}+\Re)}$. Then there exists $v_\infty(x)$ dependent on x only, such that for the solution*

$$U = \begin{pmatrix} u \\ v \end{pmatrix}$$

of

$$\frac{\partial U}{\partial t} + (\mathcal{A}+\Re)(U) = 0, \tag{27}$$

$$U(0) = \begin{pmatrix} u_0 \\ v_0 \end{pmatrix}, \tag{28}$$

the following holds

$$\begin{aligned}\|.\|_1 - \lim_{t\to\infty} u(x,t) &= 0, \\ \|.\|_1 - \lim_{t\to\infty} v(x,t) &= v_\infty(x).\end{aligned}$$

These results can be extended to possibly discontinuous generalized Prandtl-Ishlinskii operators of play type. This includes the case of possibly discontinuous Preisach operators.

Acknowledgments

The research was supported, in part, by project GD 201/03/H152 from the Czech Science Foundation and MSM4781305904 from the Czech Ministry of Education.

References

[1] Barbu V 1976 *Nonlinear Semigroups and Differential Equations in Banach Spaces* (Noordhoff International Publishing)

[2] Brokate M and Sprekels J 1996 *Hysteresis and Phase Transitions* (Springer)

[3] Crandall M G 1976 *Proc. Int. Symp.* **1** (New York: Academic Press) 131

[4] Curl R L and Keoleian G A 1984 Implicit-adsorbate model for apparent anomalies with organic adsorption on natural adsorbents *Environ. Sci. Technol.* **18** 916

[5] Everett D H 1958 *Some problems in the investigation of porosity by adsorption methods, The Structure and Properties of Porous Materials* (London: Butterworth)

[6] Ghosh D R and Keinath T M 1994 Effect of clay minerals present in aquifer soils on the adsorption and desorption of hydrophobic organic compounds *Environ. Prog.* **13** 51

[7] Gleysteen L F and Deitz V R 1945 Hysteresis in the physical adsorption of nitrogen on Bone Char and other adsorbents *J. of Research of the National Bureau of Standards* **35**

[8] Kopfová J 2005 Entropy condition for a quasilinear hyperbolic equation with hysteresis *Differential and Integral Equation* **18** 451

[9] Kopfová J 2006 Dissipative phase transitions *Series on Advances in Math. for Appl. Sciences* **71** ed P Colli (Singapore: World Scientific Publishing Co. Pte. Ltd.) 141

[10] Kordulová P 2004 An example of discontinuous solution for a quasilinear hyperbolic equation with hysteresis *Preprint* MA 49/2004

[11] Kordulová P 2006 Asymptotic behaviour of a quasilinear hyperbolic equation with hysteresis *submitted*

[12] Krasnosel'skiĭ M and Pokrovskiĭ A 1989 *Systems with Hysteresis* (New York: Springer-Verlag)

[13] Pavel M H 1987 *Nonlinear Evolution Operators and Semigroups* (Springer)

[14] Pazy A 1983 *Semigroups of Linear Operators and Applications to Partial Differential Equations* (Springer)

[15] Preisach F 1983 Über die magnetische Nachwirkung *Z. Physik* **94** 277

[16] Rhee H K, Aris R and Amundson N R 1986 *First Order Partial Differential Equations* **1** (Prentice-Hall)

[17] Ruthen D M 1984 *Principles of adsorption and adsorption processes* (New York: Wiley)

[18] Showalter R E and Peszynska M 1998 A Transport Model with Adsorption Hysteresis *Differential and Integral Equations* **11** 327

[19] Smoller J 1983 *Shock Waves and Reaction-diffusion Equations* (Springer)

[20] Visintin A 1994 *Differential Models of Hysteresis* (Berlin: Springer)

[21] Visintin A 2005 Quasilinear first order P.D.E.s with hysteresis *J. Math. Anal. Appl.* **312** 401

[22] Wittbold P 1993 *Progress in Partial Differential Equations: The Metz Surveys 2* (Pitman Res. Notes Math. Ser. **296**) (Harlow: Longman Sci. Tech.) 216

Institute of Physics Publishing
doi:10.1088/1742-6596/55/1/014
Journal of Physics: Conference Series **55** (2006) 144–154
International Workshop on Multi-Rate Processes and Hysteresis

The Kurzweil integral and hysteresis

P Krejčí

Weierstrass Institute for Applied Analysis and Stochastics, Mohrenstr. 39, D-10117 Berlin, Germany, and Mathematical Institute, Czech Academy of Sciences, Žitná 25, CZ-11567 Praha 1, Czech Republic

E-mail: krejci@wias-berlin.de and krejci@math.cas.cz

Abstract. A hysteresis operator, called *the play, with variable (possibly degenerate) characteristics*, is considered in the space of right-continuous regulated functions. The Lipschitz continuity of the input-output mapping is proved by means of a new technique based on the Kurzweil integral.

Introduction

This paper is devoted to the Kurzweil integral description of a scalar rate-independent evolution variational inequality with a moving constraint in the space $G_R(0,T)$ of right-continuous regulated functions. In the hysteresis literature, this variational inequality defines the so-called "play operator with variable characteristic", cf. [3]. We prove here that the input-output operator is Lipschitz continuous with respect to the sup-norm in $G_R(0,T)$. This fact in principle can be derived from the general Lipschitz continuity result for polyhedral sweeping processes in [7]. Our objective here is to show that the Kurzweil formalism, going back to [8], see also [9, 10], provides simple and straightforward proofs. As our main tool, we derive a Kurzweil integral characterization of monotone functions (Proposition 1.9), which we believe to be of independent interest in real analysis.

The paper is divided into two sections. In Section 1 we recall some basic facts about Kurzweil integration and give a detailed proof of Proposition 1.9; the hysteresis problem itself is treated in Section 2.

1. The Kurzweil integral

In this section, we recall the definition and some basic properties of the Kurzweil integral introduced in [8] as a framework for solving ODEs with singular right-hand sides. We cite most of the results without proof, and an interested reader can find more information in [5, 9, 10]. Proposition 1.9, however, which plays an important role in the theory of Kurzweil integral variational inequalities, seems to be new in this setting and its detailed proof is given at the end of this section.

The original definition in [8] is not suitable for an integral formulation of discontinuous evolution variational inequalities, and this is why the Young integral was used instead in [2, 6]. The extension of the Kurzweil integral in [5] contains, however, the Young integral as a special case, preserving the advantage of the easy Kurzweil formalism. Here, we deal only with *right-continuous evolution processes*, and Definition 1.1 below, which goes back to [9], turns out to be sufficient for our purposes.

The basic concept in the Kurzweil integration theory is that of a δ*-fine partition*. Consider a nondegenerate closed interval $[a,b] \subset \mathbb{R}$, and denote by $\mathcal{D}_{a,b}$ the set of all divisions of the form

$$d = \{t_0, \ldots, t_m\}, \quad a = t_0 < t_1 < \ldots < t_m = b. \tag{1.1}$$

With a division $d = \{t_0, \ldots, t_m\} \in \mathcal{D}_{a,b}$ we associate *partitions* D defined as

$$D = \{(\tau_j, [t_{j-1}, t_j]) \,;\, j = 1, \ldots, m\}\,; \qquad \tau_j \in [t_{j-1}, t_j] \quad \forall j = 1, \ldots, m. \tag{1.2}$$

We define the set

$$\Gamma(a,b) := \{\delta : [a,b] \to \mathbb{R} \,;\, \delta(t) > 0 \quad \text{for every} \quad t \in [a,b]\}. \tag{1.3}$$

An element $\delta \in \Gamma(a,b)$ is called a *gauge*. For $t \in [a,b]$ and $\delta \in \Gamma(a,b)$ we denote

$$I_\delta(t) := \,]t - \delta(t), t + \delta(t)[\,. \tag{1.4}$$

Definition 1.1 *Let $\delta \in \Gamma(a,b)$ be a given gauge. A partition D of the form (1.2) is said to be* δ-fine *if for every $j = 1, \ldots, m$ we have*

$$\tau_j \in [t_{j-1}, t_j] \subset I_\delta(\tau_j),$$

and the following implications hold:

$$\tau_j = t_{j-1} \Rightarrow j = 1, \quad \tau_j = t_j \Rightarrow j = m.$$

The set of all δ-fine partitions is denoted by $\mathcal{F}_\delta(a,b)$.

It is easy to see that $\mathcal{F}_\delta(a,b)$ is nonempty for every $\delta \in \Gamma(a,b)$; this follows e.g. from [4, Lemma 1.2].

For given functions $f, g : [a,b] \to \mathbb{R}$ and a partition D of the form (1.2) we define the Kurzweil integral sum $K_D(f,g)$ by the formula

$$K_D(f,g) = \sum_{j=1}^{m} f(\tau_j)\,(g(t_j) - g(t_{j-1}))\,. \tag{1.5}$$

Definition 1.2 *Let $f, g : [a,b] \to \mathbb{R}$ be given. We say that $J \in \mathbb{R}$ is the* Kurzweil integral *over $[a,b]$ of f with respect to g and denote*

$$J = \int_a^b f(t)\,dg(t)\,, \tag{1.6}$$

if for every $\varepsilon > 0$ there exists $\delta \in \Gamma(a,b)$ such that for every $D \in \mathcal{F}_\delta(a,b)$ we have

$$|J - K_D(f,g)| \le \varepsilon\,. \tag{1.7}$$

Using the fact that the implication

$$\delta \le \min\{\delta_1, \delta_2\} \quad \Rightarrow \quad \mathcal{F}_\delta(a,b) \subset \mathcal{F}_{\delta_1}(a,b) \cap \mathcal{F}_{\delta_2}(a,b) \tag{1.8}$$

holds for every $\delta, \delta_1, \delta_2 \in \Gamma(a,b)$, we easily check that the value of J in Definition 1.2 is uniquely determined.

We list below in Propositions 1.3, 1.4 some standard properties common to most integral concepts.

Proposition 1.3 *Let* $f, f_1, f_2, g, g_1, g_2 : [a,b] \to \mathbb{R}$ *be any functions. Then the following implications hold.*

(i) *If* $\int_a^b f_1(t)\,dg(t)$, $\int_a^b f_2(t)\,dg(t)$ *exist, then* $\int_a^b (f_1+f_2)(t)\,dg(t)$ *exists and*

$$\int_a^b (f_1+f_2)(t)\,dg(t) = \int_a^b f_1(t)\,dg(t) + \int_a^b f_2(t)\,dg(t). \tag{1.9}$$

(ii) *If* $\int_a^b f(t)\,dg_1(t)$, $\int_a^b f(t)\,dg_2(t)$ *exist, then* $\int_a^b f(t)\,d(g_1+g_2)(t)$ *exists and*

$$\int_a^b f(t)\,d(g_1+g_2)(t) = \int_a^b f(t)\,dg_1(t) + \int_a^b f(t)\,dg_2(t). \tag{1.10}$$

(iii) *If* $\int_a^b f(t)\,dg(t)$ *exist, then* $\int_a^b \lambda f(t)\,dg(t)$, $\int_a^b f(t)\,d(\lambda g)(t)$ *exist for every constant* $\lambda \in \mathbb{R}$, *and*

$$\int_a^b \lambda f(t)\,dg(t) = \int_a^b f(t)\,d(\lambda g)(t) = \lambda \int_a^b f(t)\,dg(t). \tag{1.11}$$

Proposition 1.4 *Let* $f, g : [a,b] \to \mathbb{R}$ *be given functions and let* $s \in \,]a,b[$ *be given.*

(i) *Assume that* $\int_a^b f(t)\,dg(t)$ *exists. Then* $\int_a^s f(t)\,dg(t)$, $\int_s^b f(t)\,dg(t)$ *exist.*

(ii) *Assume that* $\int_a^s f(t)\,dg(t)$, $\int_s^b f(t)\,dg(t)$ *exist. Then* $\int_a^b f(t)\,dg(t)$ *exists and*

$$\int_a^b f(t)\,dg(t) = \int_a^s f(t)\,dg(t) + \int_s^b f(t)\,dg(t). \tag{1.12}$$

In order to preserve the consistency of (1.12) also in the limit cases $s = a$ and $s = b$, we set

$$\int_s^s f(t)\,dg(t) = 0 \qquad \forall s \in [a,b],\ \forall f, g : [a,b] \to \mathbb{R}. \tag{1.13}$$

Recall that a function $f : [a,b] \to \mathbb{R}$ is said to be *regulated* if for every $t \in [a,b]$ there exist both one-sided limits $f(t+), f(t-) \in \mathbb{R}$ with the convention $f(a-) = f(a)$, $f(b+) = f(b)$, see [1]. Obviously, the set of discontinuity points of every regulated function is at most countable.

In agreement with [10], we denote by $G(a,b)$ the set of all regulated functions $f : [a,b] \to \mathbb{R}$. Let us introduce in $G(a,b)$ a system of seminorms

$$\|f\|_{[s,t]} := \sup\{|f(\tau)|\,;\ \tau \in [s,t]\} \tag{1.14}$$

for any subinterval $[s,t] \subset [a,b]$. Indeed, $\|\cdot\|_{[a,b]}$ is a norm. With this norm, $G(a,b)$ becomes a Banach space. Let us note that the space $C[a,b]$ of *continuous functions* $f : [a,b] \to \mathbb{R}$ is a closed subspace of $G(a,b)$ with respect to the norm $\|\cdot\|_{[a,b]}$. Moreover, every regulated function can be uniformly approximated by *step functions* of the form

$$w(t) = \sum_{k=0}^{m} \hat{c}_k\, \chi_{\{t_k\}}(t) + \sum_{k=1}^{m} c_k\, \chi_{]t_{k-1},t_k[}(t), \quad t \in [a,b], \tag{1.15}$$

where $d = \{t_0, \ldots, t_m\} \in \mathcal{D}_{a,b}$ is a given division, χ_A for $A \subset [a,b]$ is the characteristic function of the set A, and $\hat{c}_0, \ldots, \hat{c}_m, c_1, \ldots, c_m$ are given real numbers. We see in particular that the space $BV(a,b)$ of all functions of bounded variation on $[a,b]$ is contained as a dense subset in $G(a,b)$. In the next section, we will restrict ourselves to the spaces $G_R(a,b)$, $BV_R(a,b)$ of right-continuous functions from $G(a,b)$, $BV(a,b)$, respectively.

Remark 1.5 Proposition 1.4 needs some comment. Whenever we integrate functions f, g defined in $[a,b]$ over an interval $[r,s] \subset [a,b]$, we implicitly consider their restrictions $f|_{[r,s]}, g|_{[r,s]}$. In particular, in case of regulated functions, we have e.g. $f|_{[r,s]}(s+) = f(s)$, $f|_{[r,s]}(r-) = f(r)$.

Note that we deal here with functions that are defined *for all* $t \in [a,b]$. The concept of "almost everywhere" is meaningless here.

The following explicit formulas can easily be derived from the definition.

Proposition 1.6 *For every* $f : [a,b] \to \mathbb{R}$, $g \in G(a,b)$, $a \le r \le b$, *we have*

(i) $\displaystyle\int_a^b \chi_{\{r\}}(t)\, dg(t) \;=\; g(r+) - g(r-)\,,$

(ii) $\displaystyle\int_a^b f(t)\, d\big(\chi_{\{r\}}\big)(t) \;=\; \begin{cases} 0 & \text{if } r \in]a,b[\,, \\ -f(a) & \text{if } r = a\,, \\ f(b) & \text{if } r = b\,, \end{cases}$

(iii) $\displaystyle\int_a^b \chi_{]r,s[}(t)\, dg(t) \;=\; g(s-) - g(r+) \qquad \forall\, s \in]r,b]\,,$

(iv) $\displaystyle\int_a^b f(t)\, d\big(\chi_{]r,s[}\big)(t) \;=\; f(r) - f(s) \qquad \forall\, s \in]r,b]\,.$

We see in particular that the integral $\int_a^b f(t)\, dg(t)$ exists whenever one of the functions f, g is a step function and the other one is regulated. By a density argument, we obtain the following result.

Theorem 1.7 (Properties of the Kurzweil integral)

(i) *If* $f \in G(a,b)$ *and* $g \in BV(a,b)$, *then* $\int_a^b f(t)\, dg(t)$ *exists and satisfies the estimate*

$$\left|\int_a^b f(t)\, dg(t)\right| \;\le\; \|f\|_{[a,b]} \operatorname*{Var}_{[a,b]} g\,. \tag{1.16}$$

(ii) *If* $f \in BV(a,b)$ *and* $g \in G(a,b)$, *then* $\int_a^b f(t)\, dg(t)$ *exists and satisfies the estimate*

$$\left|f(a)\,g(a) + \int_a^b f(t)\, dg(t)\right| \;\le\; \left(|f(b)| + \operatorname*{Var}_{[a,b]} f\right) \|g\|_{[a,b]}\,. \tag{1.17}$$

(iii) *For every* $f \in G(a,b)$, $g \in BV(a,b)$ *we have the integration-by-parts formula*

$$\begin{aligned} \int_a^b f(t)\, dg(t) + \int_a^b g(t)\, df(t) \;&=\; f(b)\,g(b) - f(a)\,g(a) \\ &+ \sum_{t \in [a,b]} \Big(\big(f(t) - f(t-)\big)\big(g(t) - g(t-)\big) - \big(f(t+) - f(t)\big)\big(g(t+) - g(t)\big)\Big)\,. \end{aligned} \tag{1.18}$$

(iv) *If* $f_n \in G(a,b)$ *and* $g_n \in BV(a,b)$ *are such that* $\|f_n - f\|_{[a,b]} \to 0$, $\|g_n - g\|_{[a,b]} \to 0$ *as* $n \to \infty$, *and* $\operatorname{Var}_{[a,b]} g_n \le C$ *independently of* n, *then*

$$\lim_{n\to\infty} \int_a^b f_n(t)\, dg_n(t) \;=\; \int_a^b f(t)\, dg(t)\,. \tag{1.19}$$

Corollary 1.8 *For every $g \in BV(a,b)$ we have*

$$\int_a^b g(t+)\, dg(t) = \frac{1}{2}\left(|g(b)|^2 - |g(a)|^2\right) + \frac{1}{2}\sum_{t\in[a,b]} |g(t+) - g(t-)|^2 . \tag{1.20}$$

We conclude this section by a Kurzweil integral characterization of monotone functions, which will be referred to several times in the next section in the context of Kurzweil integral variational inequalities.

Proposition 1.9 *Let $f \in G(a,b)$ and $g \in BV(a,b)$ be such that*

(i) *$f(t) > 0$ for every $t \in [a,b]$,*

(ii) *$\int_s^t f(\tau)\, dg(\tau) \geq 0$ for every $a \leq s < t \leq b$.*

Then g is nondecreasing in $[a,b]$.

The proof of this statement will be divided into several steps. Let us start with an easy convergence result.

Lemma 1.10 *For every $f \in G(a,b)$ and $g \in BV(a,b)$ we have*

$$\lim_{h\to 0+} \int_t^{t+h} f(\tau)\, dg(\tau) = f(t)\,(g(t+) - g(t)) \qquad \forall t \in [a,b[\,, \tag{1.21}$$

$$\lim_{h\to 0+} \int_{t-h}^{t} f(\tau)\, dg(\tau) = f(t)\,(g(t) - g(t-)) \qquad \forall t \in\,]a,b]\,. \tag{1.22}$$

Proof. Let $t \in [a,b[$ be fixed. For $\tau \in [a,b]$ put $f_1(\tau) = f(t+)$, $f_2(\tau) = (f(t) - f(t+))\, \chi_{\{t\}}(\tau)$, $f_3(\tau) = f(\tau) - f_1(\tau) - f_2(\tau)$. Let $\varepsilon > 0$ be arbitrary, and let $h \in\,]0, b-t[$ be such that

$$|f(\tau) - f(t+)| \leq \varepsilon, \quad |g(\tau) - g(t+)| \leq \varepsilon \qquad \text{for } \tau \in\,]t, t+h]\,.$$

Then $|f_3(\tau)| \leq \varepsilon$ for all $\tau \in [t, t+h]$, hence, by (1.16),

$$\left|\int_t^{t+h} f_3(\tau)\, dg(\tau)\right| \leq \varepsilon \operatorname*{Var}_{[a,b]} g\,. \tag{1.23}$$

Furthemore, we have by Remark 1.5 and Proposition 1.6 that

$$\int_t^{t+h} f_1(\tau)\, dg(\tau) = f(t+)\,(g(t+h) - g(t))\,, \tag{1.24}$$

$$\int_t^{t+h} f_2(\tau)\, dg(\tau) = (f(t) - f(t+))\,(g(t+) - g(t))\,, \tag{1.25}$$

hence, as a consequence of (1.24)–(1.25),

$$\left|\int_t^{t+h} (f_1(\tau) + f_2(\tau))\, dg(\tau) - f(t)\,(g(t+) - g(t))\right| = |f(t+)\,(g(t+h) - g(t+))| \leq \varepsilon\, \|f\|_{[a,b]}\,, \tag{1.26}$$

and (1.21) follows from (1.23) and (1.26). The proof of (1.22) is similar. ■

Under the hypotheses of Proposition 1.9, we thus have in particular

$$g(t-) \leq g(t) \leq g(t+) \qquad \forall t \in [a,b]\,. \tag{1.27}$$

As a next step, we prove the following lemma.

Lemma 1.11 *Let the hypotheses of Proposition 1.9 hold. Then for every non-negative function $w \in G(a,b)$ we have*

$$\int_a^b w(\tau)\, f(\tau)\, dg(\tau) \;\geq\; 0\,. \tag{1.28}$$

Proof. It suffices to assume that w is a step function of the form (1.15). Indeed, for an arbitrary function $w \in G(a,b)$ we find a sequence $\{w_n\}$ of step functions such that $\|w - w_n\|_{[a,b]} \to 0$ as $n \to \infty$, and use Proposition 1.7 (iv).

Hence, let w be as in (1.15), and let us consider any $h > 0$ such that $t_{k-1} + h < t_k - h$ for all $k = 1, \ldots, m$. We then have

$$\int_a^b w(\tau)\, f(\tau)\, dg(\tau) \;=\; \sum_{k=1}^{m} \left(\int_{t_{k-1}}^{t_{k-1}+h} + \int_{t_{k-1}+h}^{t_k-h} + \int_{t_k-h}^{t_k} \right) w(\tau)\, f(\tau)\, dg(\tau)\,,$$

where

$$\int_{t_{k-1}+h}^{t_k-h} w(\tau)\, f(\tau)\, dg(\tau) = c_k \int_{t_{k-1}+h}^{t_k-h} f(\tau)\, dg(\tau) \;\geq\; 0$$

for all h and k by hypothesis. Letting h tend to $0+$ we obtain from Lemma 1.10 and formula (1.27) that

$$\int_a^b w(\tau)\, f(\tau)\, dg(\tau) \;\geq\; \sum_{k=1}^{m} (\hat{c}_{k-1}\, f(t_{k-1})(g(t_{k-1}+) - g(t_{k-1})) + \hat{c}_k\, f(t_k)(g(t_k) - g(t_k-))) \;\geq\; 0\,.$$

■

Lemma 1.12 *Let the hypotheses of Proposition 1.9 hold, and let in addition $f(t+) > 0$, $f(t-) > 0$ for all $t \in [a,b]$. Then g is nondecreasing in $[a,b]$.*

Proof. There exists some $r > 0$ such that $f(t) \geq r$ for all $t \in [a,b]$. Hence, we may use Lemma 1.11 with the function

$$w(\tau) = \frac{1}{f(\tau)}\, \chi_{\,]s,t[}\,(\tau)$$

for any choice of $]s,t[\, \subset [a,b]$, and the desired inequality

$$\begin{aligned} 0 \;&\leq\; \int_a^b w(\tau)\, f(\tau)\, dg(\tau) \;=\; \int_s^t 1\, dg(\tau) - \int_s^t \chi_{\{s\}}\, dg(\tau) - \int_s^t \chi_{\{t\}}\, dg(\tau) \\ &=\; g(t) - g(s) - (g(s+) - g(s)) - (g(t) - g(t-)) \\ &\leq\; g(t) - g(s)\,, \end{aligned}$$

follows easily. ■

We now pass to the proof of Proposition 1.9.

Proof of Proposition 1.9. We define the sets

$$\mathcal{N}_+ = \{t \in [a,b]\,;\, f(t+) = 0\}\,, \quad \mathcal{N}_- = \{t \in [a,b]\,;\, f(t-) = 0\}\,, \quad \mathcal{N} = \mathcal{N}_+ \cup \mathcal{N}_-\,.$$

For every $t \in \mathcal{N}$, we have either $f(t) > f(t-)$ or $f(t) > f(t+)$, hence $\mathcal{N}$ is at most countable. Furthermore, $\mathcal{N}$ is closed. Indeed, if for instance $\{t_i\}$ is a sequence in $\mathcal{N}$, $t_i \nearrow t$, then in a neighborhood of each t_i there exists $\hat{t}_i$ such that $f(\hat{t}_i) < 1/i$, $\hat{t}_i \nearrow t$, hence $f(t-) = 0$. A similar argument works for $t_i \searrow t$.

All elements of $\mathcal{N}$ can be ordered into a sequence $\{s_j\,;\ j \in \mathbb{N}\}$. Let $\varepsilon > 0$ be given. For every $j \in \mathbb{N}$ we find $h_j > 0$ such that

$$\left.\begin{array}{lcll} |g(\tau) - g(s_j+)| & \le & \varepsilon\, 2^{-j-1} & \text{for } \tau \in\,]s_j, s_j + h_j]\,, \\ |g(\tau) - g(s_j-)| & \le & \varepsilon\, 2^{-j-1} & \text{for } \tau \in [s_j - h_j, s_j[\,. \end{array}\right\} \tag{1.29}$$

The set $\mathcal{N}$ is compact, we can therefore find a finite covering

$$\mathcal{N} \subset \bigcup_{k=0}^{m} \left]s_{j_k} - h_{j_k}, s_{j_k} + h_{j_k}\right[\,, \tag{1.30}$$

with $s_{j_0} < s_{j_1} < \ldots < s_{j_m}$. We may assume that

$$s_{j_{k-1}} - h_{j_{k-1}} < s_{j_k} - h_{j_k}\,, \quad s_{j_{k-1}} + h_{j_{k-1}} < s_{j_k} + h_{j_k}$$

for all $k = 1, \ldots, m$; otherwise we remove the redundant intervals from the covering (1.30). We now set $a_0 = a$, $b_m = b$, and for $k = 1, \ldots, m$ choose

$$\left.\begin{array}{ll} b_{k-1} = s_{j_{k-1}} + h_{j_{k-1}}\,, a_k = s_{j_k} - h_{j_k} & \text{if } s_{j_{k-1}} + h_{j_{k-1}} \le s_{j_k} - h_{j_k}\,, \\ b_{k-1} = a_k \in\,]s_{j_k} - h_{j_k}, s_{j_{k-1}} + h_{j_{k-1}}[\, \cap\,]s_{j_{k-1}}, s_{j_k}[\, \setminus \mathcal{N} & \text{if } s_{j_{k-1}} + h_{j_{k-1}} > s_{j_k} - h_{j_k}\,. \end{array}\right\} \tag{1.31}$$

Then

$$\mathcal{N} \subset [a_0, b_0[\, \cup \bigcup_{k=1}^{m}\,]a_k, b_k[\, \cup\,]a_m, b_m]\,.$$

For $a_k \le s < t \le b_k$, $k = 0, \ldots, m$, we have by (1.27) and (1.29) that $g(s) \le g(t) + \varepsilon\, 2^{-j_k}$. On the other hand, for $b_{k-1} \le s < t \le a_k$, $k = 1, \ldots, m$, it follows from Lemma 1.12 that $g(s) \le g(t)$. Consequently, for all $a \le s < t \le b$ we have

$$g(s) \le g(t) + \varepsilon \sum_{k=0}^{m} 2^{-j_k} \le g(t) + \varepsilon\,.$$

Since $\varepsilon > 0$ has been chosen arbitrarily, we obtain the assertion. ∎

2. A variational inequality

We now consider a fixed interval $[0, T]$ and two (input) functions $u, r \in G_R(0, T)$ such that $r(t) \ge 0$ for all $t \in [0, T]$, and an initial condition $x_0 \in [-r(0), r(0)]$. The problem consists in finding the output $\xi \in G_R(0, T)$ such that

$$\left.\begin{array}{l} |u(t) - \xi(t)| \le r(t) \qquad\qquad \forall t \in [0, T]\,, \\ \displaystyle\int_0^T (u(t) - \xi(t) - y(t))\, d\xi(t) \ \ge\ 0 \quad \forall y \in G(0, T)\,, \quad |y(t)| \le r(t) \ \ \forall t \in [0, T]\,, \\ u(0) - \xi(0) = x_0\,. \end{array}\right\} \tag{2.1}$$

This extends the model considered in [3] in two respects: we admit discontinuous inputs, and $r(t)$ is allowed to vanish at any point. The setting of [3] however contains additional nonlinearities that we neglect here for simplicity.

We first construct a solution to Problem (2.1) if the inputs are step functions of the form

$$u(t) = \sum_{k=1}^{m} u_{k-1}\, \chi_{[t_{k-1}, t_k[}(t) + u_m\, \chi_{\{T\}}(t)\,, \quad r(t) = \sum_{k=1}^{m} r_{k-1}\, \chi_{[t_{k-1}, t_k[}(t) + r_m\, \chi_{\{T\}}(t)\,, \tag{2.2}$$

where $0 = t_0 < t_1 < \ldots < t_m = T$ is a given division, and u_k, r_k for $k = 0, \ldots m$ are given real numbers. To this end, we use the projection $Q_c(x) = \max\{-c, \min\{x, c\}\}$ for $c \geq 0$ and its complement $P_c(x) = x - Q_c(x)$ for $x \in \mathbb{R}$, and claim that the solution ξ is given by the formula

$$\xi(t) = \sum_{k=1}^{m} \xi_{k-1}\, \chi_{[t_{k-1}, t_k[}(t) + \xi_m\, \chi_{\{T\}}(t) \tag{2.3}$$

with $\xi_0 = u_0 - x_0$ and

$$\xi_k = \xi_{k-1} + P_{r_k}(u_k - \xi_{k-1}) \quad \text{for } k = 1, \ldots m. \tag{2.4}$$

Indeed, we have $u_k - \xi_k = Q_{r_k}(u_k - \xi_{k-1}) \in [-r_k, r_k]$. Using Proposition 1.6, we can evaluate the integral in (2.1) explicitly and obtain

$$\int_0^T (u(t) - \xi(t) - y(t))\, d\xi(t) = \sum_{k=1}^{m} (\xi_k - \xi_{k-1})\,(u_k - \xi_k - y(t_k)) \tag{2.5}$$

for every regulated function y such that $|y(t)| \leq r(t)$ for all t. For all $x \in \mathbb{R}$ and $|y| \leq c$ we have $P_c(x)\,(Q_c(x) - y) \geq 0$. Putting $z_k = u_k - \xi_{k-1}$, we thus can rewrite (2.5) as

$$\int_0^T (u(t) - \xi(t) - y(t))\, d\xi(t) = \sum_{k=1}^{m} P_{r_k}(z_k)\,(Q_{r_k}(z_k) - y(t_k)) \geq 0, \tag{2.6}$$

hence (2.1) is satisfied.

To extend the set admissible inputs, we state and prove the main result of this section on Lipschitz continuity of the input-output mapping defined by (2.1).

Theorem 2.1 *Let $u_i, r_i \in G_R(0,T)$ and initial conditions $x_i^0 \in [-r_i(0), r_i(0)]$ be given, $i = 1, 2$. Let ξ_1, ξ_2 be corresponding solutions to (2.1), and assume that $\xi_i \in BV_R(0,T)$ for $i = 1, 2$. Then for every $t \in [0,T]$ we have*

$$|\xi_1(t) - \xi_2(t)| \leq \max\{|\xi_1(0) - \xi_2(0)|, \|u_1 - u_2\|_{[0,t]} + \|r_1 - r_2\|_{[0,t]}\}. \tag{2.7}$$

Before passing to the proof of Theorem 2.1, we derive an easy auxiliary result.

Lemma 2.2 *Let $u, r \in G_R(0,T)$ and an initial condition $x_0 \in [-r(0), r(0)]$ be given. Let ξ be a solution to (2.1), and assume that $\xi \in BV_R(0,T)$. Then for every $0 \leq s < t \leq T$ and every regulated function y such that $|y(\tau)| \leq r(\tau)$ for all τ, we have*

$$\int_s^t (u(\tau) - \xi(\tau) - y(\tau))\, d\xi(\tau) \geq 0. \tag{2.8}$$

Furthermore, if both u and r are constant in an interval $[t_0, t_1] \subset [0,T]$, then ξ is constant in $[t_0, t_1]$.

Proof. Set $y^*(\tau) = y(\tau)\, \chi_{]s,t]}(\tau) + (u(\tau) - \xi(\tau))\,(\chi_{[0,s]} + \chi_{]t,T]})(\tau)$ for $\tau \in [0,T]$. Using Propositions 1.4, 1.6, and Remark 1.5 (note that ξ is right-continuous!), we obtain that

$$\begin{aligned} 0 &\leq \int_0^T (u(\tau) - \xi(\tau) - y^*(\tau))\, d\xi(\tau) \\ &= \int_s^t (u(\tau) - \xi(\tau) - y(\tau))\, d\xi(\tau) \\ &\quad + \int_t^T (u(\tau) - \xi(\tau) - y(\tau))\, \chi_{\{t\}}\tau)\, d\xi(\tau) - \int_s^t (u(\tau) - \xi(\tau) - y(\tau))\, \chi_{\{s\}}\tau)\, d\xi(\tau) \\ &= \int_s^t (u(\tau) - \xi(\tau) - y(\tau))\, d\xi(\tau). \end{aligned} \tag{2.9}$$

Assume now that u and r are constant in $[t_0, t_1]$. For all $t \in]t_0, t_1]$ we then have

$$\int_{t_0}^{t} (u(t_0) - \xi(\tau) - y(\tau))\, d\xi(\tau) \;\geq\; 0\,.$$

We may choose $y(\tau) = u(t_0) - \xi(t_0)$ and use (1.20) for $g(\tau) = u(t_0) - \xi(\tau)$ to obtain

$$0 \leq \int_{t_0}^{t} (\xi(t_0) - \xi(\tau))\, d\xi(\tau) = -\frac{1}{2}|\xi(t_0) - \xi(t)|^2 - \frac{1}{2} \sum_{\tau \in [t_0, t]} |\xi(\tau) - \xi(\tau-)|^2\,,$$

hence $\xi(t) = \xi(t_0)$ for all $t \in [t_0, t_1]$. ■

The above proof shows why it was convenient to choose the test functions y in Problem (2.1) in $G(0,T)$ and not in $G_R(0,T)$. More refined results in [2] however show that even the choice $y \in BV_R(0,T)$ is sufficient.

Proof of Theorem 2.1. The process defined by (2.1) is rate-independent, hence it suffices to prove that

$$|\xi_1(t) - \xi_2(t)| \;\leq\; \max\{|\xi_1(0) - \xi_2(0)|, \|u_1 - u_2\|_{[0,T]} + \|r_1 - r_2\|_{[0,T]}\} \tag{2.10}$$

for all $t \in [0,T]$. Set $\bar{u} = u_1 - u_2$, $\bar{r} = r_1 - r_2$, $\bar{\xi} = \xi_1 - \xi_2$, and assume that there exists $t_1 \in]0,T]$ such that $\bar{\xi}(t_1) > \|\bar{u}\|_{[0,T]} + \|\bar{r}\|_{[0,T]}$. Putting $y(\tau) = Q_{r_1(\tau)}(u_2(\tau) - \xi_2(\tau))$ and $y(\tau) = Q_{r_2(\tau)}(u_1(\tau) - \xi_1(\tau))$ in the variational inequalities for ξ_1 and ξ_2, respectively, we obtain for all $0 \leq s < t \leq t_1$ that

$$\int_{s}^{t} (\bar{u}(\tau) - \bar{\xi}(\tau) + P_{r_1(\tau)}(u_2(\tau) - \xi_2(\tau)))\, d\xi_1(\tau) \;\geq\; 0\,, \tag{2.11}$$

$$\int_{s}^{t} (-\bar{u}(\tau) + \bar{\xi}(\tau) + P_{r_2(\tau)}(u_1(\tau) - \xi_1(\tau)))\, d\xi_2(\tau) \;\geq\; 0\,. \tag{2.12}$$

We have for all $c, c' > 0$ and $x \in \mathbb{R}$ the implication $|x| \leq c' \Rightarrow |P_c(x)| \leq |c - c'|$, hence,

$$|P_{r_1(\tau)}(u_2(\tau) - \xi_2(\tau))| \leq \|\bar{r}\|_{[0,T]}\,, \quad |P_{r_2(\tau)}(u_1(\tau) - \xi_1(\tau))| \leq \|\bar{r}\|_{[0,T]}$$

for all $\tau \in [0,T]$. Letting $t \nearrow t_1$ in (2.11)–(2.12), and using (1.22), we obtain $\xi_1(t_1-) \geq \xi_1(t_1)$, $\xi_2(t_1-) \leq \xi_2(t_1)$, hence $\bar{\xi}(t_1-) \geq \bar{\xi}(t_1) > \|\bar{u}\|_{[0,T]} + \|\bar{r}\|_{[0,T]}$. There exists therefore an interval $[t_0, t_1] \subset [0,T]$ such that $\bar{\xi}(t) > \|\bar{u}\|_{[0,T]} + \|\bar{r}\|_{[0,T]}$ for all $t \in [t_0, t_1]$. Applying Proposition 1.9 to (2.11)–(2.12), we conclude that ξ_1 is nonincreasing, ξ_2 is nondecreasing, hence $\bar{\xi}$ is nonincreasing in $[t_0, t_1]$. Let t^* be the infimum of all $t_0 \in [0, t_1]$ such that $\bar{\xi}(t) > \|\bar{u}\|_{[0,T]} + \|\bar{r}\|_{[0,T]}$ for every $t \in [t_0, t_1]$. The above argument yields that $t^* = 0$ and that $\bar{\xi}$ is nonincreasing in $[0, t_1]$, which we wanted to prove. The case $\bar{\xi}(t_1) < -\|\bar{u}\|_{[0,T]} - \|\bar{r}\|_{[0,T]}$ is similar. The proof of Theorem 2.1 is complete. ■

Let now $u, r \in G_R(0,T)$ be given, $r(t) \geq 0$ for all t, and for some fixed interval $[t_0, t_1] \subset [0,T]$ set $u_1(t) = u_2(t) = u(t)$, $r_1(t) = r_2(t) = r(t)$ for $t \in [0, t_0]$, $u_1(t) = u(t)$, $r_1(t) = r(t)$ $u_2(t) = u(t_0)$, $r_2(t) = r(t_0)$ for $t \in [t_0, t_1]$. Then Theorem 2.1 and Lemma 2.2 yield

$$|\xi(t_1) - \xi(t_0)| \;\leq\; \|u - u(t_0)\|_{[t_0,t_1]} + \|r - r(t_0)\|_{[t_0,t_1]}\,. \tag{2.13}$$

In particular, if both u and r belong to $BV_R(0,T)$, then so does ξ and we have

$$\operatorname*{Var}_{[0,T]} \xi \;\leq\; \operatorname*{Var}_{[0,T]} u + \operatorname*{Var}_{[0,T]} r\,. \tag{2.14}$$

We have seen that Problem (2.1) has a unique solution ξ whenever u and r are step functions. Since every BV-function can be uniformly approximated by step functions with uniformly bounded variation, we obtain, as immediate consequence of Theorems 2.1 and 1.7 (iv), the following result.

Corollary 2.3 *Problem (2.1) has a unique solution $\xi \in BV_R(0,T)$ for every $u, r \in BV_R(0,T)$, $r(t) \geq 0$ for all $t \in [0,T]$, and for every $x_0 \in [-r(0), r(0)]$. Moreover, the solution operator $\mathcal{P} : (u, r, x_0) \mapsto \xi$ admits a Lipschitz continuous extension to $\mathcal{P} : \mathcal{D} \to G_R(0,T)$, where $\mathcal{D} = \{(u, r, x_0) \in G_R(0,T) \times G_R(0,T) \times \mathbb{R}\,;\, r(t) \geq 0$ for $t \in [0,T],\, x_0 \in [-r(0), r(0)]\}$, and inequality (2.7) holds for all $(u_i, r_i, x_i^0) \in \mathcal{D}$, $\xi_i = \mathcal{P}(u_i, r_i, x_i^0)$, $i = 1, 2$.*

We do not know if the extended operator $\mathcal{P} : \mathcal{D} \to G_R(0,T)$ still admits the Kurzweil integral representation (2.1), because in general, $\int_a^b f(t)\, dg(t)$ is not well defined if both f and g are only regulated. There are some indications that (2.1) is meaningful in this situation, too, but no proof is yet available. Things are different if r is bounded from below by a positive constant. Then, similarly as in [6], ξ has bounded variation and the integral formula holds. To conclude this paper, we give an elementary proof of a slightly stronger version of this fact.

Proposition 2.4 *Let $r_0 > 0$ be given, and let $(u, r, x_0) \in \mathcal{D}$ be such that $r(t) \geq r_0$ for all $t \in [0,T]$. Then $\xi = \mathcal{P}(u, r, x_0)$ is piecewise monotone and, in particular, belongs to $BV_R(0,T)$.*

Proof. Since both u and r are regulated, there exists an integer n^* with the property that for every sequence $a_1 < b_1 \leq a_2 < b_2 \leq \ldots \leq a_n < b_n$ of points in $[0,T]$, the following implication holds:

$$|u(b_k) - u(a_k)| + |r(b_k) - r(a_k)| \geq \frac{2}{3} r_0 \quad \forall k = 1, \ldots, n \implies n \leq n^* . \tag{2.15}$$

We claim that ξ has at most $(3n^* + 1)$ monotonicity intervals.

To prove this conjecture, we tacitly consider uniformly convergent BV-approximations of u and r preserving the property (2.15), so that the variational inequality (2.1) is meaningful. Since the number of monotonicity intervals (hence the total variation) of ξ remains bounded independently of the approximations, we may pass to the limit using Theorem 1.7 and obtain the assertion.

Assume that there exists $m > n^*$ and points $0 \leq t_0 < t_1 < \ldots < t_{2m} \leq T$ such that

$$\xi(t_0) < \xi(t_1) > \xi(t_2) < \ldots > \xi(t_{2m}) .$$

We make use of the following implications:

$$\left.\begin{array}{lll} u(\tau) - \xi(\tau) < r_0 & \forall \tau \in [s,t] \implies & \xi \text{ is nonincreasing in } [s,t] , \\ u(\tau) - \xi(\tau) > -r_0 & \forall \tau \in [s,t] \implies & \xi \text{ is nondecreasing in } [s,t] . \end{array}\right\} \tag{2.16}$$

Indeed, (2.16) follows from Proposition 1.9 provided we put in (2.8) $y(\tau) = \pm r_0$. Hence, the sets

$$\begin{aligned} A_{2j-1} &= \{\tau \in [t_{2j-2}, t_{2j-1}]\,;\, u(\tau) - \xi(\tau) \geq r_0\} , \\ A_{2j} &= \{\tau \in [t_{2j-1}, t_{2j}]\,;\, u(\tau) - \xi(\tau) \leq -r_0\} \end{aligned}$$

are non-empty for all $j = 1, \ldots, m$, and we may set

$$s_k = \sup A_k , \quad k = 1, \ldots, 2m . \tag{2.17}$$

By (2.16) and by the right-continuity of ξ, we have $\xi(s_{2j-1}) \geq \xi(t_{j-1})$, $\xi(s_{2j}) \leq \xi(t_{2j})$ for $j = 1, \ldots, m$, hence

$$\xi(s_1) > \xi(s_2) < \ldots > \xi(s_{2m}) , \tag{2.18}$$

and, by definition of s_k, we have

$$u(s_{2j-1}-) - \xi(s_{2j-1}-) \geq r_0\,, \quad u(s_{2j}-) - \xi(s_{2j}-) \leq -r_0 \quad \text{for } j = 1, \dots, m\,.$$

Furthermore, by (2.13), we have

$$|\xi(s_k) - \xi(s_k-)| \leq |u(s_k) - u(s_k-)| + |r(s_k) - r(s_k-)| \quad \forall k = 1, \dots, 2m\,.$$

Using (2.18), we thus obtain for $j = 1, \dots, m$ that

$$\begin{aligned} u(s_{2j-1}-) - u(s_{2j}-) &\geq 2r_0 + \xi(s_{2j-1}-) - \xi(s_{2j}-) \\ &\geq 2r_0 - |u(s_{2j-1}) - u(s_{2j-1}-)| - |r(s_{2j-1}) - r(s_{2j-1}-)| \\ &\quad - |u(s_{2j}) - u(s_{2j}-)| - |r(s_{2j}) - r(s_{2j}-)|\,, \end{aligned} \tag{2.19}$$

and similarly, for $j = 2, \dots, m$,

$$\begin{aligned} u(s_{2j-1}-) - u(s_{2j-2}-) &\geq 2r_0 - |u(s_{2j-1}) - u(s_{2j-1}-)| - |r(s_{2j-1}) - r(s_{2j-1}-)| \\ &\quad - |u(s_{2j-2}) - u(s_{2j-2}-)| - |r(s_{2j-2}) - r(s_{2j-2}-)|\,. \end{aligned} \tag{2.20}$$

The set $M := \{k = 1, \dots, 2m\,;\ |u(s_k) - u(s_k-)| + |r(s_k) - r(s_k-)| \geq (2/3)r_0\}$ contains, by (2.15), at most n^* elements. For $2m - 1 - 2n^*$ indices $k \in \{1, \dots, 2m\}$ we thus have by (2.19)–(2.20) that

$$|u(s_k-) - u(s_{k-1}-)| \geq 2r_0 - \frac{4}{3}r_0 = \frac{2}{3}r_0\,.$$

From (2.15) we conclude that $2m - 1 - 2n^* \leq n^*$, that is, $2m \leq 3n^* + 1$ as conjectured. This completes the proof of Proposition 2.4. ∎

Acknowledgments

This work has been partially supported by the University College Cork, Ireland.

Bibliography

[1] Aumann G 1954 *Reelle Funktionen* (in German) (Berlin: Springer)

[2] Brokate M and Krejčí P 2003 Duality in the space of regulated functions and the play operator *Math. Z.* **245** 667

[3] Chernorutskii V V and Krasnosel'skii M A 1992 Hysteresis systems with variable characteristics *Nonlin. Anal. TMA* **18** 543

[4] Krejčí P and Kurzweil J A nonexistence result for the Kurzweil integral *Math. Bohem.* **127** 571

[5] Krejčí P 2003 The Kurzweil integral with exclusion of negligible sets *Math. Bohem.* **128** 277

[6] Krejčí P and Laurençot Ph 2002 Generalized variational inequalities *J. Convex Anal.* **9** 159

[7] Krejčí P and Vladimirov A 2003 Polyhedral sweeping processes with oblique reflection in the space of regulated functions *Set-Valued Anal.* **11** (2003) 91

[8] Kurzweil J 1957 Generalized ordinary differential equations and continuous dependence on a parameter *Czechoslovak Math. J.* **7** (**82**) 418

[9] Schwabik Š 1973 On a modified sum integral of Stieltjes type *Časopis Pěst. Mat.* **98** 274

[10] Tvrdý M 1989 Regulated functions and the Perron-Stieltjes integral *Časopis Pěst. Mat.* **114** 187

Institute of Physics Publishing	Journal of Physics: Conference Series **55** (2006) 155–164
doi:10.1088/1742-6596/55/1/015	International Workshop on Multi-Rate Processes and Hysteresis

Some results on the stability of quasi-static paths of elastic-plastic systems with hardening

J A C Martins[1], **M D P Monteiro Marques**[2] **and A Petrov**[3]

[1] Instituto Superior Técnico, Dep. Eng. Civil and ICIST,
Av. Rovisco Pais, 1049-001 Lisboa, Portugal
[2] Centro de Matemática e Aplicações Fundamentais and F.C.U.L.,
Av. Prof. Gama Pinto, 2, 1649-003 Lisboa, Portugal
[3] Weierstraß-Institut für Angewandte Analysis und Stochastik,
Mohrenstraße 39, 10117 Berlin, Germany

E-mail: jmartins@civil.ist.utl.pt, mmarques@ptmat.fc.ul.pt, petrov@wias-berlin.de

Abstract. In this paper, a concept of stability of quasi-static paths is discussed. This takes into account the existence of fast (dynamic) and slow (quasi-static) times scales in the mechanical systems that have an elastic-plastic behavior with linear hardening. The proposed concept is essentially a continuity property relative to the size of the initial perturbations (as in Lyapunov stability) and relative to the smallness of the rate of application of the forces (which here plays the role of the small parameter in singular perturbation problems). Existence and uniqueness results for the dynamic and quasi-static problems are recalled and the stability of quasi-static paths for elastic-plastic systems with hardening is obtained.

1. Introduction

The computation of quasi-static evolutions of mechanical systems (frequently involving physical and geometric nonlinearities, plasticity, damage, etc.) is done with the presumption that (some portions of) the computed quasi-static paths are a reliable approximation of the actual paths that, in our physical world, are subjected to the universal rule of Newton's law. The common use of static equilibrium equations instead of Newton's law in the computation of the slow evolution of some mechanical systems requires a justification and an assessment of the limits of its validity. That is the purpose of the research work initiated in [6], and continued in [10, 9, 7]. To the best of our knowledge, Martins and coworkers [6, 10] were the first to recognize the distinct time scales involved in dynamic and quasi-static problems, as well as the singular perturbation nature of the problem of using static equilibrium equations to approximate governing dynamic equations, when the time rate of change of the load is decreased to zero. They performed a change of variables in the governing system of dynamic equations that consists of replacing the (fast) physical time t by a (slow) loading parameter λ whose rate of change with respect to time, $\varepsilon = d\lambda/dt$, is eventually decreased to zero. This leads to a system of singularly perturbed equations or inclusions, in which some of the highest order derivatives with respect to the loading parameter appear multiplied by the small time rate of change ε of the load.

In the present paper, the definition of stability of quasi-static paths given in [10] is adapted to a continuum (1D) case that involves an elastic-plastic material with kinematic hardening, and we shall prove that in this case the dynamic evolutions remain close to a quasi-static path when they start sufficiently close to it and the load is applied sufficiently slowly.

Johnson [3, 4] developed the variational formulation of plasticity problems with hardening. For the corresponding quasi-static problem he proved existence of a strong solution and, under some additional assumptions, a regularity result for the velocity was obtained. A similar formulation had already been obtained for perfect elastic-plastic and elastic-visco-plastic systems by Duvaut and Lions [2]. In what concerns the dynamic problems in elasto-plasticity with hardening, we address the reader to the works of Krejčí [5], Showalter and Shi [11, 12], Visintin [13], and the references therein.

The structure of the article is the following. In Section 2, the mathematical formulations for dynamic and quasi-static elastic-plastic systems with hardening are presented. In Section 3, existence and uniqueness results, which use the theory of *m-accretive* operators, are recalled [1, 2, 11, 12, 14]; some results are also shown for the elastic-visco-plastic (Yosida) regularization of the original dynamic problem followed by its finite dimensional (Galerkin) approximation. In Section 4, the definition of stability of a quasi-static path is adapted from [7, 9, 10] to the present elastic-plastic problems, and stability in that sense for these problems is proved.

2. Governing equations

We consider an elastic-plastic bar with linear kinematic hardening that has length L along the x axis. Geometrical linearity is assumed. The governing dynamic equation can be non-dimensionalized by using the non-dimensional time (τ) and load parameter (λ, $\lambda = \lambda_1 + \varepsilon\tau$), yielding

$$\varepsilon^2 u'' - \sigma_x = f(x, \lambda), \tag{1}$$

where u, r, f are the non-dimensional axial displacement, the stress in the plastic element and the applied force per unit length along the bar, respectively; σ is the stress in the elastic-plastic bar, and the subscript x denotes a derivative with respect to x. For simplicity, the mass density of the bar has been taken as unitary.

The extension e is the derivative in space of the displacement u, and it can be decomposed into elastic, e^e, and plastic, e^p, parts:

$$e = u_x = e^e + e^p. \tag{2}$$

The stress σ in the bar is related to the stress in the plastic element and the plastic strain by the linear kinematic hardening rule

$$\sigma = r + He^p, \tag{3}$$

and is related to the elastic part of the extension by means of Hooke's law

$$\sigma = Ee^e = E(e - e^p), \tag{4}$$

where the elasticity and hardening moduli E and H are positive constants. Therefore (2), (3), and (4) lead to

$$\sigma = \sigma(e, r) = D(e + H^{-1}r) \text{ where } D = (E^{-1} + H^{-1})^{-1}. \tag{5}$$

Carrying (5) and (2) into (1), we obtain

$$\varepsilon^2 u'' - Du_{xx} - DH^{-1}r_x = f. \tag{6}$$

The behavior of the plastic element is characterized by the non-dimensional inequality and flow rule:

$$|r| \leq 1, \ (e^p)' \begin{cases} \geq 0 \text{ if } r = +1, \\ = 0 \text{ if } -1 < r < +1, \\ \leq 0 \text{ if } r = -1. \end{cases} \tag{7}$$

The governing dynamic equations (6), together with the conditions (7) can be put in the form of a singular perturbation system of first order differential equations and inclusion. For that purpose, let $\mathcal{C}$ denote the following closed convex set in $L^2(0,L)$

$$\mathcal{C} = \{r \in L^2(0,L) : |r| \leq 1\}, \tag{8}$$

and let $\text{sign}^{-1}(r)$ be the normal cone to $\mathcal{C}$ at $r \in L^2(0,L)$. Then we observe that (7) can be written in the differential inclusion form:

$$(e^p)' \in \text{sign}^{-1}(r). \tag{9}$$

Relations (4) lead to

$$(e^p)' = \tilde{D}^{-1}(Eu'_x - r') = DH^{-1}u'_x - \tilde{D}^{-1}r', \text{ where } \tilde{D} = E + H. \tag{10}$$

Substituting (10) in (9), we get

$$\tilde{D}^{-1}r' - DH^{-1}u'_x + \text{sign}^{-1}(r) \ni 0. \tag{11}$$

We introduce now the following spaces

$$\mathcal{H} = L^2(0,L), \ \mathcal{V} = H^1(0,L), \ \mathcal{V}_0 = H^1_0(0,L). \tag{12}$$

We denote the norm in $\mathcal{H}$ (resp. $\mathcal{V}$) by $|\cdot|$ (resp. $\|\cdot\|$) and the scalar product in $\mathcal{H}$ by $(\cdot,\cdot)$. From (6) and (11) we finally obtain the governing dynamic system

$$\begin{cases} \varepsilon v' - Du_{xx} - DH^{-1}r_x = f, \\ \varepsilon u' - v = 0, \\ \tilde{D}^{-1}r' - DH^{-1}u'_x + \text{sign}^{-1}(r) \ni 0, \end{cases} \tag{13}$$

together with the Dirichlet boundary conditions

$$v = u = 0 \text{ on } \{0, L\} \times (\lambda_1, \lambda_2), \tag{14}$$

and the initial conditions (recall (5) and (12))

$$\big(v(\lambda_1), u(\lambda_1), r(\lambda_1)\big) = (v_1, u_1, r_1) \in \mathcal{V}_0 \times \mathcal{V}_0 \times \mathcal{C}, \text{ with } \sigma(u_{x1}, r_1) \in \mathcal{V}. \tag{15}$$

Note that the differential inclusion system (13) can be written in the variational inequality form:

$$\begin{cases} (u,r) \in \mathcal{V}_0 \times \mathcal{C} \text{ and } \forall (u^*, r^*) \in \mathcal{V}_0 \times \mathcal{C} : \\ (\varepsilon^2 u'', u^*) + a(u, u^*) + b(r, u^*) = (f, u^*), \\ (r^* - r, \tilde{D}^{-1}r') - b(r^* - r, u') \geq 0, \end{cases} \tag{16}$$

where the bilinear forms $a(\cdot,\cdot)$ and $b(\cdot,\cdot)$ are given by

$$a(u, u^*) = (Du_x, u^*_x),$$
$$b(r, u^*) = (DH^{-1}r, u^*_x),$$

for all $u, u^* \in \mathcal{V}$ and all $r \in \mathcal{H}$. Clearly there exist positive constants α_m, α_M and β_M such that, for all $u, u^* \in \mathcal{V}$ and all $r \in \mathcal{H}$:

$$\alpha_m \|u\|^2 \le a(u,u), \quad a(u,u^*) \le \alpha_M \|u\| \, \|u^*\|,$$

$$b(r,u^*) \le \beta_M |r| \, \|u^*\|.$$

These inequalities will be used throughout the rest of the paper in order to obtain all the estimates that are based on variational statements.

The corresponding quasi-static system is then (let $\varepsilon = 0$ in (13))

$$\begin{cases} -D\bar{u}_{xx} - DH^{-1}\bar{r}_x = f, \\ \tilde{D}^{-1}\bar{r}' - DH^{-1}\bar{u}_x' + \operatorname{sign}^{-1}(\bar{r}) \ni 0, \end{cases} \tag{17}$$

with the Dirichlet boundary conditions

$$\bar{u} = 0 \text{ on } \{0, L\} \times (\lambda_1, \lambda_2), \tag{18}$$

and the initial conditions (recall again (5) and (12))

$$\big(\bar{u}(\lambda_1), \bar{r}(\lambda_1)\big) = (\bar{u}_1, \bar{r}_1) \in \mathcal{V}_0 \times \mathcal{C}, \ \textit{with } \sigma(\bar{u}_{x1}, \bar{r}_1) \in \mathcal{V}. \tag{19}$$

Note that, consistent with the above, the quasi-static displacement rate with respect to the physical time vanishes ($\bar{v} \equiv 0$). On the other hand, the differential inclusion system (17) can be written in the variational form:

$$\begin{cases} (\bar{u}, \bar{r}) \in \mathcal{V}_0 \times \mathcal{C} \text{ and } \forall (u^*, r^*) \in \mathcal{V}_0 \times \mathcal{C}: \\ a(\bar{u}, u^*) + b(\bar{r}, u^*) = (f, u^*), \\ (r^* - \bar{r}, \tilde{D}^{-1}\bar{r}') - b(r^* - \bar{r}, \bar{u}') \ge 0. \end{cases} \tag{20}$$

3. Existence, uniqueness and approximation results

In this section we recall existence and uniqueness results for the dynamic and quasi-static problems that were formulated earlier. The theory of *m-accretive* operators is used with this purpose. We also recall the elastic-visco-plastic regularization of the original dynamic elastic-plastic problem, as well as the finite dimensional approximation of the regularized problem. This has the purpose of recalling or proving some properties of those approximations that will be used subsequently.

3.1. Existence and uniqueness of solution for the dynamic and quasi-static problems

The dynamic and the quasi-static systems introduced in Section 2 can be rewritten in a form that may be studied with the theory of *m-accretive* operators. Consider the differential inclusion problem that involves a multivalued operator $\mathbb{A}$ in the Hilbert space $\mathcal{Y}$, with domain $\mathcal{D}(\mathbb{A}) = \{\boldsymbol{x} \in \mathcal{Y} : \mathbb{A}\boldsymbol{x} \neq \emptyset\}$:

$$\boldsymbol{x}(\lambda) \in \mathcal{D}(\mathbb{A}), \ \forall \lambda \in [\lambda_1, \lambda_2], \tag{21a}$$

$$\boldsymbol{x}' + \mathbb{A}\boldsymbol{x} \ni \boldsymbol{g} \text{ a.e. on } (\lambda_1, \lambda_2), \tag{21b}$$

$$\boldsymbol{x}(\lambda_1) = \boldsymbol{x}_1, \tag{21c}$$

Recall that existence and uniqueness of solution to this problem can be obtained from the following (*cf.* [4] or [3]):

Proposition 3.1 *Assume that* $\mathbb{A}$ *is an m-accretive operator in the Hilbert space* $\mathcal{Y}$, $\boldsymbol{g}$ *belongs to* $W^{1,\infty}(\lambda_1, \lambda_2; \mathcal{Y})$ *and* $\boldsymbol{x}_1 \in \mathcal{D}(\mathbb{A})$. *Then there exists a unique solution* $\boldsymbol{x}$ *of* (21) *belonging to* $W^{1,\infty}(\lambda_1, \lambda_2; \mathcal{Y})$.

In what concerns the *dynamic problem*, we differentiate the second equation in the system (13) with respect to x, we use $e = u_x$ as an alternative unknown function, and we denote $\boldsymbol{x} = (v, D^{1/2}e, \tilde{D}^{-1/2}r)$. Then the governing dynamic system (13) yields the inclusion (21b) with

$$\mathbb{A} = \frac{1}{\varepsilon}\begin{pmatrix} 0 & -D^{1/2}\partial/\partial x & -E\tilde{D}^{-1/2}\partial/\partial x \\ -D^{1/2}\partial/\partial x & 0 & 0 \\ -E\tilde{D}^{-1/2}\partial/\partial x & 0 & \varepsilon\tilde{D}^{1/2}\text{sign}^{-1}(\tilde{D}^{1/2}(\cdot)) \end{pmatrix} \text{ and } \boldsymbol{g} = \frac{1}{\varepsilon}\begin{pmatrix} f \\ 0 \\ 0 \end{pmatrix}. \tag{22}$$

Letting $\mathcal{Y} = \mathcal{H}^3$, it can be proved (see [11] and [12]) that $\mathbb{A}$ is an *m-accretive* operator in $\mathcal{Y}$. Then Proposition 3.1, yields the following:

Corollary 3.2 *Assume that* f *belongs to* $W^{1,\infty}(\lambda_1, \lambda_2; \mathcal{H})$ *and that* (15) *holds. Then there exists a unique* $\boldsymbol{x} = (v, D^{1/2}e, \tilde{D}^{-1/2}r) \in W^{1,\infty}(\lambda_1, \lambda_2; \mathcal{H}^3)$ *that solves* (21) *with* $\mathbb{A}$ *and* $\boldsymbol{g}$ *given by* (22), *and with* $r(\lambda) \in \mathcal{C}$ *for all* $\lambda \in [\lambda_1, \lambda_2]$, $v \in L^\infty(\lambda_1, \lambda_2; \mathcal{V}_0)$, *and* $\sigma(e, r) \in L^\infty(\lambda_1, \lambda_2; \mathcal{V})$.

In what concerns the *quasi-static problem*, we differentiate the first identity in (17) with respect to the load parameter λ and we get

$$-D\bar{u}'_{xx} = DH^{-1}\bar{r}'_x + f', \tag{23}$$

together with the Dirichlet boundary conditions (18). Since this is an elliptic problem for $\bar{u}'$ we conclude that there exists a unique solution. For such solution $\bar{u}'_x + H^{-1}\bar{r}'$ depends linearly and continuously on f', i.e.

$$\bar{u}'_x + H^{-1}\bar{r}' = Bf', \tag{24}$$

where B is a continuous linear operator between the appropriate spaces. Inserting this in the inclusion in (17) we finally get the differential inclusion

$$\bar{r}' + Hsign^{-1}(\bar{r}) \ni DBf'. \tag{25}$$

The sub-differential $\partial\varphi(\bar{r}) = \text{sign}^{-1}(\bar{r})$ is an *m-accretive* operator since $\varphi(\bar{r})$ is a proper convex and lower semi-continuous function. For $\boldsymbol{x} = \bar{r}$, $\mathbb{A} = H\text{sign}^{-1}$, $\boldsymbol{g} = DBf'$ and $\mathcal{Y} = \mathcal{H}$, we apply Proposition 3.1 and we obtain the following Corollary:

Corollary 3.3 *Assume that* f *belongs to* $W^{1,\infty}(\lambda_1, \lambda_2; \mathcal{H})$ *and* (19) *holds. Then there exists a unique solution* $(\bar{u}, \bar{r})$ *of* (17)–(19) *such that* $(\bar{u}, \bar{r}) \in W^{1,\infty}(\lambda_1, \lambda_2; \mathcal{V}_0 \times \mathcal{H})$, *with* $\bar{r} \in \mathcal{C}$ *for all* $\lambda \in [\lambda_1, \lambda_2]$, *and* $\sigma(\bar{u}_x, \bar{r}) \in L^\infty(\lambda_1, \lambda_2; \mathcal{V})$.

3.2. Elastic-visco-plastic and Galerkin approximations

We consider first the *elastic-visco-plastic problem*:

$$\begin{cases} \varepsilon^2 u''_\mu - Du_{\mu xx} - DH^{-1}r_{\mu x} = f, \\ \tilde{D}^{-1}r'_\mu - DH^{-1}u'_{\mu x} + \mathcal{J}_\mu(r_\mu) = 0, \end{cases} \quad \text{where} \quad \mathcal{J}_\mu(r_\mu) = \frac{1}{\mu}\left(r_\mu - \text{proj}_{\mathcal{C}} r_\mu\right), \tag{26}$$

with the Dirichlet boundary conditions

$$u_\mu = 0 \text{ on } \{0, L\} \times (\lambda_1, \lambda_2), \tag{27}$$

and the initial conditions

$$\left(\varepsilon u'_\mu(\lambda_1), u_\mu(\lambda_1), r_\mu(\lambda_1)\right) = (v_1, u_1, r_1) \in \mathcal{V}_0 \times \mathcal{V}_0 \times \mathcal{C}, \text{ with } \sigma(u_{x1}, r_1) \in \mathcal{V}. \tag{28}$$

Here $\mu > 0$ is the viscosity parameter and $\text{proj}_\mathcal{C}$ denotes the projection on the convex $\mathcal{C}$.

The variational formulation of the regularized problem (26)–(28) reads

$$\begin{cases} (u_\mu, r_\mu) \in \mathcal{V}_0 \times \mathcal{H} \text{ such that } \forall (u^*, r^*) \in \mathcal{V}_0 \times \mathcal{H}, \\ (\varepsilon^2 u''_\mu, u^*) + a(u_\mu, u^*) + b(r_\mu, u^*) = (f, u^*), \\ (r^*, \tilde{D}^{-1} r'_\mu) - b(r^*, u'_\mu) + (r^*, \mathcal{J}_\mu(r_\mu)) = 0, \end{cases} \tag{29}$$

also with the initial conditions (28). Note that this elastic-visco-plastic problem is an Yosida regularization of the original elastic-plastic problem. For a similar approximation in the corresponding finite-dimensional system see [7]. Whenever convenient we shall use the notation $v_\mu = \varepsilon u'_\mu$ in the following text.

We consider now a *finite dimensional approximation* of the above elastic-visco-plastic problem, which is obtained in the following classical manner. Let $\{w_j\}_{j=1}^\infty$ be a complete orthonormal sequence in $\mathcal{H}$ whose elements belong to $H^2(0, L)$. Let $u_{\mu_n} = \sum_{i=1}^n g_{in}(\lambda) w_i(x)$ and $r_{\mu_n} = \sum_{i=1}^n h_{in}(\lambda) w_i(x)$ satisfying the following variational formulation

$$\begin{cases} \text{For all } u^* = \sum_{i=1}^n g^*_{in}(\lambda) w_i(x) \text{ and } r^* = \sum_{i=1}^n h^*_{in}(\lambda) w_i(x), \\ (\varepsilon^2 u''_{\mu_n}, u^*) + a(u_{\mu_n}, u^*) + b(r_{\mu_n}, u^*) = (f, u^*), \\ (r^*, \tilde{D}^{-1} r'_{\mu_n}) - b(r^*, u'_{\mu_n}) + (r^*, \mathcal{J}_\mu(r_{\mu_n})) = 0, \end{cases} \tag{30}$$

with

$$\varepsilon \lim_{n\to\infty} \sum_{i=1}^n g'_{in}(\lambda_1) w_i(x) = v_1, \ \lim_{n\to\infty} \sum_{i=1}^n g_{in}(\lambda_1) w_i(x) = u_1, \ \lim_{n\to\infty} \sum_{i=1}^n h_{in}(\lambda_1) w_i(x) = r_1. \tag{31}$$

The following results can be proved for the above approximations, when the dimension parameter n tends to ∞, and the viscosity parameter μ tends to 0.

Proposition 3.4 *Assume that f belongs to $W^{1,\infty}(\lambda_1, \lambda_2; \mathcal{H})$ and that (28) holds. Then there exists a unique solution (v_μ, u_μ, r_μ) of (26)–(28) such that (v_μ, u_μ, r_μ) and (v'_μ, u'_μ, r'_μ) belong respectively to $L^\infty(\lambda_1, \lambda_2; \mathcal{V}_0 \times \mathcal{V}_0 \times \mathcal{H})$ and $L^\infty(\lambda_1, \lambda_2; \mathcal{H} \times \mathcal{H} \times \mathcal{H})$ and $\sigma(u_{\mu x}, r_\mu) \in L^\infty(\lambda_1, \lambda_2; \mathcal{V})$. Moreover, as μ tends to zero, u_μ and $\sigma(u_{\mu x}, r_\mu)$ converge strongly to their limits.*

The Galerkin approximation described above together with *a priori* estimates based on the variational formulations (29), (30) can be used to prove these results. The reader can find detailed proofs in the Appendix of [8] or in [2]. This Proposition can also be proved using the theory of *m-accretive* operators.

Lemma 3.5 *Assume that (28) holds and that f belongs to $W^{1,\infty}(\lambda_1, \lambda_2; \mathcal{H})$. Then independently of $\mu > 0$, for all λ belonging to (λ_1, λ_2), $v_\mu(\lambda)$, $u_{\mu x}(\lambda)$ and $r_\mu(\lambda)$ are bounded in $\mathcal{H}$. Moreover, as μ tends to 0, it holds that, for all λ belonging to (λ_1, λ_2),*

$$v_\mu(\lambda) \to v(\lambda) \textit{ strongly in } \mathcal{H},$$
$$u_{\mu x}(\lambda) \to u_x(\lambda) \textit{ strongly in } \mathcal{H},$$
$$r_\mu(\lambda) \to r(\lambda) \textit{ strongly in } \mathcal{H},$$

and also

$$\varepsilon v'_\mu \rightharpoonup \varepsilon v' \textit{ weakly } * \textit{ in } L^\infty(\lambda_1, \lambda_2; \mathcal{V}'_0). \tag{32}$$

Energy estimation based on the variational formulation (29) leads to the above boundedness properties. The strong convergence properties are obtained by the energy estimation of the difference between the elastic-visco-plastic system (29) and the elastic-plastic system with hardening (16). The same difference and the strong convergence of $\sigma(u_{\mu x}, r) = D(u_{\mu x} + H^{-1} r_\mu)$ lead to the desired result.

Lemma 3.6 *Assume that* (28) *holds and that* f *belongs to* $W^{2,\infty}(\lambda_1, \lambda_2; \mathcal{H})$. *Then there exists a subsequence, still denoted by* v'_{μ_n}, *such that*

$$v'_{\mu_n} \rightharpoonup v'_\mu \text{ weakly } * \text{ in } L^\infty(\lambda_1, \lambda_2; \mathcal{H}). \tag{33}$$

Moreover denoting by c *a positive constant that is independent of* μ *and* n, *and letting* $v'_{\mu_n}(\lambda_1)$ *be the initial acceleration of the* μ_n *solution, the following inequality holds for all* λ *belonging to* (λ_1, λ_2)

$$|\varepsilon v'_{\mu_n}(\lambda)|^2 \leq c\big(||v_1||^2 + \varepsilon^2 |v'_{\mu_n}(\lambda_1)|^2 + \varepsilon^2 |f'(\lambda_1)|^2 + \varepsilon^2 ||f'||^2_{L^\infty(\lambda_1,\lambda_2;\mathcal{H})} + \varepsilon^2 ||f''||^2_{L^2(\lambda_1,\lambda_2;\mathcal{H})}\big). \tag{34}$$

Unlike the previous results, the proof of this Lemma requires the *differentiation of the governing system* (30) *with respect to* λ. Taking $u^* = \varepsilon^2 u''_{\mu_n}$ and $r^* = \varepsilon^2 r'_{\mu_n}$ in the resulting system, adding the two resulting identities, and using the monotonicity of $r \mapsto \mathcal{J}_\mu(r)$, yields an appropriate energy inequality. Its integration in λ, taking into account the solution values at the initial λ_1, and the application of Gronwall's Lemma, leads to the final result. The detailed proof can be found in [8].

4. Stability of quasi-static paths of elastic-plastic systems with linear hardening

In Section 4.1, we adapt the definition of stability of a quasi-static path ([7], [9], [10]) to the present elastic-plastic problem with hardening. In Section 4.2, *a priori* estimates on the elastic-visco-plastic systems are obtained, which prove that the dynamic and the quasi-static solutions remain close to each other if the dynamic solution of (13) is initially close to the quasi-static solution of (17) and the loading rate ε is sufficiently small.

4.1. Definition of stability of a quasi-static path

The mathematical definition of stability of a quasi-static path at an equilibrium point is presented in the context of the governing dynamic system (13)–(15) and the quasi-static system (17)–(19).

Definition 4.1 *The quasi-static path* $(\bar{u}(\lambda), \bar{r}(\lambda))$ *is said to be stable at* λ_1 *if there exists* $0 < \Delta\lambda \leq \lambda_2 - \lambda_1$, *such that, for all* $\delta > 0$ *there exists* $\bar{\rho}(\delta) > 0$ *and* $\bar{\varepsilon}(\delta) > 0$ *such that for all initial conditions* u_1, v_1, r_1 *and* $\bar{u}_1$, $\bar{r}_1$ *and all* $\varepsilon > 0$ *such that*

$$|v_1|^2 + ||u_1 - \bar{u}_1||^2 + |r_1 - \bar{r}_1|^2 \leq \bar{\rho}(\delta) \text{ and } \varepsilon \leq \bar{\varepsilon}(\delta),$$

the solution $(u(\lambda), v(\lambda), r(\lambda))$ *of the dynamic system* (13)–(15) *satisfies*

$$|v(\lambda)|^2 + ||u(\lambda) - \bar{u}(\lambda)||^2 + |r(\lambda) - \bar{r}(\lambda)|^2 \leq \delta,$$

for all $\lambda \in [\lambda_1, \lambda_1 + \Delta\lambda]$.

For more details, the reader is referred to [10].

4.2. Stability of a quasi-static path

In order to prove the stability of the quasi-static path, we obtain an energy estimate of the difference between the dynamic and the quasi-static solution. In order to do this we will consider an auxiliary "special dynamic solution" $(\tilde{v}, \tilde{u}, \tilde{r})$ that solves (26) with the Dirichlet boundary conditions (14) and with the initial conditions

$$\big(\tilde{v}(\lambda_1), \tilde{u}(\lambda_1), \tilde{r}(\lambda_1)\big) = (\varepsilon\bar{u}_1', \bar{u}_1, \bar{r}_1) \in \mathcal{V}_0 \times \mathcal{V}_0 \times \mathcal{C}, \text{ with } \sigma(\bar{u}_{x1}, \bar{r}_1) \in \mathcal{V}. \tag{35}$$

This means that the initial conditions for this special dynamic solution coincide with or result from the quasi-static solution. Besides these "quasi-static initial conditions" (35) the special dynamic solution thus satisfies the variational statements:

$$\begin{cases} (\tilde{u}, \tilde{r}) \in \mathcal{V}_0 \times \mathcal{C} \text{ and } \forall (u^*, r^*) \in \mathcal{V}_0 \times \mathcal{C}, \\ (\varepsilon^2 \tilde{u}'', u^*) + a(\tilde{u}, u^*) + b(\tilde{r}, u^*) = (f, u^*), \\ (r^* - \tilde{r}, \tilde{D}^{-1}\tilde{r}') - b(r^* - \tilde{r}, \tilde{u}') \geq 0. \end{cases} \tag{36}$$

This special dynamic solution is used in the stability proof in the following manner: the difference between the dynamic and the quasi-static solution is estimated by evaluating separately the difference between the dynamic solution and the special dynamic solution, and the difference between the latter and the quasi-static solution:

$$\begin{aligned} &|v(\lambda)|^2 + \|u(\lambda) - \bar{u}(\lambda)\|^2 + |r(\lambda) - \bar{r}(\lambda)|^2 \\ &\leq 2\big(|v(\lambda) - \tilde{v}(\lambda)|^2 + \|u(\lambda) - \tilde{u}(\lambda)\|^2 + |r(\lambda) - \tilde{r}(\lambda)|^2 \\ &+ |\tilde{v}(\lambda)|^2 + \|\tilde{u}(\lambda) - \bar{u}(\lambda)\|^2 + |\tilde{r}(\lambda) - \bar{r}(\lambda)|^2\big), \end{aligned} \tag{37}$$

In what concerns the difference between the dynamic solution and the special dynamic solution, the desired estimate is nothing but a continuity property of the solutions of the dynamic problem relatively to the initial conditions. The corresponding energy estimate is obtained in Lemma 4.2. A similar "well-posedness" result was recently obtained by Visintin [13].

In Lemma 4.3 an energy estimate of the difference between the special dynamic solution and the quasi-static one is obtained which makes use of the results of the Yosida and the Galerkin approximations of the previous section.

Lemma 4.2 *Assume that* (15) *and* (35) *hold and that* f *belongs to* $W^{1,\infty}(\lambda_1, \lambda_2; \mathcal{H})$. *Then there exists a positive constant* c *such that:*

$$\begin{aligned} &|v(\lambda) - \tilde{v}(\lambda)|^2 + \|u(\lambda) - \tilde{u}(\lambda)\|^2 + |r(\lambda) - \tilde{r}(\lambda)|^2 \\ &\leq c\big(|v_1 - \varepsilon\bar{u}'(\lambda_1)|^2 + \|u_1 - \bar{u}_1\|^2 + |r_1 - \bar{r}_1|^2\big). \end{aligned} \tag{38}$$

Proof. Subtracting the equalities in (16) and (36), and choosing $u^* = u' - \tilde{u}'$ in both, we have

$$\big(\varepsilon^2(u'' - \tilde{u}''), u' - \tilde{u}'\big) + a(u - \tilde{u}, u' - \tilde{u}') + b(r - \tilde{r}, u' - \tilde{u}') = 0. \tag{39}$$

On the other hand, taking $r^* = \tilde{r}$ in the inequality of (16) and $r^* = r$ in the inequality of (36), we get, after adding the resulting inequalities:

$$(r - \tilde{r}, \tilde{D}^{-1}(r' - \tilde{r}')) \leq b(r - \tilde{r}, u' - \tilde{u}'). \tag{40}$$

Carrying (40) into (39) and since $v = \varepsilon u'$ and $\tilde{v} = \varepsilon\tilde{u}'$, we obtain

$$\frac{d}{d\lambda}\left(|v - \tilde{v}|^2 + a(u - \tilde{u}, u - \tilde{u}) + |\tilde{D}^{-1/2}(r - \tilde{r})|^2\right) \leq 0. \tag{41}$$

Integrating (41) over (λ_1, λ), $\lambda \in [\lambda_1, \lambda_2]$, and using the initial conditions (15) and (35) leads to the desired result.

Lemma 4.3 *Assume that f belongs to $W^{1,\infty}(\lambda_1, \lambda_2; \mathcal{H})$ and that (19) and (35) hold. Then there exists a positive constant c, such that*

$$|\tilde{v}(\lambda)|^2 + \|\tilde{u}(\lambda) - \bar{u}(\lambda)\|^2 + |\tilde{r}(\lambda) - \bar{r}(\lambda)|^2 \le c\big[\varepsilon^2|\bar{u}'(\lambda_1)|^2 + \int_{\lambda_1}^{\lambda} (\varepsilon\tilde{v}', \bar{u}')\, d\xi\big] \tag{42}$$

Proof. Proceeding as in the previous Lemma, but dealing now with the difference between the "special dynamic solution" and the quasi-static one, we obtain successively

$$\varepsilon^2(\tilde{u}'', \tilde{u}') + a(\tilde{u} - \bar{u}, \tilde{u}' - \bar{u}') + b(\tilde{r} - \bar{r}, \tilde{u}' - \bar{u}') = \varepsilon^2(\tilde{u}'', \bar{u}'), \tag{43}$$

$$(\tilde{r} - \bar{r}, \tilde{D}^{-1}(\tilde{r}' - \bar{r}')) \le b(\tilde{r} - \bar{r}, \tilde{u}' - \bar{u}'), \tag{44}$$

$$\frac{d}{d\lambda}\left(|\tilde{v}|^2 + a(\tilde{u} - \bar{u}, \tilde{u} - \bar{u}) + |\tilde{D}^{-1/2}(\tilde{r} - \bar{r})|^2\right) \le 2\varepsilon^2(\tilde{u}'', \bar{u}'). \tag{45}$$

The desired result follows from integrating (45) over (λ_1, λ) with $\lambda \in [\lambda_1, \lambda_2]$, since the initial conditions (35) for the special dynamic solution are such that $\tilde{v}(\lambda_1) = \varepsilon\bar{u}_1'$, $\tilde{u}(\lambda_1) = \bar{u}_1$, and $\tilde{r}(\lambda_1) = \bar{r}_1$.

In order to obtain the final stability result, the estimates (38) and (42) of the previous lemmas are inserted in the inequality (37). We have

$$\begin{aligned} &|v(\lambda)|^2 + \|u(\lambda) - \bar{u}(\lambda)\|^2 + |r(\lambda) - \bar{r}(\lambda)|^2 \\ &\le c\big[|v_1 - \varepsilon\bar{u}'(\lambda_1)|^2 + \|u_1 - \bar{u}_1\|^2 + |r_1 - \bar{r}_1|^2 \\ &+ \varepsilon^2|\bar{u}'(\lambda_1)|^2 + \int_{\lambda_1}^{\lambda} (\varepsilon\tilde{v}', \bar{u}')\, d\xi\big], \end{aligned} \tag{46}$$

where, again, c is some positive constant that is independent of ε. In order to estimate the last integral, we consider the finite dimensional (Galerkin) approximation of the elastic-visco-plastic (Yosida) regularization of the inertia term $\varepsilon\tilde{v}'$:

$$\int_{\lambda_1}^{\lambda} (\varepsilon\tilde{v}', \bar{u}')\, d\xi = \int_{\lambda_1}^{\lambda} (\varepsilon(\tilde{v}' - \tilde{v}_\mu'), \bar{u}')\, d\xi + \int_{\lambda_1}^{\lambda} (\varepsilon(\tilde{v}_\mu' - \tilde{v}_{\mu n}'), \bar{u}')\, d\xi + \int_{\lambda_1}^{\lambda} (\varepsilon\tilde{v}_{\mu n}', \bar{u}')\, d\xi. \tag{47}$$

The conclusion follows then from Lemma 3.6: the estimate (34) of $\varepsilon\tilde{v}_{\mu n}'$ and the weak $*$ convergence (32) and (33) of the terms that involve differences of inertia terms. Note, in particular, that the initial terms involved in the estimate (34) of $\varepsilon\tilde{v}_{\mu n}'$ have the following simplifications: $\|\tilde{v}_1\|^2 = \varepsilon^2\|\bar{u}'(\lambda_1)\|^2$ and $\tilde{v}_{\mu n}'(\lambda_1) = 0$. It is because of these simplifications that we use the special dynamic solution in our proof and, in Lemma 4.3, we estimate the distance between the quasi-static solution and the special dynamic solution (rather than directly estimate the distance between the quasi-static solution and an arbitrary dynamic solution).

Before stating the desired stability result, we remark that the proof given here differs from the one presented in [8], namely, on the use made here of the weak $*$ convergence (32) of $\varepsilon\tilde{v}_\mu'$ to $\varepsilon\tilde{v}'$, and on some aspects of the notation and of the order of the presentation, which hopefully will improve its clarity.

Proposition 4.4 *(Stability). Assume that f belongs to $W^{2,\infty}(\lambda_1, \lambda_2; \mathcal{H})$ and that (15) and (19) hold. Then there exists $\gamma > 0$ such that for $0 < \varepsilon < 1$,*

$$\begin{aligned} &|v(\lambda)|^2 + \|u(\lambda) - \bar{u}(\lambda)\|^2 + |r(\lambda) - \bar{r}(\lambda)|^2 \\ &\le \gamma\big(|v_1|^2 + \|u_1 - \bar{u}_1\|^2 + |r_1 - \bar{r}_1|^2 + \varepsilon\big). \end{aligned}$$

Acknowledgments
This work is part of the project "New materials, adaptive systems and their nonlinearities; modelling control and numerical simulation" carried out in the framework of the European Community Program "Improving the human research potential and socio-economic knowledge base" (contract n°HPRN-CT-2002-00284). J.A.C. Martins, M.D.P. Monteiro Marques and A. Petrov were partially supported by F.C.T. (Fundação para a Ciência e a Tecnologia) /POCTI/FEDER and by Project POCTI/MAT/40867/2001 "Estabilidade de Trajectórias Quase-Estáticas e Problemas de Perturbação Singular". A. Petrov thanks also the Research Center "MATHEON" (project C18) for generous support.

References

[1] Brézis H 1973 *Opérateurs maximaux monotones et semi-groupes de contractions dans les espaces de Hilbert.* North-Holland Mathematics Studies, no 5. Notas de Matemática (50). (Amsterdam: North-Holland Publishing Co)
[2] Duvaut G and Lions J-L 1976 *Inequalities in mechanics and physics.* (Berlin: Springer-Verlag) Translated from the French by C W John, Grundlehren der Mathematischen Wissenschaften 219p.
[3] Johnson C 1976 Existence theorems for plasticity problems *J. Math. Pures Appl. (9)* **55** 431
[4] Johnson C 1978 On plasticity with hardening *J. Math. Anal. Appl.* **62** 325
[5] Krejčí P 1996 *Hysteresis, Convexity and Dissipation in Hyperbolic Equations* (Tokyo: Gakkotosho)
[6] Loret B, Simões F M F and Martins J A C 2000 Flutter instability and ill-posedness in solids and fluid-saturated porous media *Material Instabilities in Elastic and Plastic Solids (Udine, 1999)* ed Petryk H *CISM Courses and Lectures* **414** pp 109–207 (Vienna: Springer)
[7] Martins J A C, Monteiro Marques M D P and Petrov A On the stability of quasi-static paths for finite dimensional elastic-plastic systems with hardening (submitted)
[8] Martins J A C, Monteiro Marques M D P and Petrov A On the stability of elastic-plastic systems with hardening (submitted)
[9] Martins J A C, Rebrova N V and Sobolev V A 2006 On the (in)stability of quasi-static paths of smooth systems: definitions and sufficient conditions. *Math. Methods in the Appl. Sciences* **29** 741
[10] Martins J A C, Simões F M F A, Pinto da Costa A and Coelho I 2004 Three examples on "(in)stability of quasi-static paths" (submitted)
[11] Showalter R E and Shi P 1997 Plasticity models and nonlinear semigroups. *J. Math. Anal. Appl.* **216** 218
[12] Showalter R E and Shi P 1998 Dynamic plasticity models. *Comput. Methods Appl. Engrg.* **151** 501
[13] Visintin A 2005 On homogenization of elasto-plasticity. *J. Phys.: Conf. Series* **22** 222
[14] Zeidler E 1985 *Nonlinear functional analysis and its applications. III* (New York: Springer-Verlag) Variational methods and optimization, translated from the German by Leo F Boron

Institute of Physics Publishing
doi:10.1088/1742-6596/55/1/016

Journal of Physics: Conference Series **55** (2006) 165–170
International Workshop on Multi-Rate Processes and Hysteresis

Complex nonlinear ordinary differential equations and geometry

B McKay
Mathematics Department, University College, Cork, Ireland
E-mail: b.mckey@ucc.ie

Abstract. This paper is a study of global properties of complex nonlinear second order differential equations. We uncover a purely local expression, known since the late 19th Century, and easy to compute in examples, and prove that it determines whether the solutions of such an equation close up as topological spheres.

1. Introduction

1.1. Differential equations and curves

This paper belongs to the mathematics of the 17th-19th centuries. More specifically, I want to think about ordinary differential equations in complex variables. As the principal example which we will frequently return to, consider

$$\frac{d^2w}{dz^2} = 0,$$

a differential equation in two complex variables z and w. The solutions are of course the linear functions, $w = a + bz$. Lets keep this example in back of our minds, while we consider the general theory of second order ordinary differential equations

$$\frac{d^2w}{dz^2} = f\left(z, w, \frac{dw}{dz}\right).$$

All solutions of any complex ordinary differential equation are complex analytic functions. But I want to picture them geometrically, as curves in the z, w-plane. In my example, those curves will be the straight lines. Of course, the graph of a complex analytic function is a surface in a 4-dimensional space, not a curve in a plane. But it is convenient to think of the graph a complex analytic function as a "complex curve."

1.2. Projective geometry as analogue

Two further subtleties arise when we think about our example more carefully. First, if we think about lines in the plane, we include among them the vertical lines. These are not solutions of our differential equation. However, geometrically the lines which are neither horizontal nor vertical are solutions to either of the equations

$$\frac{d^2w}{dz^2} = 0 \text{ or } \frac{d^2z}{dw^2} = 0.$$

So we can imagine that an ordinary differential equation might have the same generic solution as some other equation in perhaps a different choice of variables, i.e. after a coordinate change. In our example, this is the coordinate change $(z, w) \to (w, z)$. We want to consider such a pair of equations to be equivalent, and "glue them together", just as we would "glue together" the various branches of a multivalued analytic function.

There is a further subtlety, of the same kind, occuring in our example. Consider the coordinate change taking a point (z, w) to the point

$$(\zeta, \omega) = \left(\frac{1}{z}, \frac{w}{z}\right).$$

The lines still become lines:

$$w = a + bz \to \omega = a\zeta + b.$$

This allows us to include on each line all of the usual points, as well as the point $\zeta = 0, \omega = b$, which does not correspond to a point of the line in the z, w-plane. It is the point at infinity. Moreover, in the ζ, ω-plane, there is an additional line: $\zeta = 0$, which does not correspond to any line in the z, w-plane. Putting this all together, we have added a point "at infinity" on each line, and these points at infinity form a line at infinity.

We should certainly not expect to be able to attach points "at infinity" to the solutions of arbitrary ordinary differential equations. Indeed, it seems to be a rare phenomenon that one could sensibly attach such points, via some clever coordinate change that took us to a new ordinary differential equation near infinity. However, the main observation here is that the solutions of our example equation are smooth at infinity, and the equation continues to make sense across infinity, as long as we make the right choice of coordinate change. We want to classify differential equations exhibiting this phenomenon. The complex plane with a point at infinity attached is called the *Riemann sphere*, and is topologically a sphere.

1.3. The problem

Our problem: classify the second order ordinary differential equations whose solutions are Riemann spheres. More precisely, we want to determine which equations admit analytic continuation along each solution, so that (after possibly some coordinate changes) the solutions close up to become smooth complex curves, which are topologically spheres. We will call these equations *straight*.

The reader might well wonder whether there could be a better choice of complex curve to build a theory on than the Riemann sphere. Perhaps there is something simpler. Rest assured that the function theory of the Riemann sphere is the simplest possible, by far simpler than even the complex plane, or any other complex curve. The curious reader is strongly encouraged to read the beautiful books of Clemens [3] and Forster [5]. Analysts speak of the Riemann sphere (or the "extended complex plane"), but geometers always refer to rational curves to mean complex curves which are topologically spheres. It turns out that rational curves can always be identified with the Riemann sphere by some homeomorphism which matches up the local and global function theory.

1.4. The solution

Definition 1 *For a second order ordinary differential equation*

$$\frac{d^2 w}{dz^2} = f\left(z, w, \frac{dw}{dz}\right),$$

(with f complex analytic) define the torsion *[8, 2, 1]:*

$$\frac{d^2}{dz^2}\frac{\partial^2 f}{\partial \dot{w}^2} - 4\frac{d}{dz}\frac{\partial^2 f}{\partial w \partial \dot{w}} + \frac{\partial f}{\partial \dot{w}}\left(4\frac{\partial^2 f}{\partial w \partial \dot{w}} - \frac{d}{dz}\frac{\partial^2 f}{\partial \dot{w}^2}\right) - 3\frac{\partial f}{\partial w}\frac{\partial^2 f}{\partial \dot{w}^2} + 6\frac{\partial^2 f}{\partial w^2}.$$

The torsion depends on derivatives of fourth order of the function f.

Example 1 *The Lienard equation*

$$\frac{d^2w}{dz^2} + w\frac{dw}{dz} + \frac{1}{9}w^3 = 0$$

has vanishing torsion.

Theorem 1 (McKay [6]) *A second order ordinary differential equation is torsion-free (i.e. the torsion vanishes) if and only if it is straight.*

2. Torsion

Tresse [8] discovered his torsion expression by asking a simpler question: what are the invariants of an ordinary differential equation of second order, under change of variables? In fact, he considered a somewhat more general notion of changing variables, on the phase space of an ordinary differential equation.

To motivate this idea, let us return once again to our example of lines in the plane. We can consider the plane itself as the "configuration space," in the terminology of physicists. The "phase space" is the space of pointed lines, i.e. of choices of a line in the plane, and a point on that line. This phase space is clearly three dimensional, since there is a two-dimensional space of points (the plane), and through each point, a one-dimensional space of lines (determined by angle). Of course, once again we are being cagey about whether we are working with real or complex variables, in using a term like "angle". But since the intuition comes from real variables, we can work with real variables for the moment.

So lets imagine that we have (x, y) coordinates on the plane, and then (x, y, θ) coordinates on the phase space. Tresse's geometric picture is that the phase space is a three-dimensional block, but sliced into ribbons by the curves we obtain by varying θ and fixing x and y. These curves are in 1-1 correspondence with the points of the plane. There is however another collection of curves slicing up the three-dimensional block, in another direction: the curves we obtain by fixing a choice of line, and moving (x, y) along that line. This, obviously, fixes the angle θ. The reader might wish to try to draw for her own understanding the two different slicings of the three-dimensional phase space.

This is only for our example, so far. But every ordinary differential equation

$$\frac{d^2w}{dz^2} = f\left(z, w, \frac{dw}{dz}\right).$$

has a phase space, which for us is parameterized locally by variables (z, w, p), say, so that z and w parameterize the z, w-plane, and p is some additional variable. We can view the first collection of curves slicing up the phase space as the curves of constant z, w-values, varying only p. We can view the other collection of curves as the solutions of the system

$$\begin{aligned} \frac{dw}{dz} &= p, \\ \frac{dp}{dz} &= f(z, w, p). \end{aligned}$$

This is an old trick, which goes by many names, and is usually presented to students in their first course in differential equations.

Tresse considers the general problem of when a three-dimensional space with two families of curves slicing it up comes about from an ordinary differential equation in this way. His first observation is that (1) the curves from each of the two slicings-up must smoothly cut up the three-dimensional space, with no two curves from the same family passing through the same point. His second is that (2) if a curve from one family strikes the same point as a curve from the other family, then the two curves cannot be tangent. Finally, he notes that (3) any vector field perpendicular to the curves of both families must have nonzero curl. These conditions can be checked explicitly. Tresse then uses the theorem of Frobenius (see Spivak [7]) to prove that these three conditions precisely characterize the geometry of a second order ordinary differential equation. One can recover the equation from the two families of curves directly.

Tresse made a monumental computation to find which combinations of derivatives of the function f above will be invariant under coordinate changes. In fact, he does not find the torsion this way, but rather discovers in his amazing long calculations that the torsion changes from one coordinate system to another, but in the process only gets multiplied by a nonzero factor. Hence the torsion can change under coordinate changes, but vanishing of the torsion holds in one choice of coordinates just when it vanishes in any.

3. Rationality of solutions

3.1. Global function theory of rational curves

Theorem 2 *Every holomorphic function on the Riemann sphere is constant.*

Proof. Such a function f must have a maximum modulus $|f|$ at some point of the Riemann sphere, by compactness. But by the maximum modulus theorem, only constant functions have interior maxima of modulus. Every point of the sphere is interior.

3.2. Vanishing torsion from rationality

Unfortunately, torsion is *not* a function. One can compute it as a function in one coordinate system, but then we get different values if we change coordinates. So the elementary argument of theorem 2 won't apply to torsion. Indeed the argument really requires that we have a function defined all the way around the Riemann sphere, right out to infinity.

What is the torsion, if not a function? Lets consider a simpler problem: what is an integral in complex analysis? We can integrate things that look like $f(z)\,dz$. Integrands transform under a change of variables in the only way they can: by the change of variables formula for integrals. So, for example, if we make a new variable $\zeta = \frac{1}{z}$, then we change the integrand according to

$$f(z)\,dz = -\frac{1}{\zeta^2} f\left(\frac{1}{\zeta}\right)\,d\zeta.$$

This is just undergraduate calculus substitution in an integral. If we have an integrand defined on the entire Riemann sphere, then

$$f\left(\frac{1}{\zeta}\right) \cong \zeta^2,$$

so

$$f(z) \cong \frac{1}{z^2},$$

decays quadratically at infinity. Therefore $f(z)$ must extend to the entire Riemann sphere continuously, setting $f(\infty) = 0$. But by the theorem above, $f(z)$ must be constant, and since $f(\infty) = 0$, we must have $f(z) = 0$ everywhere.

Torsion behaves similarly to an integrand under coordinate changes: if the solution of an ordinary differential equation closes up to a rational curve, then torsion is forced to vanish everywhere on that curve. Therefore if the generic solution is rational, then the torsion vanishes everywhere.

4. Duality

Tresse's geometric picture of second order ordinary differential equations has a natural symmetry to it: there are two families of curves in phase space, and if we change our minds about which is which, we still get a second order ordinary differential equation, by Tresse's characterization. This trick interchanges the configuration space with the space of solutions, yielding an ordinary differential equation on the space of solutions. For example (a simple, but not very serious, example): if we start with

$$\frac{d^2w}{dz^2} = -\left(\frac{dw}{dz}\right)^2,$$

the solutions are

$$w = \log(z+a) + b.$$

Now fix z and w, and think of a and b as variables. (They parameterize the space of solutions.) Differentiate once in a:

$$0 = \frac{1}{z+a} + \frac{db}{da},$$

and once again:

$$0 = -\frac{1}{(z+a)^2} + \frac{d^2b}{da^2}.$$

Solve for all z and w variables appearing:

$$0 = -\left(\frac{db}{da}\right)^2 + \frac{d^2b}{da^2}.$$

This is the *dual* ordinary differential equation, obtaining by switching the choices of families of curves in phase space.

As an easier example, our standard example equation

$$\frac{d^2w}{dz^2} = 0$$

is self-dual in this sense.

Duality shows us that the invariants under coordinate changes of an equation and of its dual must get interchanged by duality. Cartan [2] creates a complicated dictionary explaining how to compute the invariants of the dual from the invariants of the original equation. The power of duality comes from the fact that Tresse's complicated nonlinear torsion is interchanged under duality with the simple linear expression

$$\frac{d^4f}{d\dot{w}^4}.$$

In particular, an equation has vanishing torsion just when its dual equation has the form

$$\frac{d^2w}{dz^2} = \sum_{k=0}^{3} a_k(z)\left(\frac{dw}{dz}\right)^k.$$

4.1. Rationality from vanishing torsion

The vanishing of torsion implies that the dual equation has this very simple form. We now want to see why this forces rationality of the solutions of the original equation. But in terms of the dual equation, we are asking about rationality of the curves obtained by varying p and fixing z, w. The curves represent the various directions (angles) at the point z, w in configuration space. Since the equation has this simple form, we can check that the equation extends to all directions, by analytic continuation (just changing variables from (z, w) to (w, z)). Therefore there is a unique solution in each direction. The phase space above the configuration point (z, w) is identified analytically with the "circle" of directions through the point (z, w), which is a rational curve.

5. Conclusions and directions for further research

Cartan [2] proves that the generic torsion-free equation is an integrable system. The examples that have arisen are (perhaps not surprisingly) rather dull. All of the torsion-free ordinary differential equations found so far turn out to be integrable, and always by some elementary trick. The author has checked all of the equations given in all of the large handbooks in his library, coming up with about a dozen examples that are torsion-free.

The same ideas work in the context of 3rd order equations [6], and for higher order as well [4]. It seems possibly that they might work for complex curves somewhat more complicated than rational curves. For equations whose solutions are more complicated complex curves, the moduli ([3]) of the curves provide conservation laws, so I anticipate that the equations will be integrable, but through a more exciting mechanism.

A few advanced remarks: The requirement that the phase space of a second order complex ordinary differential equation be a closed Kähler manifold turns out to produce topological constraints. Since Mori theory (after Siu) tells us so much about three-dimensional closed Kähler manifolds, it should be possible to classify the phase spaces, and the equations. These would be very special equations, with a kind of nonlinear Painlevé property.

References

[1] Arnol'd V I 1988 *Geometrical methods in the theory of ordinary differential equations*, second ed., Grundlehren der Mathematischen Wissenschaften [Fundamental Principles of Mathematical Sciences], **250** (New York: Springer-Verlag) Translated from the Russian by Joseph Szücs [József M. Szűcs] MR MR947141 (89h:58049)

[2] Cartan É 1938 Les espaces généralisés et l'integration de certaines classes d'équations differentielles *C. R. Acad. Sci. Paris Sér. I Math.* no. 206, 1689 also in Œuvres Completes vol III partie 2 1621

[3] Herbert Clemens C 2003 *A scrapbook of complex curve theory*, second ed. Graduate Studies in Mathematics **55** (Providence: American Mathematical Society) MR MR1946768 (2003m:14001)

[4] Dunajski M and Tod P February 2005 *Paraconformal geometry of nth order ODEs, and exotic holonomy in dimension four* DAMTP-2005-20, math.DG/0502524

[5] Forster O 1991 *Lectures on Riemann surfaces*, Graduate Texts in Mathematics **81**, (New York: Springer-Verlag) Translated from the 1977 German original by Bruce Gilligan, Reprint of the 1981 English translation. MR MR1185074 (93h:30061)

[6] McKay B 2005 Rational curves and ordinary differential equations math.DG/0507087

[7] Spivak M 1979 *A comprehensive introduction to differential geometry* vol I, second ed. (Wilmington, Del.: Publish or Perish Inc.) MR 82g:53003a

[8] Tresse A 1896 *Détermination des invariants ponctuels de l'équation différentielle ordinaire du second ordre* $y'' = \omega(x, y, y')$, Ph.D. thesis, Universtät Leipzig, Leipzig, Presented to the Prince Jablonowski Society of Leipzig; unpublished

Institute of Physics Publishing
doi:10.1088/1742-6596/55/1/017

Differentiability of evolution operators for dynamical systems with hysteresis

A Pokrovskii[1], **T Power**[2], **D Rachinskii**[3] **and A Zhezherun**[4]
[1,3]Department of Applied Mathematics, University College Cork, Ireland
[2]Department of Computing, Mathematics & Physics, Waterford Institute of Technology, Ireland
[4]Department of Civil & Environmental Engineering, University College Cork, Ireland

E-mail: [1]`a.pokrovskii.ucc.ie`, [2]`tpower@wit.ie`, [3]`d.rachinskii@ucc.ie`, [4]`a.zhezherun@mars.ucc.ie`

Abstract. Hysteresis is a collective name for strongly nonlinear phenomena which occur in engineering, physical and economic systems. The wording 'strongly nonlinear' means that linearization cannot encapsulate the observed effects. In particular, standard linearization techniques are not applicable in the analysis of dynamical systems with hysteresis nonlinearities. Nevertheless, evolution operators of such dynamical systems typically have derivatives at many "important" points, which is enough to develop efficient modifications of gradient methods, linear stability analysis methods, etc. In this paper, we suggest a linearization algorithm for dynamical systems with the Preisach hysteresis operator. We present and justify an algorithm for calculation of the derivative of the Poincare map for scalar equations with a periodic solution of a simple form. We also suggest a version of this algorithm for a general class of non-periodic solutions of multi-dimensional equations with the Preisach nonlinearity.

1. Introduction

The mathematical theory of systems with hysteresis is a new chapter of applied mathematics, which is being actively developed worldwide by synergized efforts of mathematicians and engineers on the basis of methods of nonlinear analysis, dynamical systems and control theory, see [1, 2, 3, 4, 5, 7, 10]. A review of the current state of the art and further references may be found in the fundamental three volume set [11].

Following the paradigm suggested in [4], hysteresis nonlinearities are treated as nonlinear operators in functional spaces. An important feature of all hysteresis nonlinearities is that they are intrinsically non-smooth: there always exist points at which the operators do not have a derivative in a natural (e.g. Fréchet) sense.

There is a rapidly growing interest in the theoretical analysis and computer modeling of closed loop dynamical systems with hysteresis nonlinearities. The aforementioned non-smoothness of hysteresis operators causes non-smoothness of the corresponding evolution operators at least at some points of the phase space. Thus, most efficient methods (standard in other problems) based on linear approximations, such as gradient methods for solving boundary value problems, linear stability analysis, etc., are not yet developed for systems with hysteresis nonlinearities.

There are strong indications that the evolution operators generated by dynamical systems with hysteresis are in fact linearizable on a subset of the phase space of the dynamical system, and the "size" of this subset is sufficient to develop efficient modifications of classical linearization techniques for a range of boundary-value and stability problems. Moreover, the derivatives of the evolution operator can be found from differential-operator equations which are analogues to classical first variational equations.

The dynamical systems with hysteresis studied in this paper have the form of the equation

$$\frac{dz}{dt} = f(t, z, y) \tag{1}$$

coupled with the hysteresis relation between the variables $x = \ell(z)$ and y, the scalar input and output, respectively, of the Preisach hysteresis operator with the internal state ω. Here $x = \ell(z)$ is a scalar component of the vector z in an appropriate basis. The evolution of the system is thus described by the evolutions of the phase variable z and the state ω of the hysteresis nonlinearity, while the output y of this nonlinearity is a given function of ω. Correspondingly, the evolution operator for any time interval τ is a mapping in the space of couples (z, ω). Equations (1), and their further modifications, are important in various application areas, for example, ferromagnetic oscillators, control systems, power electronics and hydrology. The famous Preisach operator [4, 5], relating the variables y and x, is used as a model for many real world phenomena, see [11, 12, 13] and the references therein.

This paper is organized as follows. In section 2 we consider a simple example: a scalar differential equation (1), which has a periodic solution of a simple form. We formulate an algorithm for the calculation of the derivative of the Poincare map for this equation at the initial value of the periodic solution and discuss the application of this algorithm to the stability analysis of such solutions. The results of section 2 are proved in section 3. In section 4 we consider a more general class of non-periodic solutions of vector equations (1) and suggest a generalization of the linearization algorithm of section 2 for this class. However, we formulate this algorithm as a conjecture without proving it: we suggest a set of conditions which we believe ensure that our linearization algorithm is correct for the setting of section 4.

2. Stability of a periodic solution

2.1. Preisach model

In applications the Preisach model naturally arises as a superposition of simple two-state hysteresis nonlinearities, non-ideal relays (switches), with a common input. We use an equivalent alternative definition of the model due to [4]. The output of the model is a scalar function defined by the formula

$$y(t) = \mathrm{mes}_\mu\, \Omega(t) := \iint_{\Omega(t)} \mu(\alpha, \beta)\, d\alpha\, d\beta \tag{2}$$

where $\mu : \Pi \to \mathbb{R}$ is a nonnegative continuously differentiable integrable function, $\Pi = \{(\alpha, \beta) : 0 \le \beta - \alpha \le d\}$ is a strip in the (α, β) plane, and the domain $\Omega(t) \subset \Pi$ of integration changes in time. For each t, this domain has the form

$$\Omega(t) = \{(\alpha, \beta) \in \Pi : \alpha + \beta \le 2x(t) + \omega(t; \beta - \alpha)\}, \tag{3}$$

where $x(t)$ is a scalar continuous input of the model and the function $\omega(t; \cdot) : [0, d] \to \mathbb{R}$, called the state of the model, satisfies

$$\omega(t; 0) = 0; \qquad |\omega(t; \xi_1) - \omega(t; \xi_1)| \le |\xi_1 - \xi_2|, \qquad 0 \le \xi_1, \xi_2 \le d,$$

for all t. The evolution of the state $\omega = \omega(t; \cdot)$ (and hence, the evolution of Ω) is determined by the input with the following simple rules. If the continuous input is monotone on a segment

$[t_1, t_2]$, then on this segment

$$\omega(t;\xi) = \begin{cases} \max\{-\xi, \omega(t_1;\xi) - 2x(t) + 2x(t_1)\} & \text{if } x \text{ increases,} \\ \min\{\xi, \omega(t_1;\xi) - 2x(t) + 2x(t_1)\} & \text{if } x \text{ decreases;} \end{cases} \qquad 0 \le \xi \le d. \tag{4}$$

Now, a standard two-step procedure is used to define the state and output for any continuous input $x : \mathbb{R}_+ \to \mathbb{R}$ (see, e.g. [4, 5, 6]). First, formula (4) is combined with the semigroup identity to define the state $\omega(t;\cdot)$ at every moment $t \ge 0$ for any continuous piecewise monotone input $x : \mathbb{R}_+ \to \mathbb{R}$ and any initial state $\omega(0;\cdot) = \omega_0(\cdot)$, while formulas (2), (3) define the corresponding output. We denote the output by

$$y(t) = \mathcal{P}[\omega_0]x(t), \qquad t \ge 0, \tag{5}$$

reflecting the fact that it depends both on input and initial state, and remark that $\omega = \omega(t;\xi)$ is continuous in both arguments due to (4), hence the output $y : \mathbb{R}_+ \to \mathbb{R}$ of the Preisach model is also continuous. As the second step, one extends the input-output operator (5) to the whole class $C(\mathbb{R}_+)$ of continuous inputs from the subset of piecewise monotone ones by continuity, based on the Lipschitz estimate that holds for every $t \ge 0$:

$$\|\mathcal{P}[\omega_0^1]x_1 - \mathcal{P}[\omega_0^2]x_2\|_{C[0,t]} \le M\big(\|\omega_0^1 - \omega_0^2\|_C + \|x_1 - x_2\|_{C[0,t]}\big),$$

where $\|\omega_0^1 - \omega_0^2\|_C = \max_{\xi\in[0,d]} |\omega_0^1(\xi) - \omega_0^2(\xi)|$ measures the distance between initial states and the uniform norm in the space of inputs and outputs is used.

In section 4 we continue the discussion of the evolution of states of the Preisach model.

2.2. Assumptions

We consider the equation

$$\begin{aligned} x'(t) &= f(t, x, y), \\ y(t) &= \mathcal{P}[\omega_0]x(t), \end{aligned} \tag{6}$$

where x, y are the scalar input and scalar output of the Preisach nonlinearity, and ω_0 is its initial state. The function f is assumed to be T-periodic in the variable t:

$$f(t, x, y) = f(t + T, x, y),$$

and continuously differentiable with respect to the set of its arguments.

Solutions of (6) are pairs $(x(t), \omega(t))$, $t \ge 0$, where $\omega(t)$ is the state of the Preisach nonlinearity. Differentiability of f ensures that any initial data

$$x(0) = x_0, \qquad \omega(0) = \omega_0 \tag{7}$$

define a unique solution

$$(x(t), \omega(t)) = (x(t; x_0, \omega_0), \omega(t; x_0, \omega_0)) \tag{8}$$

of (6), at least on some interval $0 \le t < \delta$ (see, e.g. [1]).

In this section we suppose that system (6) has a periodic solution $(x_*(t), \omega_*(t))$ such that the following conditions hold (see figure 1):

1. The function $x_*(t)$ has one local maximum and one local minimum in the segment $[0, T]$, the minimum x_*^m is reached at the moment $t_1 > 0$, and the maximum x_*^M is reached at the moment t_2, $t_1 < t_2 < T$.

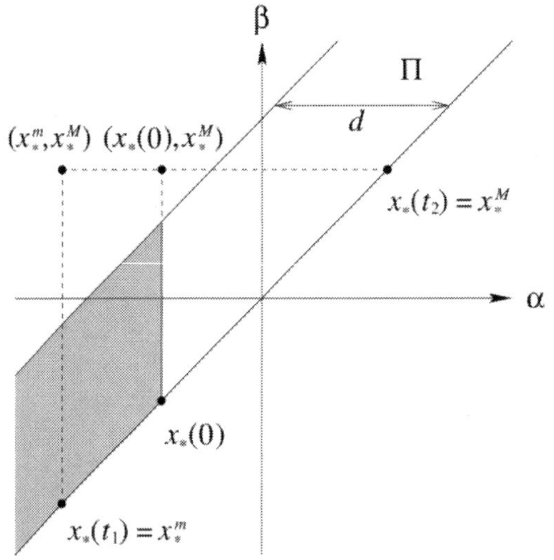

Figure 1. Diagram of $x_*(t)$ on the Preisach plane

2. The derivative of the function $x_*(t)$ is strictly negative on the intervals $[0, t_1)$ and $(t_2, T]$, and strictly positive on the interval (t_1, t_2).
3. The relations $x_*''(t_1) > 0$, $x_*''(t_2) < 0$ hold.
4. The point $(x_*(0), x_*^M)$ lies outside the strip Π on the Preisach plane (α, β).

These conditions imply that the variable state $\omega_*(t; \cdot)$ is a *vertical segment* for all t from the interval $[0, t_1]$ and a *horizontal segment* at the moment t_2, i.e.,

$$\omega_*(t) = \omega_v \quad \text{for } t \in [0, t_1]; \quad \omega_*(t_2) = \omega_h$$

with

$$\omega_v(\xi) = \xi, \quad \omega_h(\xi) = -\xi, \quad \xi \in [0, d].$$

2.3. Main result

Denote

$$\begin{aligned}
\varphi(x, v) &= \operatorname{mes}_\mu\{(\alpha, \beta) \in \Pi\colon \alpha \le v\} + \operatorname{mes}_\mu\{(\alpha, \beta) \in \Pi\colon v \le \alpha \le \beta \le x\}, \quad x \ge v, \\
\psi(x, v) &= \operatorname{mes}_\mu\{(\alpha, \beta) \in \Pi\colon \beta \le v\} - \operatorname{mes}_\mu\{(\alpha, \beta) \in \Pi\colon x \le \alpha \le \beta \le v\}, \quad x \le v,
\end{aligned}$$

(see figure 2), and

$$\begin{aligned}
J_\varphi^x(x, v) &= \frac{\partial \varphi}{\partial x}(x, v) = \int_{\max\{v, x-d\}}^{x} \mu(\alpha, x)\, d\alpha, \quad J_\varphi^v(x, v) = \frac{\partial \varphi}{\partial v}(x, v) = \int_{\min\{x, v+d\}}^{v+d} \mu(v, \beta)\, d\beta, \\
J_\psi^x(x, v) &= \frac{\partial \psi}{\partial x}(x, v) = \int_{x}^{\min\{v, x+d\}} \mu(x, \beta)\, d\beta, \quad J_\psi^v(x, v) = \frac{\partial \psi}{\partial v}(x, v) = \int_{\min\{v-d, x\}}^{x} \mu(\alpha, x)\, d\alpha.
\end{aligned} \tag{9}$$

Consider the following two auxiliary initial value problems

$$x' = f(t, x, \varphi(x, v)), \quad x(t_1) = v, \tag{10}$$

$$x' = f(t, x, \psi(x, v)), \quad x(t_2) = v, \tag{11}$$

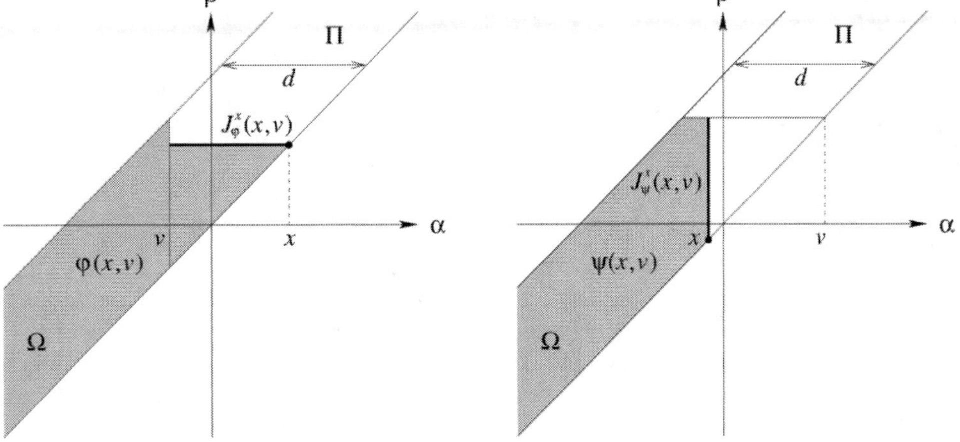

Figure 2. Functions $\varphi(x,v)$ and $\psi(x,v)$ are the measures of the shaded sets

with a scalar parameter v.

From Conditions 1–4 above it follows that $x_*(t)$ is a solution of problem (10) on the interval $[t_1, t_2]$ with $v = x_*^m$, and a solution of problem (11) on the interval $[t_2, t_1 + T]$ with $v = x_*^M$.

Denote by y_* the output of the Preisach nonlinearity for the periodic solution: $y_*(t) = \mathcal{P}[\omega_*(0)]x_*(t)$. Consider the linearization (first variational equation) of problem (10) in a vicinity of its solution $x_*(t)$:

$$\begin{aligned} \tilde{x}_1'(t) &= F_1(t)\tilde{x}_1(t) + G_1(t), \\ \tilde{x}_1(\theta) &= 1 \end{aligned} \tag{12}$$

with $\theta = t_1$, and

$$\begin{aligned} F_1(t) &= \frac{\partial f}{\partial x}(t, x_*(t), y_*(t)) + \frac{\partial f}{\partial y}(t, x_*(t), y_*(t)) J_\varphi^x(x_*(t), x_*^m), \\ G_1(t) &= \frac{\partial f}{\partial y}(t, x_*(t), y_*(t)) J_\varphi^v(x_*(t), x_*^m), \end{aligned} \tag{13}$$

where $\tilde{x}_1 = \frac{\partial x}{\partial v}$. Similarly, consider the linearization of problem (11):

$$\begin{aligned} \tilde{x}_2'(t) &= F_2(t)\tilde{x}_2(t) + G_2(t), \\ \tilde{x}_2(\theta) &= 1, \end{aligned} \tag{14}$$

with $\theta = t_2$, and

$$\begin{aligned} F_2(t) &= \frac{\partial f}{\partial x}(t, x_*(t), y_*(t)) + \frac{\partial f}{\partial y}(t, x_*(t), y_*(t)) J_\psi^x(x_*(t), x_*^M), \\ G_2(t) &= \frac{\partial f}{\partial y}(t, x_*(t), y_*(t)) J_\psi^v(x_*(t), x_*^M). \end{aligned} \tag{15}$$

We denote the solutions of (12) and (14) by $\tilde{x}_1(t;\theta)$ and $\tilde{x}_2(t;\theta)$.

Note that if $|x_0 - x_*(0)|$ is sufficiently small, and the initial state of the Preisach nonlinearity is the vertical segment ω_v, then the solution $(x(t), \omega(t))$ of equation (6) with initial values (x_0, ω_v) is defined on the segment $[0, T]$.

Theorem 1. *Let $|x_0 - x_*(0)|$ be sufficiently small, and let the initial state of the Preisach nonlinearity be the vertical segment ω_v. Then the function $x(T; x_0, \omega_v)$ of the argument x_0 is differentiable at the point $x_0 = x_*(0)$, and its derivative is equal to*

$$\frac{\partial x}{\partial x_0}(T; x_*(0), \omega_v) = \tilde{x}_1(t_2; t_1)\tilde{x}_2(t_1 + T; t_2), \tag{16}$$

where $\tilde{x}_1(t;\theta)$ and $\tilde{x}_2(t;\theta)$ are solutions of problems (12) *and* (14).

2.4. Remarks

a. If initial data (7) are close to the initial values $x_*(0)$, $\omega_*(0) = \omega_v$ of the periodic solution (i.e., $|x_0 - x_*(0)|$, $\|w_0 - w_v\|_C$ are sufficiently small), then Condition 4 implies that $\omega(T) = \omega_v$ due to the continuous dependence of solution of (6) on initial data. Therefore, Theorem 1 implies that the periodic solution $(x_*(t), \omega_*(t))$ is exponentially stable if $\left|\frac{\partial x}{\partial x_0}(T; x_*(0), \omega_v)\right| < 1$, and unstable if $\left|\frac{\partial x}{\partial x_0}(T; x_*(0), \omega_v)\right| > 1$, where the derivative is given by (16). More precisely, if $\left|\frac{\partial x}{\partial x_0}(T; x_*(0), \omega_v)\right| < 1$, then for any initial data (7) that are sufficiently close to $x_*(0)$, $\omega_*(0)$ solution (8) is defined for all $t \geq 0$ and

$$|x(t) - x_*(t)| + |y(t) - y_*(t)| + \|w(t) - w_*(t)\|_C \leq$$
$$\leq Ce^{-\gamma t}\big(|x_0 - x_*(0)| + \|w_0 - w_*(0)\|_C\big),$$

for each t with some $C, \gamma > 0$ independent of x_0, ω_0. If $\left|\frac{\partial x}{\partial x_0}(T; x_*(0), \omega_v)\right| > 1$, then $x(t)$ leaves a fixed vicinity of $x_*(t)$ at some moment $t > 0$ for any initial data (7) close enough to, but different from, $x_*(0)$, $\omega_*(0)$.

b. Condition 4 can be replaced with the condition $(x_*^m, x_*^M) \notin \Pi$ (we chose to use Condition 4 in order to simplify the proof). If the point (x_*^m, x_*^M) lies in the interior of the strip Π, then generically equation (6) has a connected continuum of periodic solutions whenever it has one such solution (see [2]), consequently periodic solutions are not asymptotically stable in this case.

c. Statements similar to Theorem 1 are valid for multidimensional systems with the Preisach nonlinearity and for periodic solutions with the x_*-component having more than two local extrema on the period. Although we do not consider them here, a straightforward modification of the above linearization algorithm works for such more complicated solutions if the local extrema of $x_*(t)$ satisfy some additional non-degeneracy conditions. These are either a modification of Condition 4 or, alternatively, another non-degeneracy condition, which is considered in the next section for non-periodic solutions.

Analogues of Theorem 1 can be formulated also for equations with the Preisach operator on the left hand side, like for example $y' = f(t,x)$, $y(t) = \mathcal{P}[\omega_0]x(t)$. For these, Condition 4 is unnecessary. Such equations naturally arise in hydrological models [8, 9]; further examples of equations with the time derivative of the Preisach operator can be found in [10].

d. Formula (16) can be easily understood for an equation (6) where f has jumps in t. More precisely, assume that f is discontinuous at $t = t_1, t_2$ and has finite left and right limits on these lines. Suppose the component x_* of a periodic solution satisfies Conditions 1, 2 and 4 above and the following assumption, which replaces Condition 3:

3*. The left and right derivatives of the function x_* at its extremum points t_1, t_2 are non-zero.

Then Lemma 1 of the next section implies that the x-component of a solution $(x(t), \omega(t))$ with initial values $x(0) = x_0$, $\omega(0) = \omega_*(0) = \omega_v$ is C^1-close to x_* on each of the segments $[0, t_1]$, $[t_1, t_2]$ and $[t_2, T]$ whenever x_0 is close to $x_*(0)$. Consequently, x satisfies the same Conditions 1, 2, 3* and 4. Hence both x_* and x strictly decrease on the intervals $[0, t_1]$, $[t_2, T]$, strictly increase on $[t_1, t_2]$, and reach minimum and maximum values on the segment $[0, T]$ at the same points t_1 and t_2, respectively; also, the point $(x(0), x^M)$ lies outside the strip Π on the Preisach plane (α, β), like the point $(x_*(0), x_*^M)$, where we use the notation $x^m = x(t_1)$, $x^M = x(t_2)$ for the minimum and maximum of x.

Now, from the definition of the Preisach nonlinearity it follows that because x decreases on $[0, t_1]$ and the initial state is $\omega(0) = w_v$, the state of the Preisach nonlinearity at every moment $t \in [0, t_1]$ is w_v and the output y on the segment $[0, t_1]$ equals

$$y(t) = \mathcal{P}[\omega_v]x(t) = \chi(x(t)) \quad \text{with} \quad \chi(v) := \text{mes}_\mu\{(\alpha, \beta) \in \Pi\colon \alpha \le v\}.$$

By the same argument, $y_*(t) = \chi(x_*(t))$ for $t \in [0, t_1]$ and hence on this time segment x and x_* are solutions of the same ordinary differential equation

$$x' = f(t, x, \chi(x)). \tag{17}$$

Similarly, the relations $\omega(t_1) = \omega_*(t_1) = \omega_v$ and the fact that x and x_* increase on the segment $[t_1, t_2]$ imply the equalities

$$y(t) = \varphi(x(t), x(t_1)), \qquad y_*(t) = \varphi(x_*(t), x_*(t_1)) \quad \text{for} \quad t \in [t_1, t_2]$$

and one infers that on the segment $[t_1, t_2]$ the function x solves problem (10) for the parameter value $v = x(t_1) = x^m$, while x_* solves the same problem for the parameter value $v = x_*(t_1) = x_*^m$. Finally, because $\omega(t_2) = \omega_*(t_2) = \omega_h$ (which follows from the relations $(x^m, x^M) \notin \Pi$, $(x_*^m, x_*^M) \notin \Pi$) and x and x_* decrease on the segment $[t_2, T]$, one concludes that

$$y(t) = \psi(x(t), x(t_2)), \qquad y_*(t) = \psi(x_*(t), x_*(t_2)) \quad \text{for} \quad t \in [t_2, T]$$

and consequently x solves problem (11) on the segment $[t_2, T]$ for the parameter value $v = x(t_2) = x^M$, and x_* solves this problem on the same segment for the parameter value $v = x_*(t_2) = x_*^M$.

Thus, for the initial values x_0 close to $x_*(0)$ and $\omega(0) = \omega_v$, system (6) on the segment $[0, T]$ is equivalent to the sequence of three initial value problems for ordinary differential equations on the time intervals $[0, t_1]$, $[t_1, t_2]$ and $[t_2, T]$. Therefore, one calculates the derivative $\partial x / \partial x_0(T; x_*(0), \omega_v)$ of the Poincare map of (6) simply by the standard linearization of ODEs. It should be just noted that the value $x(t_1)$ of a solution of equation (17) at the end point of the first segment $[0, t_1]$ is used both as an initial value and as a value of the parameter v in the right-hand side for problem (10) on the second segment $[t_1, t_2]$ (and similarly the end value of the solution of (10) is used as an initial value and the parameter value for problem (11) on the last segment $[t_2, T]$). Therefore, one uses theorems on differentiation of solutions of ODE both with respect to initial values and parameters. In this manner, one arrives at the formula

$$\frac{\partial x}{\partial x_0}(T; x_*(0), \omega_v) = \tilde{x}_3(t_1; 0)\tilde{x}_1(t_2; t_1)\tilde{x}_2(T; t_2), \tag{18}$$

where $\tilde{x}_3(\cdot; 0)$ is the solution of the linearization

$$\tilde{x}_3'(t) = D(t)\tilde{x}_3(t), \qquad D(t) = \frac{\partial f}{\partial x}(t, x_*(t), y_*(t)) + \frac{\partial f}{\partial y}(t, x_*(t), y_*(t))\frac{\partial \chi}{\partial x}(x_*(t))$$
$$\tilde{x}_3(0) = 1,$$

of (17) around its solution x_*, and $\tilde{x}_1$, $\tilde{x}_2$ are as in Theorem 1. It is easy to see then that Condition 4 implies $\tilde{x}_3(t_1; 0) = \tilde{x}_2(t_1; 0)$ and that $\tilde{x}_2(t_1; 0)\tilde{x}_2(T; t_2) = \tilde{x}_2(T + t_1; t_2)$ (details are in the next section), hence (18) is equivalent to formula (16) of Theorem 1.

The proof of Theorem 1 below is somewhat more involved than the above argument. The reason is that under the Conditions 1–4 a solution x close to x_* can have more than one minimum and more than one maximum in the interval $[0, T]$; also generally, the perturbed solution x reaches its global minimum and maximum not precisely at the same moments t_1 and t_2 as x_* does. However, all local extrema of x are located close to the extremum points t_1, t_2 of x_* for small perturbations of the periodic solution of (6). As we show in the next section, this leads to the same formula (16) under the conditions of Theorem 1 as in the simpler case of Condition 3*.

Example.

Consider the equation

$$x'(t) = x(t) + \mathcal{P}[\omega_0]x(t) + a + b \sin \frac{2\pi t}{T}, \tag{19}$$

where the Preisach operator $\mathcal{P}$ has a uniform measure $\mu(\alpha, \beta) = \mu$ defined on the set

$$\{(\alpha, \beta) \colon |\alpha|, |\beta| \leq \gamma \text{ and } 0 \leq \beta - \alpha \leq d\}.$$

Note that we can always consider a measure with a bounded support, as long as the solution of the system does not leave the region where the measure support is located.

For the purposes of this example we choose the constants a, b, T, d, μ and γ such that (19) has a periodic solution that has an explicit form. Let

$$a = -\frac{1}{2}, \quad b = \frac{27}{28}, \quad T = \frac{2}{5}(4 + 7 \ln 5), \quad d = 1, \quad \mu = \frac{2}{7}, \quad \gamma = 2.$$

Hence, the functions $\varphi(x, v)$ and $\psi(x, v)$ take the form

$$\varphi(x, v) = \begin{cases} \frac{1}{7}(3 + 2x), & x > 1 + v \text{ and } v \leq 1, \\ \frac{1}{7}(3 - 2v(x - 2) + x^2, & v > 1, \\ \frac{1}{7}(4 + x^2 + v^2 - 2v(x - 1), & \text{otherwise;} \end{cases}$$

$$\psi(x, v) = \begin{cases} \frac{2}{7}(2 + x), & x < v - 1 \text{ and } v \geq -1, \\ \frac{1}{7}(2 + 2v - x)(2 + x), & v < -1, \\ \frac{1}{7}(3 - x^2 - v^2 + 2v(x + 1), & \text{otherwise.} \end{cases}$$

Consider the initial value problem for (19) with $x_*(0) = -1$ and $\omega_0 = \omega_v$. The solution $x_*(t)$ of this problem is T-periodic, reaches its maximum $x_*^M = 1$ at the moment $\frac{1}{2}T$ and minimum $x_*^m = -1$ at the moment T, and has the following explicit form on the interval $[0, T]$:

$$x_*(t) = \begin{cases} \frac{5t-4}{2(t+2)}, & 0 \leq t \leq \frac{4}{5}, \\ \frac{5}{4}\left(1 - e^{\frac{1}{7}(4-5t)}\right), & \frac{4}{5} < t \leq \frac{1}{2}T, \\ -\frac{5(5t-8-7\ln 5)}{2(5t+6-7\ln 5)}, & \frac{1}{2}T < t \leq \frac{1}{2}T + \frac{4}{5}, \\ \frac{5}{4}\left(-1 + e^{\frac{1}{7}(8-5t)}\right), & \frac{1}{2}T + \frac{4}{5} < t \leq T. \end{cases}$$

To apply Theorem 1 we need to solve the problems (12) and (14), which take the form

$$\tilde{x}_1(t) = -\frac{5}{7}\tilde{x}_1(t) - \min\left\{0, \frac{2}{7}x_*(t)\right\}(1 - \tilde{x}_1(t)),$$

$$\tilde{x}_2(t) = -\frac{5}{7}\tilde{x}_2(t) + \max\left\{0, \frac{2}{7}x_*(t)\right\}(1 - \tilde{x}_2(t)).$$

These problems can also be solved explicitely. With straightforward calculations, for the equality (16) we obtain

$$\frac{\partial x}{\partial x_0}(T;-1,\omega_v) = \frac{351649}{26471025} \approx 0.01328.$$

Hence, the periodic solution of (19) is asymptotically stable.

3. Proof of Theorem 1

3.1. Auxiliary lemmas

We start with a rough estimation of the deviation of the perturbed solution $(x(t),\omega(t))$ with initial values $x(0)=x_0$, $\omega(0)=\omega_v$ from the periodic solution. Denote

$$r(t)=x(t)-x_*(t), \quad r_0=r(0)=x_0-x_*(0).$$

Lemma 1. *If $|r_0|$ is sufficiently small, then*

$$|r(t)| \le e^{L_1 t}|r_0|, \quad 0 \le t \le T.$$

Proof. Fix a vicinity U of the solution $x_*(t)$. In this vicinity function f is a Lipschitz function:

$$|f(x_2,y_2)-f(x_1,y_1)| \le c_1|x_2-x_1|+c_2|y_2-y_1|.$$

If x_0 is close to $x_*(0)$, then $x(t)\in U$ for $t\in[0,T]$ by the theorem on continuous dependence of solutions on initial conditions. Therefore, the Lipschitz estimate implies

$$|x'(t)-x'_*(t)| \le c_1|x(t)-x_*(t)|+c_2|y(t)-y_*(t)|.$$

Now, from the Lipschitz continuity of the Preisach operator in $C[0,T]$ it follows that

$$|x'(t)-x'_*(t)| \le L_1\|x(\xi)-x_*(\xi)\|_{C[0,t]}. \tag{20}$$

Denote

$$u(t)=|r_0|+\int_0^t |r'(s)|ds = |x_0-x_*(0)|+\int_0^t |x'(s)-x'_*(s)|ds.$$

Then (20) implies $u'(t)\le L_1 u(t)$, which implies

$$u(t)\le u(0)e^{L_1 t},$$

and the conclusion of the Lemma follows from $|r(t)|\le u(t)$, $r(0)=u(0)$. □

Lemma 1 together with relation (20) implies that for sufficiently small $|r_0|$

$$\|r(t)\|_{C^1[0,T]} \le L|r_0|. \tag{21}$$

Let $k=\min\{|x''_*(t_1)|,|x''_*(t_2)|\}$ and $K=\max\{|x''_*(t_2)|,|x''_*(t_2)|\}$. The continuity of x''_* implies that there exist $\Delta_1>0$ and $\Delta_2>0$ such that

$$2K \ge x''_*(\xi_1) \ge \frac{k}{2} \text{ for } \xi_1\in[t_1-\Delta_1,t_1+\Delta_1],$$
$$-2K \le x''_*(\xi_2) \le -\frac{k}{2} \text{ for } \xi_2\in[t_2-\Delta_2,t_2+\Delta_2].$$

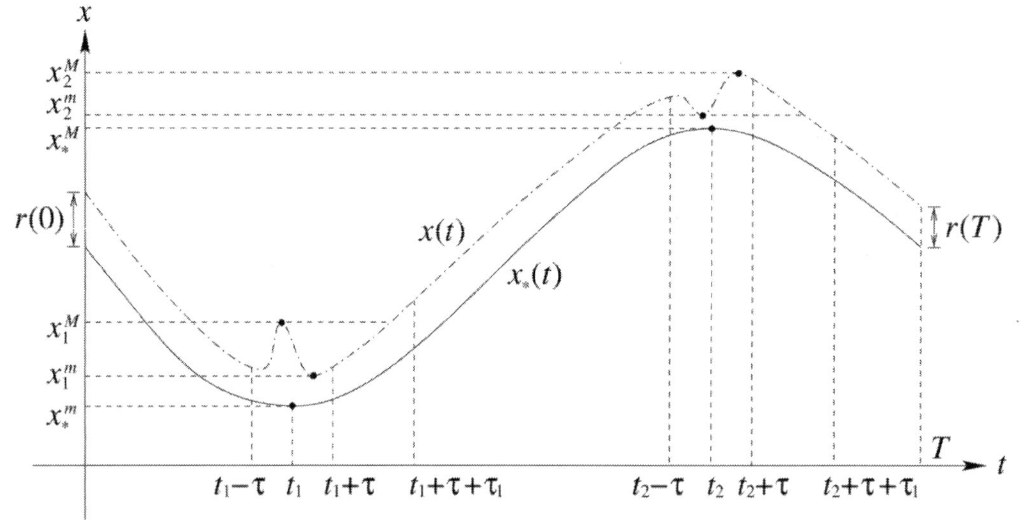

Figure 3. Solutions $x_*(t)$ and $x(t)$

Let $\Delta = \min\{\Delta_1, \Delta_2\}$. Then, because $x'_*(t_1) = x'_*(t_2) = 0$, we have the following relations:

$$2K(t - t_1) \le x'_*(t) \le \frac{k}{2}(t - t_1) \text{ for } t \in [t_1 - \Delta, t_1], \tag{22a}$$

$$2K(t - t_1) \ge x'_*(t) \ge \frac{k}{2}(t - t_1) \text{ for } t \in [t_1, t_1 + \Delta], \tag{22b}$$

$$2K(t_2 - t) \ge x'_*(t) \ge \frac{k}{2}(t_2 - t) \text{ for } t \in [t_2 - \Delta, t_2], \tag{22c}$$

$$2K(t_2 - t) \le x'_*(t) \le \frac{k}{2}(t_2 - t) \text{ for } t \in [t_2, t_2 + \Delta]. \tag{22d}$$

Now consider the derivative $x'_*(t)$ on the intervals $[0, t_1 - \Delta]$, $[t_1 + \Delta, t_2 - \Delta]$, and $[t_2 + \Delta, T]$. According to Condition 2, $x'_*(t)$ is strictly negative on the first and the third of these intervals, and strictly positive on the second one. Therefore there exists a constant $a > 0$ such that $|x'_*(t)| \ge a$ on all the three intervals.

Let

$$\tau = r_0 \frac{4L}{k}, \quad \tau_1 = 20\tau \frac{K}{k}, \tag{23}$$

where L is defined in (21). Consider a sufficiently small r_0 such that (21) holds, and

$$\tau < \Delta, \quad \tau < t_1, \quad 2\tau + \tau_1 < t_2 - t_1, \quad \tau + \tau_1 < T - t_2, \quad \frac{k\tau}{2} < a. \tag{24}$$

Consider the solutions $x(t)$ and $x_*(t)$ on the intervals $[0, t_1 - \tau]$, $[t_1 - \tau, t_1 + \tau]$, $[t_1 + \tau, t_2 - \tau]$, $[t_2 - \tau, t_2 + \tau]$ and $[t_2 + \tau, T]$. Conditions 1 and 2 imply that $x'_*(t) < 0$ on the interval $[0, t_1 - \tau]$, then on the interval $[t_1 - \tau, t_1 + \tau]$ the function $x_*(t)$ reaches its minimum x_*^m at the moment t_1, and $x'_*(t) > 0$ on the interval $[t_1 + \tau, t_2 - \tau]$. The following lemma states that the perturbed solution $x(t)$ is strictly monotone outside the two small intervals around the points t_1, t_2 (see figure 3) and gives a lower estimate of $|x'(t)|$.

Lemma 2. *The function $x(t)$ strictly increases on the interval $[t_1+\tau, t_2-\tau]$ and strictly decreases on the intervals $[0, t_1-\tau]$ and $[t_2+\tau, T]$. Moreover, on these intervals*

$$|x'_*(t)| \geq \frac{k}{2}\tau, \quad |x'(t)| \geq \frac{k}{4}\tau. \tag{25}$$

Proof. First consider the interval $[0, t_1-\tau]$. Relation (22a) implies that on $[t_1-\Delta, t_1-\tau]$

$$x'_*(t) \leq \frac{k}{2}(t-t_1) \leq -\frac{k\tau}{2}.$$

According to (24), $x'_*(t) \leq -a \leq -\frac{k\tau}{2}$ on $[0, t_1-\Delta]$. Hence,

$$x'_*(t) \leq -\frac{k\tau}{2} \quad \text{for } t \in [0, t_1-\tau].$$

Therefore, (21) and (23) imply that on the interval $[0, t_1-\tau]$

$$x'(t) \leq x'_*(t) + |r'(t)| \leq -\frac{k\tau}{2} + Lr_0 = -\frac{k\tau}{4} < 0. \tag{26}$$

Thus the function $x(t)$ decreases on the interval $[0, t_1-\tau]$, and the derivatives $x'_*(t)$ and $x'(t)$ on this interval satisfy (25). The other intervals can be considered in the same way. □

3.2. Calculation of the derivative of evolution map

The conclusion of Theorem 1 follows from the three lemmas below.

Lemma 3. *Denote by x_1^m the absolute minimum of the function $x(t)$ on the interval $[t_1-\tau, t_1+\tau]$, and by x_1^M the absolute maximum of x on the same interval. Then the following relations are valid:*

$$r(t_1+\tau+\tau_1) = \tilde{x}_2(t_1;0)r_0 + o(r_0), \tag{27}$$
$$x_1^m - x_*^m = \tilde{x}_2(t_1;0)r_0 + o(r_0), \tag{28}$$

and the output of the Preisach operator for the preturbed solution at the moment $t_1+\tau+\tau_1$ equals

$$y(t_1+\tau+\tau_1) = \varphi(x(t_1+\tau+\tau_1), x_1^m). \tag{29}$$

Lemma 4. *Denote by x_2^m the absolute minimum of function $x(t)$ on the interval $[t_2-\tau, t_2+\tau]$, and by x_2^M the absolute maximum of x on the same interval. Then the following relations hold:*

$$r(t_2+\tau+\tau_1) = \tilde{x}_2(t_1;0)\tilde{x}_1(t_2;t_1)r_0 + o(r_0), \tag{30}$$
$$x_2^M - x_*^M = \tilde{x}_2(t_1;0)\tilde{x}_1(t_2;t_1)r_0 + o(r_0), \tag{31}$$

and the output of the Preisach operator for the perturbed solution at the moment $t_2+\tau+\tau_1$ equals

$$y(t_2+\tau+\tau_1) = \psi(x(t_2+\tau+\tau_1), x_2^M).$$

Lemma 5. *The following equality holds:*

$$r(T) = \tilde{x}_2(t_1;0)\tilde{x}_1(t_2;t_1)\tilde{x}_2(T;t_2)r_0 + o(r_0). \tag{32}$$

Lemmas 3 and 4 are proved in the next two subsections. The proof of Lemma 5 is similar to that of Lemma 4 and we omit it.

To complete the proof, it remains to show that

$$\tilde{x}_2(t_1;0)\tilde{x}_2(T;t_2) = \tilde{x}_2(T+t_1;t_2), \tag{33}$$

then (32) implies (16). First of all, because the right-hand side of (14) is T-periodic in t, $\tilde{x}_2(t_1;0) = \tilde{x}_2(T+t_1;T)$. As we show in the proof of Lemma 3 below, $\frac{\partial\psi}{\partial v}(x_*(t), x_*^M) = 0$ on the interval $[T;T+t_1]$, hence on this interval $\tilde{x}_2$ is a solution of the homogeneous linear equation $\tilde{x}_2'(t) = F_2(t)\tilde{x}_2(t)$. This implies that if we denote by $\tilde{X}_2(t;\theta;\tilde{x}_0)$ the solution of (14) with an initial value $\tilde{X}_2(\theta;\theta;\tilde{x}_0) = \tilde{x}_0$, then

$$\tilde{X}_2(T+t_1;T;\tilde{x}_0) = \tilde{x}_0 \cdot \tilde{X}_2(T+t_1;T;1) = \tilde{x}_0\tilde{x}_2(T+t_1;T),$$

since $\tilde{X}_2(\cdot;\cdot;1) = \tilde{x}_2(\cdot;\cdot)$ by definition. Finally, according to the semi-group property of solutions,

$$\tilde{X}_2(T+t_1;t_2;1) = \tilde{X}_2(T+t_1;T;\tilde{X}_2(T;t_2;1)) = \tilde{X}_2(T;t_2;1)\tilde{X}_2(T+t_1;T;1),$$

i.e.

$$\tilde{x}_2(T+t_1;t_2) = \tilde{x}_2(T;t_2)\tilde{x}_2(T+t_1;T),$$

which is equivalent to (33). Thus, (32) is equivalent to

$$r(T) = \tilde{x}_2(T+t_1;t_2)\tilde{x}_1(t_2;t_1)r_0 + o(r_0),$$

and after dividing by r_0 and passing to the limit as $r_0 \to 0$, we get equation (16), so the theorem is proved. □

3.3. Proof of Lemma 3

The functions x and x_* are both decreasing on the interval $[0, t_1 - \tau]$, therefore the outputs y and y_* of the Preisach nonlinearity on this interval are determined by the equalities

$$y = \psi(x(t), x_*^M), \quad y_* = \psi(x_*(t), x_*^M).$$

Hence, on the interval $[0, t_1 - \tau]$ the functions x and x_* are solutions of the equation

$$x' = f(t, x, \psi(x, x_*^M))$$

with initial values x_0 and $x_*(0)$, respectively. Therefore, to estimate $r(t_1 - \tau)$ we use the linearization of this equation along the solution $x_*(t)$:

$$\tilde{x}_2'(t) = F_2(t)\tilde{x}_2(t), \quad \tilde{x}_2(0) = 1, \tag{34}$$

where $F_2(t)$ is defined in (15). Note that $J_\psi^v(x_*(t), x_*^M) = \frac{\partial\psi}{\partial v}(x_*(t), x_*^M) = 0$ on the interval $[0, t_1]$, because $(x_*(t), x_*^M) \notin \Pi$ on this interval. Therefore, the solution $\tilde{x}_2$ of (34) on this interval is equal to the solution $\tilde{x}_2(t;0)$ of (14). Hence

$$r(t_1 - \tau) = \tilde{x}_2(t_1 - \tau; 0)r_0 + o(r_0). \tag{35}$$

Furthermore,

$$r(t_1 + \tau + \tau_1) = r(t_1 - \tau) + r'(\xi)(2\tau + \tau_1)$$

for some $\xi \in [t_1-\tau, t_1+\tau+\tau_1]$. Combining the last two equations and taking into account that $\|r'\|_{C[0,T]}, \tau, \tau_1 = O(r_0)$, we obtain

$$r(t_1+\tau+\tau_1) = \tilde{x}_2(t_1-\tau;0)r_0 + o(r_0).$$

To prove (27) it remains to note that $\tau = O(r_0)$ implies

$$\tilde{x}_2(t_1-\tau;0) = \tilde{x}_2(t_1;0) + O(r_0). \tag{36}$$

Now consider the absolute maximum x_1^M of the function $x(t)$ on the interval $[t_1-\tau, t_1+\tau]$. Equations (22b) and (22c) imply that $|x_*'(t)| \le 2K\tau$ on this interval, therefore

$$|x'(t)| \le |x_*'(t)| + |r'(t)| \le 2K\tau + Lr_0 \le \left(2K + \frac{k}{2}\right)\tau \le \frac{5K}{2}\tau. \tag{37}$$

Hence,

$$x_1^M - x(t_1+\tau) \le |x'(\xi)| \cdot 2\tau \le 5K\tau^2.$$

Similarly to (26), we get from (25):

$$x'(t) \ge x_*'(t) - |r'(t)| \ge \frac{k\tau}{2} - Lr_0 \ge \frac{k\tau}{4} > 0$$

for $t \in [t_1+\tau, t_2-\tau]$, so $x(t)$ increases on this interval. Then, taking into account that $\tau_1 = 20\tau K/k$,

$$x(t_1+\tau+\tau_1) - x(t_1+\tau) \ge \frac{k\tau}{4}\tau_1 = 5K\tau^2 \ge x_1^M - x(t_1+\tau),$$

which implies $x(t_1+\tau+\tau_1) \ge x_1^M$. Therefore, by the wiping out property of the Preisach nonlinearity, the state $\omega(t_1+\tau+\tau_1)$ consists of two segments, which implies that $y(t_1+\tau+\tau_1) = \varphi(x(t_1+\tau+\tau_1), x_1^m)$ and thus (29) holds.

The absolute minimum x_1^m is estimated as follows:

$$\begin{aligned} x_1^m - x_*^m &= (x_1^m - x(t_1-\tau)) + (x(t_1-\tau) - x_*(t_1-\tau)) + (x_*(t_1-\tau) - x_*^m) \\ &= o(r_0) + r(t_1-\tau) + o(r_0) = \tilde{x}_2(t_1-\tau;0)r_0 + o(r_0), \end{aligned}$$

where we use (35) and (37). This together with (36) implies (28). □

3.4. Proof of Lemma 4

Consider the interval $[t_1+\tau+\tau_1, t_2-\tau]$. The functions x_* and x are both increasing on this interval. Therefore x_* solves the equation

$$x' = f(t, x, \varphi(x, v)) \tag{38}$$

for the parameter value $v = x_*^m$; the function x solves the same equation for $v = x_1^m$. The initial values $x(t_1+\tau+\tau_1)$ and $x_*(t_1+\tau+\tau_1)$ are related by (27), while the parameter values x_1^m and x_*^m are related by (28). Differentiating (38) with respect to the initial value of x and the parameter v along the solution x_*, we obtain

$$\begin{aligned} r(t_2-\tau) = \alpha(t_2-\tau)r(t_1+\tau+\tau_1) + \beta(t_2-\tau)(x^m - x_*^m) \\ + o(r(t_1+\tau+\tau_1) + |x^m - x_*^m|), \end{aligned}$$

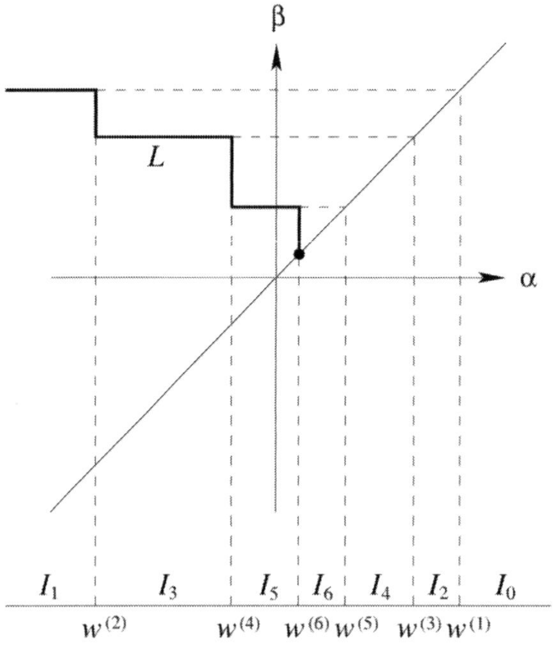

Figure 4. State of the Preisach operator defined by a step-like piecewise linear curve

where α and β solve the problems

$$\begin{aligned} \alpha'(t) &= F_1(t)\alpha(t), & \alpha(t_1+\tau+\tau_1) &= 1, \\ \beta'(t) &= F_1(t)\beta(t) + G_1(t), & \beta(t_1+\tau+\tau_1) &= 0, \end{aligned}$$

with $F_1(t)$ and $G_1(t)$ defined by (13). Now, applying (27) and (28) and noticing that $\alpha(t)+\beta(t) = \tilde{x}_1(t; t_1+\tau+\tau_1)$ solves the problem (12), we obtain

$$r(t_2-\tau) = \tilde{x}_1(t_2-\tau; t_1+\tau+\tau_1)\tilde{x}_2(t_1; 0)r_0 + o(r_0).$$

Combining this equation with

$$\begin{aligned} r(t_2+\tau+\tau_1) &= r(t_2-\tau) + o(r_0), \\ \tilde{x}_1(t_2-\tau; t_1+\tau+\tau_1) &= \tilde{x}_1(t_2; t_1) + O(r_0), \end{aligned}$$

we arrive at (30). The other conclusions of the Lemma can be proved in the same way as that of Lemma 3. □

4. General linearization algorithm

4.1. Step-like states of the Preisach operator

In this section we consider a Preisach operator with the domain $\Omega(t)$, see (3), restricted to the sets $S \in \Pi$ bounded by lines L with an infinite horizontal link, and with a finite number of vertical and horizontal links, see figure 4.

We denote the height of the infinite horizontal link by $w^{(1)}$. Let $w^{(2k)}$, $1 < 2k \le K$, be the ascending sequence of the horizontal coordinates of the vertical links. That is, the vertical links belong to the straight lines $L_{2k} = (w^{(2k)}, \beta)$, $1 < 2k \le K$. Similarly, let $w^{(2k+1)}$, $1 < 2k+1 \le K$, be the descending sequence of the vertical coordinates of the finite horizontal links. By definition, the numbers $w^{(k)}$, $1 \le k \le K$, satisfy the inequalities

$$\begin{array}{ll} w^{(1)} > \ldots > w^{(2k_0-1)} > w^{(2k_0)} > \ldots > w^{(2)} & \text{if } K = 2k_0, \\ w^{(1)} > \ldots > w^{(2k_0+1)} > w^{(2k_0)} > \ldots > w^{(2)} & \text{if } K = 2k_0 + 1. \end{array} \tag{39}$$

The totality W of all states ω described by such step-like curves may be naturally identified with the family of all finite vectors

$$w = \{w^{(k)}\}_1^K$$

satisfying the inequalities (39). The dimension of such a vector will be denoted by $\#w = K$. The totality of all K-dimensional vectors will be denoted by W^K. It is convenient to define also $w^{(0)} = -\infty$, $w^{(-1)} = \infty$.

We will identify the variable state $\omega(t)$ of the Preisach operator with a variable vector $w(t)$ of variable dimension K^t. In these notations, the variable state $\omega(t;\xi)$ of the Preisach model is given by the following equalities:

$$\omega(t;\xi) = \begin{cases} (-1)^{K^t}\xi, & \xi < (-1)^{K^t}(w^{(K^t-1)} - w^{(K^t)}), \\ 2(w^{(j)} - w^{(K^t)}) + (-1)^j\xi, & 2 \le j \le K^t - 1, \\ \quad (-1)^j(w^{(j+1)} - w^{(j)}) \le \xi < (-1)^j(w^{(j-1)} - w^{(j)}), & \\ 2(w^{(1)} - w^{(K)}) - \xi, & \xi \ge w^{(1)} - w^{(2)}. \end{cases} \tag{40}$$

We also introduce semi-closed intervals $I_k = I_k(w) = [w^{(k-1)}, w^{(k+1)})$, $0 \le k < K$, and an open interval $I_K = I_K(w) = (w^{(K)}, w^{(K-1)})$, which form a partition of the whole real line. For any sequence $w \in W^K$ we denote by $J_k(w)$, $i = 1, \ldots, K$, the following expressions:

$$\begin{array}{ll} J_{2j+1}(w) = \displaystyle\int_{I_{2j+1}(w)} \mu(\alpha, w_{2j+1})\, d\alpha, & 1 \le 2j+1 \le K, \\ J_{2j}(w) = \displaystyle\int_{I_{2j}(w)} \mu(w_{2j}, \beta)\, d\beta, & 1 < 2j \le K. \end{array} \tag{41}$$

In other words, (41) are 'μ-weights' of the horizontal and vertical links of L, and I_k are their projections on the vertical and horizontal axes, see figure 4. Comparing this notation with (9), we note that $J_K(w) = J_\varphi^x(w^{(K)}, w^{(K-1)})$ for an odd K, and $J_K(w) = J_\psi^x(w^{(K)}, w^{(K-1)})$ for an even K.

We also need the following 'complementary' notation:

$$\begin{array}{ll} \overline{J}_K(w) = \displaystyle\int_{w^{(K-1)}}^{w^{(K)}} \mu(w^{(K)}, \beta)\, d\beta, & \text{if } K \text{ is odd}, \\ \overline{J}_K(w) = -\displaystyle\int_{w^{(K-1)}}^{w^{(K)}} \mu(\alpha, w^{(K)})\, d\alpha, & \text{if } K \text{ is even}. \end{array} \tag{42}$$

4.2. Assumptions

Consider the system

$$\begin{array}{ll} z'(t) = f(t, z(t), y(t)), & z \in \mathbb{R}^n, \\ y(t) = \mathcal{P}[\omega_0]x(t), & \\ x(t) = \langle c, z(t) \rangle. & \end{array} \tag{43}$$

Here the function $f : \mathbb{R}^{n+2} \to \mathbb{R}^n$ is continuously differentiable with respect to the set of its arguments, $y(t)$ denotes the output of the Preisach operator with the input $x(t)$, $c \in \mathbb{R}^n$ is a constant vector, and $\langle \cdot, \cdot \rangle$ is a scalar product in $\mathbb{R}^n$.

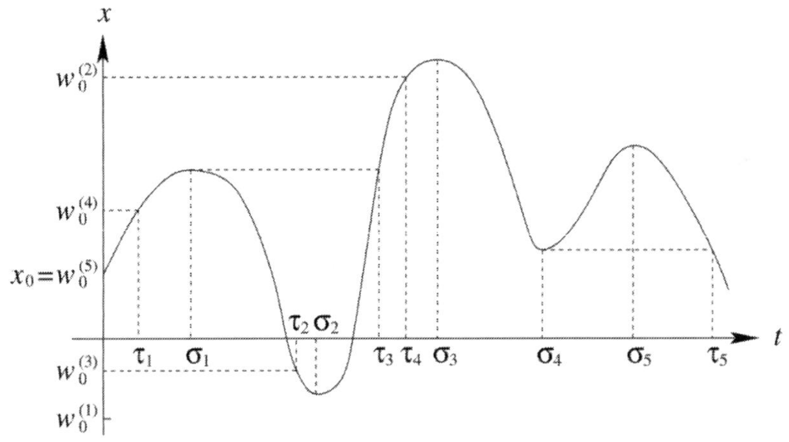

Figure 5. Numbers τ_i and σ_i

Let $(z(t), \omega(t))$, $0 \le t \le T$, be a solution of system (43). Let the following three assumptions hold:

(i) The initial state $\omega_0 = w_0$ belongs to W^{K^0} and the following natural condition holds:

$$x(0) = \langle c, z_0 \rangle = w_0^{(K^0)},$$

where $K^0 = \# w_0$.

(ii) The function $x(t) = \langle c, z(t) \rangle$, $0 \le t \le T$ is piecewise monotone with the extremum points σ_i:

$$0 < \sigma_1 < \ldots < \sigma_m < T.$$

Moreover, $x'(t) \neq 0$ within each interval (σ_{i-1}, σ_i), $i = 1, \ldots, m$ with $\sigma_0 = 0$. We denote the set of points σ_i, $i = 1, \ldots, m$, by $\mathcal{S}$.

(iii) All numbers $x_j = x(\sigma_j)$ are pairwise different, and also different from the numbers $w_0^{(i)}$, $i = 1, \ldots, K^0$.

Assumptions (i) and (ii) ensure that $\omega(t) \in W$ for each t. Moreover, for every t

$$x(t) = \langle c, z(t) \rangle = w^{(K^t)}(t).$$

To avoid some clumsy notation we also assume that either

$$K^0 \text{ is odd, and } x'(0) = \langle c, f(0, z_0, y_0) \rangle > 0,$$

or

$$K^0 \text{ is even, and } x'(0) = \langle c, f(0, z_0, y_0) \rangle < 0,$$

where $y_0 = y(0)$.

We denote by $\mathcal{T}$ the set of numbers τ such that $x(\tau)$ equals for the first time either $w_0^{(i)}$ for some $1 \le i < K^0$, or some $x(\sigma_i)$ with $\sigma_i < \tau$, see figure 5. Namely,

$$\mathcal{T} = \bigcup_{i=1}^{K^0-1} \{\inf\{t \colon t > 0 \text{ and } x(t) = w_0^{(i)}\}\} \cup \bigcup_{i=1}^{m} \{\inf\{t \colon t > \sigma_i \text{ and } x(t) = x(\sigma_i)\}\}.$$

4.3. Evolution operator and its derivative

Denote by $\mathcal{E}_{[0,t]}$ the evolution operator along the trajectory of a solution of (43), which acts on the subset of $\mathbb{R}^n \times W$ of pairs (z, w) satisfying $\langle c, z\rangle = w^{(K)}$:

$$\mathcal{E}_{[0,t]}(z_0, w_0) := (z(t), w(t)).$$

Here $(z(t), w(t))$ is a solution of (43) with initial conditions (z_0, w_0).

We will be interested in small perturbations of initial conditions (z_0, w_0), for which the perturbed Preisach state w_0^δ belongs to the same class $W^{K^0} \subset W$ as w_0 does. Since the admissible initial conditions are pairs $(z, w) \in \mathbb{R}^n \times W^{K^0}$, which satisfy the restriction $x = \langle c, z\rangle = w^{(K^0)}$, they can be naturally identified with vectors

$$(z, \mathrm{Ch}\, w) = (z_1, \ldots, z_n, w^{(1)}, \ldots, w^{(K^0-1)}),$$

where $z = (z_1, \ldots, z_n)$, $\mathrm{Ch}\, w = (w^{(1)}, \ldots, w^{(K-1)})$, and $w^{(i)}$ satisfy (39). The totality of such vectors is denoted by

$$\mathcal{H}(K^0) \subset \mathbb{R}^n \times \mathbb{R}^{K^0-1} = \mathbb{R}^{n+K^0-1}.$$

Thus we can locally interpret the evolution operator as a mapping from $\mathcal{H}(K^0)$ to $\mathcal{H}(K^t)$.

We now consider the question about a derivative of the evolution operator with respect to initial data. Let assumptions (i)–(iii) hold. Consider a perturbation $(z_0^\delta, w_0^\delta) \in \mathbb{R}^n \times W^{K^0}$ of the initial data $(z_0, w_0) \in \mathbb{R}^n \times W^{K^0}$ of the form

$$z_0^\delta = z_0 + \delta \tilde{z}_0, \quad w_0^\delta = w^0 + \delta \tilde{w}_0, \tag{44}$$

where $\tilde{z}_0 \in \mathbb{R}^n$ and $\tilde{w}_0 \in \mathbb{R}^{K^0}$ are given vectors satisfying $\tilde{w}_0^{(K^0)} = \langle c, \tilde{z}_0\rangle$, and δ is a small positive parameter.

The next proposition follows from the continuous dependence of solutions of system (43) on initial data in C^1-norm.

Proposition 1. *For any closed set $\mathcal{R} \subset [0, T]$ such that $\mathcal{R} \cap (\mathcal{S} \cup \mathcal{T}) = \varnothing$ and any pair $(\tilde{z}_0, \tilde{w}_0) \in \mathbb{R}^n \times W^{K^0}$ with $\tilde{w}_0^{(K^0)} = \langle c, \tilde{z}_0\rangle$ there exists a $\bar{\delta} > 0$ such that the perturbed solution $(z^\delta(t), w^\delta(t))$ with the initial data (44) and $0 < \delta < \bar{\delta}$ is defined on $[0, T]$, and the dimension $\#w^\delta(t)$ of $w^\delta(t)$ is equal to the dimension $\#w(t)$ of $w(t)$ for each $t \in \mathcal{R}$.*

Proposition 1 implies that for any $t \notin \mathcal{S} \cup \mathcal{T}$ there exists a sufficiently small $\bar{\delta}(t) > 0$ such that the difference

$$r(t, \delta) = (z^\delta(t), \mathrm{Ch}\, w^\delta(t)) - (z(t), \mathrm{Ch}\, w(t)) = \mathcal{E}_{[0,t]}(z_0^\delta, \mathrm{Ch}\, w_0^\delta) - \mathcal{E}_{[0,t]}(z_0, \mathrm{Ch}\, w_0)$$

is well defined for any $0 < \delta < \bar{\delta}(t)$.

We would like to construct a linear mapping $A_t : \mathcal{H}(K^0) \to \mathcal{H}(K^t)$ for $t \notin \mathcal{S} \cup \mathcal{T}$ such that for $0 < \delta < \bar{\delta}(t)$ the following relationship holds:

$$\mathcal{E}_{[0,t]}(z_0^\delta, \mathrm{Ch}\, w_0^\delta) - \mathcal{E}_{[0,t]}(z_0, \mathrm{Ch}\, w_0) = \delta A_t(\tilde{z}_0, \mathrm{Ch}\, \tilde{w}_0) + o(\delta). \tag{45}$$

In coordinate form, it means that we are going to construct a matrix A_t of variable size $(n + K^0 - 1) \times (n + K^t - 1)$.

Below we construct the functions $\tilde{z}(t)$ and $\tilde{w}(t)$ such that

$$r(t, \delta) = \delta(\tilde{z}(t), \mathrm{Ch}\, \tilde{w}(t)) + o(\delta), \quad t \notin \mathcal{S} \cup \mathcal{T}. \tag{46}$$

Let $0 < t_1 < \ldots < t_p < T$ be points from the set $\mathcal{S} \cup \mathcal{T}$, and denote $t_0 = 0$, $t_{p+1} = T$. Then the following iterative procedure is used to construct $\tilde{z}(t)$ and $\tilde{w}(t)$:

1. Let $\tilde{z}(0) = \tilde{z}_0$, $\tilde{w}(0) = \tilde{w}_0$, and $i = 0$.
2. Consider the interval (t_i, t_{i+1}). On this interval let

$$\operatorname{Ch}\tilde{w}(t) = \operatorname{Ch}\tilde{w}(t_i).$$

Hence, the vector $\operatorname{Ch}\tilde{w}(t)$ is constant on each interval (t_i, t_{i+1}). Note that the interval (t_i, t_{i+1}) does not contain points from $\mathcal{S} \cup \mathcal{T}$, thus the vector $\operatorname{Ch} w(t)$ is also constant on this interval due to the properties of the Preisach operator, and $\#\tilde{w}(t) = \#w(t) = K^{t_i}$ for $t \in (t_i, t_{i+1})$. The last component equals $\tilde{w}^{(K^t)} = \langle c, \tilde{z}(t)\rangle$ for every t.
3. The function $\tilde{z}(t)$ is defined on this interval from the system of linear differential equations

$$\begin{aligned}\tilde{z}' &= f_z(t, z(t), y(t))\tilde{z} \\ &\quad + f_y(t, z(t), y(t)) H(\operatorname{Ch} w(t); \operatorname{Ch}\tilde{w}(t)) \\ &\quad + f_y(t, z(t), y(t)) \left(-\tilde{w}_{K^t-1}\overline{J}_{K^t}(w(t)) + \langle c, \tilde{z}\rangle J_{K^t}(w(t))\right),\end{aligned} \tag{47}$$

where

$$H(\operatorname{Ch} w(t); \operatorname{Ch}\tilde{w}(t)) = \sum_{i=1}^{K^t-1} \tilde{w}^{(i)}(t) J_i(\operatorname{Ch} w(t)).$$

4. At the moment t_{i+1} the dimension and the components of the vector $\tilde{w}(t)$ are updated according to the following switch rules:
If $t_{i+1} \in \mathcal{T}$, then

$$\begin{aligned}K^{t_{i+1}} &= K^{t_i} - 2, \\ \tilde{w}^{(k)}(t_{i+1}) &= \tilde{w}^{(k)}(t_i), \quad 1 \le k < K^{t_{i+1}}.\end{aligned} \tag{R1}$$

If $t_{i+1} \in \mathcal{S}$, then

$$\begin{aligned}K^{t_{i+1}} &= K^{t_i} + 1, \\ \tilde{w}^{(k)}(t_{i+1}) &= \tilde{w}^{(k)}(t_i), \quad 1 \le k < K^{t_{i+1}} - 1, \\ \tilde{w}^{(K^{t_{i+1}}-1)}(t_{i+1}) &= \langle c, \tilde{z}(t_{i+1})\rangle.\end{aligned} \tag{R2}$$

If $t_{i+1} = T$, then the procedure ends.
5. Let $i := i + 1$, and go to step number 2.

Hypothesis 1. *Suppose that $x''(\sigma_i) \neq 0$, $i = 1, \ldots, m$. Let $(\tilde{z}(t), \operatorname{Ch}\tilde{w}(t))$ be the functions constructed by repeating steps 1–5 above, satisfying system* (47)*, switch rules* (R1)–(R2) *and the initial conditions $\tilde{z}(0) = \tilde{z}_0$, $\operatorname{Ch}\tilde{w}(0) = \operatorname{Ch}\tilde{w}_0$. Then equation* (46) *holds for $t \notin \mathcal{S} \cup \mathcal{T}$ and sufficiently small $\delta > 0$.*

Now we can define the linear mapping A_t. Equation (47) may be rewritten equivalently as

$$\begin{aligned}\tilde{z}' &= \left(f_z(t, z(t), y(t)) + J_K(x(t)) f_y(t, z(t), y(t)) c^T\right)\tilde{z} \\ &\quad + f_y(t, z(t), y(t)) \left(H(\operatorname{Ch} w(t); \tilde{w}(t)) - \tilde{w}_{K^t-1}(t)\overline{J}_K(w(t))\right),\end{aligned} \tag{48}$$

where the vectors are treated as columns, and c^T is the transposition of c. Equation (48) is a linear nonhomogeneous system in $\tilde{z}$. Note that the nonhomogeneous term

$$f_y(t, z(t), y(t)) \left(H(\operatorname{Ch} w(t); \operatorname{Ch}\tilde{w}(t)) - \tilde{w}_{K^t-1}(t)\overline{J}_K(w(t))\right)$$

is linear with respect to $\operatorname{Ch}\tilde{w}$. Also the switch rule (R1) is linear with respect $\operatorname{Ch}\tilde{w}$, and the switch rule (R2) is linear with respect to $\tilde{z}$ and $\tilde{w}$. Thus, we conclude that the functions $\tilde{z}(t)$ and $\operatorname{Ch}\tilde{w}(t)$ depend linearly on $(\tilde{z}_0, \operatorname{Ch}\tilde{w}_0)$. Hence, the mapping

$$A_t : (\tilde{z}_0, \operatorname{Ch}\tilde{w}_0) \to (\tilde{z}(t), \operatorname{Ch}\tilde{w}(t)) \tag{49}$$

is linear. Hypothesis 1 implies that (45) holds for $t \notin \mathcal{S} \cup \mathcal{T}$ and sufficiently small $\delta > 0$. Therefore, the following corollary holds.

Corollary 1. *Suppose that $x''(\sigma_i) \neq 0$, $i = 1, \dots, m$. Then for each $t \notin \mathcal{S} \cup \mathcal{T}$ the evolution operator $\mathcal{E}_{[0,t]}$ is differentiable at the point $(z_0, \operatorname{Ch} w_0)$, and its derivative is defined by* (49).

4.4. Discontinuous measure densities

The weight function $\mu(\alpha, \beta)$ in the definition of the output of the Preisach nonlinearity was assumed to be smooth everywhere above. In this final section we consider the weight functions that have some discontinuities (which are natural in many examples and applications [1, 7, 8, 9]). Let us restrict ourselves to the situation where the only possible discontinuities are along the straight lines $\alpha = \alpha_*$ and $\beta = \beta_*$, and moreover, let the limits $\mu(\alpha_*, v \pm 0)$ and $\mu(v \pm 0, \beta_*)$ be finite. Denote

$$\begin{aligned} \gamma_-(v) &= \mu(\alpha_*, v + 0) - \mu(\alpha_*, v - 0), \\ \gamma_+(v) &= \mu(v + 0, \beta_*) - \mu(v - 0, \beta_*) \end{aligned} \tag{50}$$

and

$$\Gamma_-(v) = \int_{\alpha_*}^{v} \gamma_-(s)\, ds, \quad \Gamma_+(v) = \int_{v}^{\beta_*} \gamma_+(s)\, ds. \tag{51}$$

Let us introduce the sets

$$\begin{aligned} \mathcal{U}_- &= \{t \geq 0\colon x(t) = \alpha_* \text{ and } x'(t) < 0\}, \\ \mathcal{U}_+ &= \{t \geq 0\colon x(t) = \beta_* \text{ and } x'(t) > 0\}, \\ \mathcal{U}_0 &= \{t \geq 0\colon (x(t) = \alpha_* \text{ or } x(t) = \beta_*) \text{ and } x'(t) = 0\}. \end{aligned}$$

We suppose that the set $\mathcal{U}_0$ is empty, and the sets $\mathcal{U}_\pm$ are finite (or empty), and that they do not intersect with $\mathcal{S} \cup \mathcal{T}$.

Let the evolution of $(\tilde{z}, \tilde{w})$ be defined by (47), (R1), (R2) and the additional switch rule: if $t_{i+1} \in \mathcal{U}_+ \cup \mathcal{U}_-$, then

$$\begin{aligned} K^{t_{i+1}} &= K^{t_i}, \\ \tilde{w}(t_{i+1}) &= \tilde{w}(t_i), \\ \tilde{z}(t_{i+1}) = \tilde{z}(t_{i+1} - 0) + \langle c, \tilde{z}(t_{i+1} - 0)\rangle &\frac{\Gamma_\pm\left(w^{(K^{t_{i+1}}-1)}(t_{i+1})\right)}{x'(t_{i+1})} f_y(t_{i+1}, z(t_{i+1}), y(t_{i+1})) \end{aligned} \tag{R3}$$

with $x(t) = \langle c, z(t)\rangle$ and $x'(t) = \langle c, z'(t)\rangle = \langle c, f(t, z(t), y(t))\rangle$, where t_i inlcude points from $\mathcal{U}_\pm$, and Γ_+ is used if $t_{i+1} \in \mathcal{U}_+$, and Γ_- if $t_{i+1} \in \mathcal{U}_-$. We note that (R3) may be rewritten equivalently as

$$\tilde{z}(t_{i+1}) = \left(I + \frac{\Gamma_\pm\left(w^{(K^{t_{i+1}}-1)}(t_{i+1})\right)}{x'(t_{i+1})} f_y(t_{i+1}, z(t_{i+1}), y(t_{i+1})) c^T \right) \tilde{z}(t_{i+1} - 0),$$

where I denotes the identity $n \times n$ matrix. Hence, the switch rule (R3) is linear with respect to $\tilde{z}$.

Hypothesis 2. *Let $\mu(\alpha, \beta)$ be smooth, except for discontinuities at the straight lines $\alpha = \alpha_*$ and $\beta = \beta_*$. Suppose that $x''(\sigma_i) \neq 0$, $i = 1, \dots, m$. Let $(\tilde{z}, \operatorname{Ch} \tilde{w})$ be the functions constructed by repeating steps 1–5 from section 4.3, satisfying system* (47), *switch rules* (R1)–(R3) *and the initial conditions $\tilde{z}(0) = \tilde{z}_0$, $\operatorname{Ch} \tilde{w}(0) = \operatorname{Ch} \tilde{w}_0$. Then for $t \notin \mathcal{S} \cup \mathcal{T} \cup \mathcal{U}_+ \cup \mathcal{U}_-$ the equality*

$$r(t, \delta) = \delta(\tilde{z}(t), \operatorname{Ch} \tilde{w}(t)) + o(\delta) \tag{52}$$

holds for sufficiently small $\delta > 0$.

Consider the linear mapping

$$A_t : (\tilde{z}_0, \mathrm{Ch}\,\tilde{w}_0) \to (\tilde{z}(t), \mathrm{Ch}\,\tilde{w}(t)), \tag{53}$$

with functions $\tilde{z}$ and $\mathrm{Ch}\,\tilde{w}$ defined as in Hypothesis 2.

Corollary 2. *Let $\mu(\alpha, \beta)$ be smooth, except for discontinuities at the straight lines $\alpha = \alpha_*$ and $\beta = \beta_*$. Suppose that $x''(\sigma_i) \neq 0$, $i = 1, \ldots, m$. Then for each $t \notin \mathcal{S} \cup \mathcal{T} \cup \mathcal{U}_+ \cup \mathcal{U}_-$ the evolution operator $\mathcal{E}_{[0,t]}$ is differentiable at the point $(z_0, \mathrm{Ch}\,w_0)$, and its derivative is defined by (53).*

Acknowledgments

This work was supported by grant 06/RFP/MAT048 of the Science Foundation Ireland, by the Boole Centre for Research in Informatics (BCRI), and by the Russian Foundation for Basic Research Grant 04-01-00330.

References

[1] Brokate M, Pokrovskiĭ A, Rachinskiĭ D and Rasskazov O 2006 Differential equations with hysteresis via a canonical example *The Science of Hysteresis* vol 1, ed G Bertotti and I Mayergoyz (Elsevier, Academic Press) pp 125–291

[2] Brokate M, Pokrovskiĭ A and Rachinskiĭ D 2006 Asymptotic stability of continual sets of periodic solutions to systems with hysteresis *J. Math. Anal. Appl.* **319** 94–109

[3] Krejčí P 1996 *Hysteresis, Convexity and Dissipation in Hyperbolic Equations* (Tokyo: Gakkōtosho)

[4] Krasnosel'skiĭ M and Pokrovskiĭ A 1989 *Systems with Hysteresis* (New York: Springer-Verlag)

[5] Brokate M and Sprekels J 1996 *Hysteresis and Phase Transitions*, Appl. Math. Sci., vol 121 (New York: Springer)

[6] Mayergoyz I 1991 *Mathematical Models for Hysteresis* (New York: Springer)

[7] Mayergoyz I 2003 *Mathematical Models of Hysteresis and Their Applications: Electro-magnetism* (Academic Press)

[8] O'Kane J P 2005 The FEST model—a test bed for hysteresis in hydrology and soil physics, *J. Phys.: Conf. Series* ed M P Mortell, R E O'Malley, A Pokrovskiĭ and V Sobolev **22** 148–63

[9] Flynn D, McNamara H, O'Kane J P and Pokrovskiĭ A 2006 Application of the Preisach model to soil-moisture hysteresis *The Science of Hysteresis* vol 3, ed G Bertotti and I Mayergoyz (Elsevier, Academic Press) pp 689–744

[10] Visintin A 1994 *Differential Models of Hysteresis*, Appl. Math. Sci., vol 111 (New York: Springer)

[11] 2005 *The Science of Hysteresis* vol I, II, III, ed G Bertotti and I Mayergoyz (Elsevier, Academic Press)

[12] 2005 *Preisach Memorial Book: Hysteresis Models in Mathematics, Physics and Engineering* ed A Ivanyi (Budapest: Akadémiai Kiadó)

[13] 2006 Proceedings of the 5th International Symposium on Hysteresis Modeling and Micromagnetics *Physica* B **372**

[14] Flynn D and Rasskazov O 2005 On the integration of an ODE involving the derivative of a Preisach nonlinearity *J. Phys.: Conf. Series* ed M P Mortell, R E O'Malley, A Pokrovskiĭ and V Sobolev **22** 43–55

[15] Krasnosel'skiĭ M A, Chernorutskiĭ V V, Pokrovskiĭ A and Tronel G 1992 On the dynamics of controlled systems described by equations of parabolic type with hysteresis nonlinearities *Automation and Remote Control* **53** no 11, part 1, pp 1705–11

[16] Pokrovskiĭ A 1993 Shuttle algorithm in analysis of systems with hysteresis nonlinearities *Pitman Research Notes in Math. Series* **286** (Longman Sci. Tech.) pp 124–42

Institute of Physics Publishing
doi:10.1088/1742-6596/55/1/018
Journal of Physics: Conference Series **55** (2006) 191–202
International Workshop on Multi-Rate Processes and Hysteresis

Multiple scales in spray modelling

S Sazhin

School of Engineering, Faculty of Science and Engineering, The University of Brighton, Brighton BN2 8EJ, UK

E-mail: S.Sazhin@brighton.ac.uk

Abstract. A summary of the results of modelling of several processes in diesel engine sprays is presented. These include spray penetration, droplet heating and droplet evaporation. Typical scales of these processes differ by more than 7 orders of magnitude from about 3×10^{-2} m (spray penetration) to about 10^{-9} m (diameters of individual molecules). It has been confirmed that the dynamic decomposition techniques for solving stiff systems of ordinary differential equations, reported earlier, can be an effective tool in spray modelling applications for CFD codes.

1. Introduction

The problem of spray modelling have been discussed in numerous monographs and review papers (e.g. [1]-[9]). A number of reviews have been focused on various specific aspects of this modelling, including heating and evaporation of individual droplets [10, 11]. The main objective of this paper is to present an overview of the results, not covered in the previous paper [11] with particular emphasis on the multiscale nature of the models.

A spray can be defined as one of the types of two phase flow, involving liquid as the dispersed phase in the form of droplets or ligaments and a gas as the continuous phase (cf. the definition given in [1]). In the environment, sprays are easily observed; rain, waterfalls and volcanic eruptions are just a few examples. They are widely used in medical and engineering applications, including those in direct injection internal combustion engines. This paper will focus mainly on the latter, although the generalisation of the results to other types of sprays is straightforward.

Modelling of sprays involves analysis of a number of processes which can take place simultaneously. These processes include spray formation, break-up of the liquid jet, ligaments and individual droplets, cavitation inside the nozzle, droplet heating and evaporation and, in the case of internal combustion engines, ignition of the fuel vapour/ air mixture. These processes take place over vastly different scales in space and time. For example, a typical spray penetration length in diesel engines used in passenger cars is 3×10^{-2} m. The diameter of the nozzle in these engines is about 10^{-4} m; a typical droplet diameter is about 10^{-5} m. The thickness of the Knudsen layer around droplets is about 10^{-7} m, while the diameter of individual molecules is about 10^{-9} m.

This review will focus on: modelling spray penetration, when the effect of droplet heating and evaporation can be ignored (Section 2); modelling of the heating of individual droplets when the effects of evaporation can be ignored (Section 3); and the modelling of droplet evaporation (Section 4). In Section 5 some results of further developments of the dynamic decomposition technique, already discussed in [11], are summarised.

2. Spray penetration

Detailed comparisons between different models for spray penetration and experimental results were given by Dent [12] and Hay and Jones [13]. They concluded that models predicting the penetration distance s to be proportional to the square root of time $\sqrt{t}$ are in good agreement with observations. This was confirmed by Sendyka [14]. A more detailed analysis of the penetration length shows that the proportionality of s to $\sqrt{t}$ only occurs at times exceeding the so called jet break-up time. At shorter times, s is proportional to t [15]. There is some ambiguity, however, regarding the definition of the jet break-up time.

The problem of modelling spray penetration is closely linked to the problem of modelling induced air velocity within droplet driven sprays. The latter was studied by Ghosh and Hunt [17], and some of their results were incorporated into the spray penetration model described in [16]. In [17] 3 spray zones were considered: zone 1, where the initial velocities of droplets are much greater than that of the air stream and are not much affected by it; zone 2, where the droplets slow down and their velocities become comparable with the air velocity; zone 3, where the droplets' velocities decrease so much that they becomes less than the terminal velocity. These zones, introduced in [17], have the same physical meanings as mixing, transition and fully developed regions considered in [15].

Firstly, the focus of this section will be on the analysis of a simple model referring to zone 3, which was suggested in [16], and its possible further developments. Secondly, a simplified model referring to zone 1, developed in [16, 18], will be considered.

2.1. Two-phase flow

When the distance from the nozzle becomes large enough, the velocity of the entrained air approaches the velocity of droplets. As a result, the dynamics of both droplets and entrained air can be described in terms of a two phase flow with a zero relative velocity between air and droplets. The basic properties of this flow can be derived from the equations of conservation of mass and momentum.

From the equation of conservation of mass of droplets we obtain [16]:

$$\rho_l A_0 v_{\text{in}} = \rho_m A_m v_m - (1 - \alpha_d) A_m \rho_g v_m, \tag{1}$$

where A_0 is the cross-sectional area of the nozzle, v_{in} is the initial velocity of droplets, ρ_l is the density of liquid (droplets), ρ_m is the density of a mixture of droplets and gas, A_m is the cross-sectional area of a spray, v_m is the velocity of a mixture, α_d is the volume fraction of droplets. The left hand side of Equation (1) is the mass flow rate of fuel at the nozzle. The second term in the right hand side of Equation (1) takes into account the contribution of the mass flow rate of entrained air. The first term in the right hand side of this equation gives the mass flow rate of the mixture of fuel and air. The relation between A_m and A_0 can be presented in the form [16]:

$$A_m = A_0 + \pi D_0 s \tan\theta + \pi s^2 \tan^2\theta, \tag{2}$$

where θ is the half angle of the spray cone.

Equation (1) should be supplemented by the equation of conservation of momentum:

$$\rho_l A_0 v_{\text{in}}^2 = \rho_m A_m v_m^2. \tag{3}$$

The combination of Equations (1) – (3) leads us to the following equation for the velocity of spray penetration [16]:

$$\frac{\mathrm{d}s}{\mathrm{d}t}_{|m} = \frac{2v_{\text{in}}}{1 + \sqrt{a + bs + cs^2}}, \tag{4}$$

where

$$a = 1 + 4(1-\alpha_d)\tilde{\rho_a}; \qquad b = \frac{16(1-\alpha_d)\tilde{\rho_a}\tan\theta}{D_0}; \qquad c = \frac{16(1-\alpha_d)\tilde{\rho_a}\tan^2\theta}{D_0^2},$$

$$\tilde{A} = A_m/A_0 = 1 + \frac{4s\tan\theta}{D_0} + \frac{4s^2\tan^2\theta}{D_0^2}; \qquad \tilde{\rho_r} = \rho_m/\rho_l; \qquad \tilde{\rho_a} = \rho_g/\rho_l; \qquad \tilde{v} = v_{\text{in}}/v_m.$$

subscript $_m$ indicates that $\frac{ds}{dt}_{|m}$ is the velocity of the mixture.

As follows from (4), in the case of no entrained air ($\alpha_d = 1$) we have $\tilde{v} = 1$, which means that $v_m = v_{\text{in}}$. This solution, however, does not have a physical meaning since the formation of spray always includes the entrained air. In a realistic spray environment $\alpha_d \ll 1$.

Integration of (4) gives [16]:

$$s + \frac{2cs+b}{4c}\sqrt{a+bs+cs^2} - \frac{b\sqrt{a}}{4c} + + \frac{4ac-b^2}{8c^{3/2}}\ln\left[\frac{2\sqrt{c(a+bs+cs^2)}+2cs+b}{2\sqrt{ac}+b}\right] = 2v_{\text{in}}t. \quad (5)$$

In the case when $a \gg bs \gg cs^2$ (the immediate vicinity of the nozzle), Equation (5) is simplified to [16]:

$$s = \frac{2v_{\text{in}}t}{1+\sqrt{a+\frac{b}{8c\sqrt{a}}(a-1)}} \approx v_{\text{in}}t. \quad (6)$$

When deriving (6) the fact that $\tilde{\rho_a} \ll 1$ and $a - 1 \ll 1$ is taken into account. If $a \ll bs \ll cs^2$ (s is large and/or D_0 is small) then $\frac{b}{2\sqrt{ac}} \approx 2\sqrt{\tilde{\rho_a}} \ll 1$ and the logarithmic term can be ignored. Thus:

$$s = \frac{\sqrt{v_{\text{in}}D_0t}}{(1-\alpha_d)^{1/4}\tilde{\rho_a}^{1/4}\sqrt{\tan\theta}}\left(1 - \frac{\sqrt{D_0}}{4\sqrt{v_{\text{in}}}(1-\alpha_d)^{1/4}\tilde{\rho_a}^{1/4}\sqrt{\tan\theta}\sqrt{t}}\right). \quad (7)$$

Equation (7) can be further simplified if the second term in the right hand side is ignored:

$$s = \frac{\sqrt{v_{\text{in}}D_0t}}{(1-\alpha_d)^{1/4}\tilde{\rho_a}^{1/4}\sqrt{\tan\theta}}. \quad (8)$$

Equation (8) is identical to the one suggested by a number of authors (see [12]). Equations (7) and (6) can be considered as generalisations of previously discussed formulae. The combination of Equations (6) and (8) gives essentially the same expression for spray penetration as discussed in [2]. The main advantage of Equation (5) is that it can accurately predict a smooth transition from the immediate vicinity of the nozzle to the two-phase flow in the region where the spray is fully formed. Separate solutions for the near zone and the far zone, discussed in [15], inevitably lead to a physically unrealistic jump in the velocity between these zones. The range of applicability and accuracy of Equation (8) can be determined based on direct comparison between the predictions of Equations (5), (6) and (7) and the available experimental data. In [19] the predictions of the above mentioned equations were compared with the results of experimental studies of spray penetration, obtained using a rapid compression diesel spray rig based at Brighton University (UK) and a high-pressure dimethyl ether spray chamber based at Chungbuk National University (Korea). In both cases the experimental results are shown to be in agreement with the prediction of theoretical models, even though the equipment used and the range of ambient densities tested varied significantly between the two studies. In most cases, however, the errors of measurement seem to be comparable with the improved accuracy offered by Eq. (7). The model suggested in [16] is based on the assumption that the effects of turbulence on spray dynamics can be ignored. The latter effects lead to spray diffusion in directions perpendicular to the spray penetration. Taking into account this effect leads to a new spray penetration model which predicts that $s \propto t^{3/2}$[20]. Although this dependence has occasionally been observed in our experiments, it cannot be considered as typical.

More advanced models of spray penetration, based on mass and momentum conservation, but taking into account the non-zero velocity between droplets and gas, have been developed in numerous papers including [21]-[23]. In this case the effect of droplet evaporation also needs to be taken into account. In gasoline sprays, the spray penetration process is sometimes accompanied by the formation of vortex rings [24]. In this case a different approach to the analysis would be required (cf. [24]).

The overall spray penetration length, discussed so far, is the largest measure used for the analysis of sprays. In the next section the focus will be on the initial stage of spray penetration.

2.2. The initial stage

In this section, the results reported earlier in [16, 18] are briefly summarised. The contribution of the jet in the spray penetration process can be ignored as the first approximation, and the initial sizes of droplets can be assumed equal to the nozzle diameter [16]. This can be justified by the fact that for typical diesel engines the expected break-up regime lies between the first and second wind induced break-ups [25].

At the initial stage of spray penetration the analysis can be focused on the equation describing the dynamics of an individual droplet:

$$\frac{\mathrm{d}^2 s}{\mathrm{d}t^2} = -\frac{3}{8R_d} C_D \frac{\rho_g}{\rho_l} \left(\frac{\mathrm{d}s}{\mathrm{d}t} - v_g \right) \left| \frac{\mathrm{d}s}{\mathrm{d}t} - v_g \right|, \tag{9}$$

where C_D is the drag coefficient, which depends on the Reynolds number $\mathrm{Re} = 2\rho_g |v_d - v_g| R_d / \mu_g$, R_d is the droplet's radius, $v_d = \frac{\mathrm{d}s}{\mathrm{d}t}$, v_g is the gas velocity, and the droplet is assumed to be a perfect sphere.

Equation (9) refers to the smallest scale considered so far which is comparable with the diameter of the nozzle. The solution of this equation requires knowledge of C_D and v_g. A number of approximations for C_D have been suggested ([15], [26]-[30]). The one given in [31] was used in our analysis. In this approximation, three ranges of Reynolds numbers were considered: $\mathrm{Re} \le 2$ (Stokes flow), $2 < \mathrm{Re} \le 500$ (Allen flow) and $500 < \mathrm{Re} \le 10^5$ (Newton flow). The functions $C_D(\mathrm{Re})$ for these flows are given by the following expressions: $C_D = 24/\mathrm{Re}$ (Stokes flow), $C_D = 18.5/\mathrm{Re}^{0.6}$ (Allen flow) $C_D = 0.44$ (Newton flow). The condition $\mathrm{Re} < 10^5$ is always satisfied for realistic droplets in diesel engines. Note that there is a printing mistake in [31] referring to the value of Re when the transition from Stokes to Allen flow takes place. This mistake was overlooked in the original paper [18].

The drag force acting on the droplets leads not only to their deceleration, but also to momentum transfer from the droplets to the gas entrained into the spray. In [17] the following equation for gas velocity in zone 1 of the spray was derived:

$$\frac{\mathrm{d}}{\mathrm{d}s} \left(\pi R_j^2 v_g^2 \right) = \frac{3\pi R_s^2 \alpha_d}{8R_d} C_D \left(v_d - v_g \right)^2, \tag{10}$$

where R_j and R_s are radii of the gas/air jet and the spray itself respectively at a certain position s, R_j is usually slightly larger than R_s, but as a first approximation we will assume that they are both equal to $s \tan \theta$.

We assume that v_g is so small compared with v_d that its contribution in the right hand side of Eq. (9) can be ignored altogether. Remembering the definitions of C_D for the three types of flow, Equation (9) can be simplified to [18]:

$$\frac{\mathrm{d}^2 s}{\mathrm{d}t^2} = -\frac{9}{2} \frac{\mu_g}{\rho_l R_d^2} \frac{\mathrm{d}s}{\mathrm{d}t} \quad \text{(Stokes flow)}, \tag{11}$$

$$\frac{d^2s}{dt^2} = -\frac{4.577\mu_g^{0.6}\rho_g^{0.4}}{\rho_l R_d^{1.6}}\left(\frac{ds}{dt}\right)^{1.4} \quad \text{(Allen flow)}, \tag{12}$$

$$\frac{d^2s}{dt^2} = -\frac{0.165\rho_g}{\rho_l r}\left(\frac{ds}{dt}\right)^{2} \quad \text{(Newton flow)}. \tag{13}$$

Assuming that $R_d = \text{const}$ (no evaporation and break-up) the solutions of Equations (11) – (13) can easily be found [16]. The effect of air entrainment can be considered as the perturbation to the solutions of Equations (11) - (13) [16]. This leads us to the following expressions for s in the limit of small t:

$$s = v_{d0}t - \frac{1}{2}v_{d0}\alpha_s t^2 + \frac{4}{15}\alpha\kappa_e\sqrt{v_{d0}}t^{5/2} \quad \text{(Stokes flow)}, \tag{14}$$

$$s = v_{d0}t - 0.5v_{d0}^{1.4}\beta t^2 + 0.373\kappa_e v_{d0}^{0.9}\beta t^{5/2} \quad \text{(Allen flow)}, \tag{15}$$

$$s = v_{d0}t - 0.5v_{d0}^{2}\gamma t^2 + 0.533\gamma\kappa_e v_{d0}^{3/2}t^{5/2} \quad \text{(Newton flow)}, \tag{16}$$

where v_{d0} is the initial droplet velocity,

$$\kappa_e = \frac{3R_s^2\alpha_d}{8R_d R_{g0}}C_D v_{d0}^2, \quad \alpha_s = \frac{9\nu_g\rho_g}{2R_d^2\rho_l}, \quad \beta = \frac{4.577\nu_g^{0.6}\rho_g}{R_d^{1.6}\rho_l}, \quad \gamma = \frac{0.165\rho_g}{R_d\rho_l}.$$

The main novelty of Equations (14)-(16) is the contribution of the terms taking into account air entrainment (terms proportional to $t^{5/2}$). The main limitation of these equations is that they do not consider the contribution of droplet break-up. Among numerous droplet break-up models suggested so far we mention the bag and stripping break-up models [32, 33], WAVE break-up model [34], Taylor Analogy Break-up (TAB) model [35] and the stochastic break-up model developed in [36, 37]. The bag and stripping break-up models turned out to be particularly important for the analytical study of the effects of break-up on spray penetration. They were used in [16, 18]. Other models have been used in numerical simulations of spray dynamics [38, 37]. The analysis of these models and their implementation into computational fluid dynamics codes is beyond the scope of this paper.

Also this approach does not take into account the effects at scales less that the diameter of the nozzle in the immediate vicinity of the nozzle. The most important of these effects is cavitation inside the nozzle [39, 40].

The scales considered so far refer mainly to the range from about 10^{-4} m (typical diameter of the nozzle) to about $10^{-2} - 10^{-1}$ m (typical penetration length). In the following section we move to scales of about 10^{-5} m (typical diameter of droplets away from the nozzle).

3. Droplet heating

The problem of the transient heating of semitransparent spherical droplets immersed in a stationary hot gas, taking into account the effect of thermal radiation, has been considered by a number of authors (see e.g. [41]-[45]). One of the main assumptions made in all these papers is that the boundary layer around droplets has developed and steady state correlations for the convection heat transfer coefficient (Nusselt number) can be used. A brief review of some of the results reported in these papers is given in [11].

In the general case, the process of droplet heating takes place simultaneously with the formation of the boundary layer. For moving droplets this is a complicated problem, requiring a numerical solution. The heating of stagnant droplets by the ambient gas reduces to a conduction problem. Assuming the spherical symmetry of the problem, its mathematical formulation is based on the solution of the equation:

$$\frac{\partial T}{\partial t} = \kappa \left(\frac{\partial^2 T}{\partial R^2} + \frac{2}{R}\frac{\partial T}{\partial R} \right) + P(R,t), \tag{17}$$

where

$$\kappa = \begin{cases} \kappa_l = k_l/(c_l \rho_l) & \text{when} \quad R \leq R_d \\ \kappa_g = k_g/(c_{pg}\rho_g) & \text{when} \quad R_d < R \leq \infty, \end{cases} \tag{18}$$

$\kappa_{l(g)}$, $k_{l(g)}$ and $c_{l(pg)}$ are the liquid (gas) thermal diffusivity, thermal conductivity and specific heat capacity respectively, R is the distance from the centre of the sphere.

This equation needs to be solved subject to the following initial and boundary conditions:

$$T|_{t=0} = \begin{cases} T_{d0}(R) & \text{when} \quad R \leq R_d \\ T_{g0}(R) & \text{when} \quad R_d < R \leq R_g, \end{cases} \tag{19}$$

$$T|_{R=R_d-0} = T|_{R=R_d+0}; \quad k_l \left.\frac{\partial T}{\partial R}\right|_{R=R_d-0} = k_g \left.\frac{\partial T}{\partial R}\right|_{R=R_d+0}; \quad T|_{R_g} = T_{g0}. \tag{20}$$

Assuming that $T_{d0}(R) = T_{d0} = \text{const}_1$, and $T_{g0}(R) = T_{g0} = \text{const}_2$, $R_g = \infty$, and ignoring the effect of thermal radiation, Cooper [46] has solved this problem analytically:

$$T(R \leq R_d) = T_{g\infty} + \frac{2k_l}{\pi k_g}\sqrt{\frac{\kappa_g}{\kappa_l}}\left(T_{g\infty} - T_{d0}\right)\frac{R_d}{R}$$

$$\times \int_0^\infty \mathrm{d}u \Phi(u) \exp\left(-u^2 \mathrm{Fo}\frac{\kappa_l}{\kappa_g}\right) \sin\left(u\frac{R}{R_d}\right), \tag{21}$$

$$T(R \geq R_d) = T_{g\infty} + \frac{2k_l}{\pi k_g}\sqrt{\frac{\kappa_g}{\kappa_l}}\left(T_{g\infty} - T_{d0}\right)\frac{R_d}{R}\int_0^\infty \frac{\mathrm{d}u}{u}\Phi(u)\exp\left(-u^2\mathrm{Fo}\frac{\kappa_l}{\kappa_g}\right)$$

$$\times \left\{ \cos(\gamma_c u)\sin u + \sqrt{\frac{\kappa_g}{\kappa_l}}\sin\gamma_c \left[\frac{k_l}{k_g}(u\cos u - \sin u) + \sin u\right]\right\}, \tag{22}$$

where

$$\Phi(u) = \frac{(u\cos u - \sin u)}{u^2 \sin^2 u + \frac{\kappa_g}{\kappa_l}\left[\frac{k_l}{k_g}(u\cos u - \sin u) + \sin u\right]^2},$$

$\mathrm{Fo} = t\kappa_g/R_d^2$ (Fourier number),

$$\gamma_c = \sqrt{\frac{\kappa_l}{\kappa_g}}\left[\frac{R - R_d}{R}\right]u.$$

As expected, this solution predicts diffusion of heat from gas to droplets, if $T_{g\infty} > T_{d0}$. Initially, the heat flux from gas to droplets predicted by the equation

$$\dot{q} = -k_g \left.\frac{\partial T}{\partial R}\right|_{R=R_d+0} \tag{23}$$

is infinitely large. It approaches zero at $t \to \infty$.

An alternative solution of Equation (17), taking into account the absorption of thermal radiation inside the droplet, finite R_g and the dependence of T_{g0} on R was suggested in [47, 48]:

$$T(R,t) = T_{g0} + \frac{1}{R}\sum_{n=1}^{\infty}\left[\exp\left(-\lambda_n^2 t\right)\frac{1}{||v_n||^2}\int_0^{R_d}(-(T_{g0} - T_{d0}(R))Rv_n(R)c_l\rho_l \mathrm{d}R\right.$$

$$\left. + \int_0^t \exp\left(-\lambda_n^2(t-\tau)\right)p_n(\tau)\mathrm{d}\tau\right]v_n(R), \tag{24}$$

where

$$v_n(R) = \begin{cases} \frac{\sin(\lambda_n a_d R)}{\sin(\lambda_n a_d R_d)} & \text{when} \quad R < R_d \\ \frac{\sin(\lambda_n a_g (R-R_g))}{\sin(\lambda_n a_g (R_d-R_g))} & \text{when} \quad R_d \le R \le R_g, \end{cases} \tag{25}$$

$$||v_n||^2 = \frac{c_l \rho_l R_d}{2\sin^2(\lambda_n a_d R_d)} + \frac{c_{pg}\rho_g(R_g - R_d)}{2\sin^2(\lambda_n a_g(R_d - R_g))} - \frac{k_l - k_g}{2R_d\lambda_n^2},$$

$$p_n(t) = \frac{c_l\rho_l}{||v_n||^2}\int_0^{R_b} RP(t,R)v_n(R)\mathrm{d}R.$$

A countable set of positive eigenvalues λ_n is found from the solution of the equation:

$$\sqrt{k_l c_l \rho_l}\cot(\lambda a_b R_b) - \sqrt{k_g c_{pg}\rho_g}\cot(\lambda a_g(R_d - R_g)) = \frac{k_l - k_g}{R_d \lambda}. \tag{26}$$

These are arranged in ascending order $0 < \lambda_1 < \lambda_2 < \ldots$. $a_d = \sqrt{\frac{c_l\rho_l}{k_l}}$, $a_g = \sqrt{\frac{c_{pg}\rho_g}{k_g}}$.

The practical implementation of any of these solutions into computational fluid dynamics codes remains unclear, since these codes are based on the separation of the processes in the liquid and gas phases. The required separation between the solutions could be achieved based on the comparison between the thermal diffusivities of gas and liquid. Following [10], let us consider typical values of parameters for diesel fuel spray droplets and assume that these droplets have initial temperature 300 K and are injected into a gas at temperature 800 K and pressure 30 atm [49]:

$$\rho_l = 600 \quad \mathrm{kg/m^3}, \quad k_l = 0.145 \quad \mathrm{W/(mK)}, \quad c_l = 2830 \quad \mathrm{J/(kgK)},$$

$$\rho_g = 23.8 \quad \mathrm{kg/m^3}, \quad k_g = 0.061 \quad \mathrm{W/(mK)}, \quad c_{pg} = 1120 \quad \mathrm{J/(kgK)}.$$

For these parameter values we obtain $\kappa_l = 8.53 \times 10^{-8}$ m^2/s and $\kappa_g = 2.28 \times 10^{-6}$ m^2/s. This allows us to assume:

$$\kappa_l \ll \kappa_g. \tag{27}$$

This condition tells us that gas responds much more quickly to changes in the thermal environment than liquid. As a zeroth approximation we can ignore the changes in liquid temperature altogether, and assume that the droplet surface temperature remains constant in time. This immediately allows us to decouple the solution of Equation (17) from the trivial solution of this equation for the liquid phase ($T(R \le R_d)$ =const). The former solution can be presented as:

$$T(R > R_d) = T_{g\infty} + \frac{R_d}{R}(T_{d0} - T_{g\infty})\left[1 - \mathrm{erf}\left(\frac{R - R_d}{2\sqrt{\kappa_g t}}\right)\right], \tag{28}$$

where

$$\mathrm{erf}(x) = \frac{2}{\sqrt{\pi}}\int_0^x \exp(-t^2)\mathrm{d}t.$$

In the limit $R = R_d$, Equation (28) gives $T = T_{d0}$. In the limit $t \to 0$, but $R \neq R_d$, this equation gives $T = T_{g\infty}$. From (28) we obtain the following equation for the heat flux from gas to droplets [50]:

$$\dot{q} = \frac{k_g(T_{d0} - T_{g\infty})}{R_d}\left(1 + \frac{R_d}{\sqrt{\pi\kappa_g t}}\right). \tag{29}$$

(cf. [51, 52]). For $t \gg t_d \equiv R_d^2/(\pi\kappa_g)$, Equation (29) can be further simplified to:

$$|\dot{q}| = h(T_{g\infty} - T_{d0}), \tag{30}$$

where $h = k_g/R_d$ is the convection heat transfer coefficient. Remembering the definition of the Nusselt number Nu$= 2R_d h/k_g$, this value of h corresponds to Nu$= 2$.

The equation similar to (30) could be obtained from Solutions (21) and (24) if we replace h in it with $\tilde{h} = \frac{k_g}{R_d}\chi_t$, where in the case of Solution (21) the correction factor χ_t is defined as [48]:

$$\chi_t = \frac{-\frac{k_l}{k_g}\int_0^\infty (u\cos u - \sin u)\Phi(u)\exp(-u^2\mathrm{Fo}\frac{\kappa_l}{\kappa_g})\mathrm{d}u}{\int_0^\infty \sin u\Phi(u)\exp(-u^2\mathrm{Fo}\frac{\kappa_l}{\kappa_g})\mathrm{d}u}, \tag{31}$$

$\Phi(u)$ is the same as in Equations (21) and (22).

In the case of Solution (24), χ_t is defined as [48]:

$$\chi_t = 1 + \frac{\Delta T_c + \Delta T_r}{T_{g0} - T_s}, \tag{32}$$

where

$$\Delta T_c = (T_{g0} - T_{d0})k_l a_d a_g \sum_{n=1}^{\infty} A_n \exp(-\lambda_n^2 t), \tag{33}$$

$$\Delta T_r = -\frac{3a\sigma R_d^{b-1}\theta_R^4 a_g}{a_d}\sum_{n=1}^{\infty}\frac{A_n}{\lambda_n^2}\left(1 - \exp(-\lambda_n^2 t\right), \tag{34}$$

$$A_n = \frac{1}{||v_n||^2}\left[R_d\cot(\lambda_n a_d R_d) - \frac{1}{\lambda_n a_d}\right]\cot(\lambda_n a_g(R_d - R_g)).$$

When deriving Equation (34) the radiation term was presented in an explicit form [48]:

$$P(R) = 3a\sigma R_d^{b-1}\theta_R^4/c_l\rho_l\,, \tag{35}$$

where θ_R is the radiation temperature,

$$a = a_0 + a_1\theta_R/10^3 + a_2\left(\theta_R/10^3\right)^2,$$

$$b = b_0 + b_1\theta_R/10^3 + b_2\left(\theta_R/10^3\right)^2.$$

θ_R can be assumed equal to the external temperature in the case of an optically thin gas in the whole domain.

It was shown in [47, 48] that for $R_g \geq 25R_d$, the values of χ_t defined by Eqs. (31) and (32) coincide for $0.1 \leq \mathrm{Fo} \leq 100$. In both cases they decrease from about 2.8 at Fo $= 0.1$ to about 0.6 at Fo $= 100$. It remains unclear how the multiscale nature of the processes in gas and liquid phases can be taken into account or described by the above mentioned solutions for temperature and heat fluxes.

4. Droplet evaporation

Most of the currently used 'hydrodynamic' models for droplet evaporation implicitly assume that the rate of detachment of molecules of fuel is such that the concentration of fuel vapour in the vicinity of the droplet surface is maintained at saturation level [10]. The applicability of this assumption to the problem of modelling droplet evaporation into a high pressure (above 1 atm) gas is not at first evident. Rigorous analysis of the dynamics of vapour molecules in the immediate vicinity of the droplets should be based on the analysis of the distribution function of background gas (air) $f_a \equiv f_a(\mathbf{r}, t, \mathbf{v})$ and vapour $f_v \equiv f_v(\mathbf{r}, t, \mathbf{v})$ using the Boltzmann kinetic equations. The general form of these equations, taking into account the interaction of vapour molecules with the background gas (air) can be presented as [53]:

$$\left.\begin{array}{l} \frac{\partial f_a}{\partial t} + \mathbf{v}_a \frac{\partial f_a}{\partial \mathbf{r}} = J_{aa} + J_{av} \\ \frac{\partial f_v}{\partial t} + \mathbf{v}_v \frac{\partial f_v}{\partial \mathbf{r}} = J_{va} + J_{vv} \end{array}\right\}, \tag{36}$$

where $J_{\alpha\beta}$ ($\alpha = a, v$; $\beta = a, v$) are collision integrals defined as:

$$J_{\alpha\beta} = \frac{\sigma_{\alpha\beta}^2}{2} \int_{-\infty}^{+\infty} \mathrm{d}\mathbf{v_1} \int_0^{\pi} \sin\theta \mathrm{d}\theta \int_0^{2\pi} \mathrm{d}\phi \left(f_\alpha^{'} f_{\beta 1}^{'} - f_\alpha f_{\beta 1} \right) \left| \mathbf{v}_\alpha - \mathbf{v}_{\beta 1} \right|, \tag{37}$$

$\sigma_{\alpha\beta} = (\sigma_\alpha + \sigma_\beta)/2$, σ_α and σ_β are the corresponding diameters of molecules, θ and ϕ are angular coordinates of molecules β relative to molecules α after the collision, superscript $'$ indicates the velocities and the distribution functions after collisions. Subscript $_1$ indicates that molecules of type β collide with molecules of type α and as a result of this interaction the function f_α is modified. When deriving (36) and (37) it was assumed that molecules are rigid elastic spheres and body forces acting on them are negligible. These equations are solved subject to appropriate boundary and initial conditions, including the conditions at the droplet surface [53].

In the case of evaporation into a high pressure gas, Equations (36) are solved in a relatively thin layer in the immediate vicinity of the droplet surface (Knudsen layer). In the case of fuel droplet evaporation into a diesel engine combustion chamber with air pressure at about 30 atm, the thickness of this layer is less than 10^{-6} m [53]. This is a new scale of the problem. The solution of Equations (36) within this layer is performed at even smaller scales comparable with the diameter of individual molecules (about 10^{-9} m). This is the smallest scale of the problem.

The solutions of Equations (36) are matched with the solutions of the hydrodynamic equations at the outer boundary of the Knudsen layer. Since Equation (36) predicts the finite rate of detachment of vapour molecules from the surface of the droplet (in contrast to the infinite rate of this detachment predicted by the hydrodynamic models), the evaporation rate predicted by hydrodynamic models is always larger when compared with the rate predicted by kinetic models. Although the kinetic processes take place in a thin Knudsen layer, a noticeable reduction in the density of vapour is observed between the surface of the droplet and the outer boundary of this layer. In this case, the flux density of vapour, as predicted by the kinetic model, is controlled by the vapour density at the outer boundary of the Knudsen layer. The latter is less than the vapour density at the droplet surface, which controls the vapour flux predicted by the hydrodynamic model. This is consistent with the above mentioned conclusion regarding the evaporation rates predicted by kinetic and hydrodynamic models.

The solution of Equation (36) can be based on numerical methods only [53]. Asymptotic analysis of these equations is possible in some limiting cases. The analysis in [54] is based on the method previously reported in [55] and a number of simplified assumptions including ignoring the contribution of background gas in the Knudsen layer. Both rigorous numerical and asymptotic analyses led to essentially the same conclusions; even in the case of fuel droplet evaporation into a very high pressure gas (30 atm; the situation typical for diesel engines), the kinetic effects can lead to a noticeable reduction in droplet evaporation time (about 5-10 % for droplets with radii 5-10 μm).

5. Coupled solutions

Although the processes considered so far have scales which differ by more than 7 orders of magnitude, this does not bring any additional complexity to their analysis. This is related to the fact that the rate of change of these scales with time is generally small, and the hierarchy of scales remains unchanged. This, however, is not always the case, especially when chemical reactions (autoignition) become involved. This leads us to the problem of dynamic decomposition of the system of ordinary differential equations describing the processes in individual computational cells [56]. A summary of the main results of the latter paper was reported in [11]. This approach is based on comparing the values of the right hand sides of the equations, leading to the separation of the equations into 'fast' and 'slow' variables. The hierarchy of the decomposition is allowed to vary with time. Equations for fast variables are solved by a stiff ODE system solver with the slow variables taken at the beginning of the time step. The solution of the equations for the slow variables is presented in a simplified form, assuming linearised variation of these variables for the known time evolution of the fast variables. This can be considered as the first order approximation for the fast manifold. This technique is applied to analyse the explosion of a polydisperse spray of diesel fuel. Clear advantages were demonstrated from the point of view of accuracy and CPU efficiency when compared with the conventional approach widely used in CFD codes. The difference between the solution of the full system of equations and the solution of the decomposed system of equations was shown to be negligibly small for practical applications.

The model described in [56] has been further developed in [57, 58]. In [57], the description of the autoignition process based on the one-step Arrhenius reaction was replaced by its description in terms of the so called Shell model [59]-[62]. The main conclusions of this paper were essentially the same as those of [56]. In [58], a new approach to dynamic decomposition of the systems of ODEs to fast and slow subsystems is suggested. The suggested approach is based on the comparative analysis of the magnitudes of the eigenvalues of the matrix $\mathbf{JJ}^*$, where $\mathbf{J}$ is the local Jacobi matrix of the system under consideration. The eigenvectors of this matrix are used for construction of the transformation matrix $\mathbf{Q}$. The hierarchy of the decomposition is allowed to vary with time as in the original method described in [56]. As in the original approach [56], equations for fast variables were solved by a stiff ODE system solver with the slow variables taken at the beginning of the time step. The solution of equations for slow variables was presented in a simplified form, assuming linear variations of these variables during the time evolution of the fast variables. The results showed advantages of the new approach when compared with the one discussed in [56], both from the point of view of accuracy and CPU efficiency.

6. Conclusions

Modelling of sprays involves modelling of a number of processes which can take place simultaneously at scales which differ by more than 7 orders of magnitude. A summary of the results of the modelling of several such processes is presented, with particular emphasis on sprays in diesel engines used in passenger cars. These include spray penetration, droplet heating and droplet evaporation. The modelling of spray penetration (scales of the order of 3×10^{-2} m) is based on equations of mass and momentum conservation of the mixture of droplets and entrained air. It is shown that the well known correlation that the penetration length is proportional to the root square of time, can be generalised to include the corrections to this relationship. At the initial stage of spray penetration the analysis of the process can be based on calculating the trajectories of individual droplets, taking entrainment into account.

The modelling of droplet heating involves scales comparable with a typical droplet diameter 10^{-5} m. In the general case, this modelling involves the solution of the heat conduction equation inside droplets and in surrounding gas. Remembering that for typical diesel engine parameters the gas diffusivity is much greater than liquid diffusivity, this solution can be simplified and based on the assumption that the liquid surface temperature is constant during the time step used in computational fluid dynamics (CFD) codes. The modelling of droplet evaporation, even in a high pressure background gas, requires the solution of the Boltzmann kinetic equation inside the Knudsen layer (thickness about 10^{-7} m) with appropriate boundary conditions at the droplet surface and the outer boundary of the Knudsen layer. Recent numerical and asymptotic studies of this problem are reviewed. These studies involve scales comparable with the diameters of individual molecules (about 10^{-9} m).

It is confirmed that the dynamic decomposition techniques for solving the stiff system of ordinary differential equations, reported earlier, can be an effective tool in spray modelling applications for CFD codes. Further developments of this technique, based on the comparative analysis of the magnitudes of the eigenvalues of the matrix $\mathbf{JJ}^*$, where $\mathbf{J}$ is the local Jacobi matrix of the system under consideration, are discussed.

Acknowledgments
The author is grateful to European Regional Development Fund Franco-British INTERREG IIIa (Project Ref 162/025/247) for partial financial support of the work on this review, his colleagues W Abdelghaffar, V Bykov, C Crua, G Feng, I Goldfarb, V Gol'dshtein, M Heikal, D Kennaird, P Krutitskii, A Kryukov, V Levashov , S Martynov, E Sazhina and I Shishkova in collaboration with whom some results presented in this review were originally obtained.

References

[1] Sirignano W A 1999 *Fluid Dynamics and Transport of Droplets and Sprays* (Cambridge University Press)
[2] Michaelides E E 2003 *ASME J Fluid Engineering* **125** 209
[3] Fritsching U 2004 *Spray Simulation* (Cambridge University Press).
[4] Chiu H H 2000 *Prog Energy Combustion Science* **26** 381
[5] Bellan J 2000 *Prog Energy Combustion Science* **26** 329
[6] Givler S D and Abraham J 1996 *Prog Energy Combustion Science* **22** 1
[7] Babinsky E and Sojka P E 2002 *Prog Energy Combustion Science* **28** 303
[8] Loth E 2000 *Prog Energy Combustion Science* **26** 161
[9] Orme M 1997 *Prog Energy Combustion Science* **23** 65
[10] Sazhin S S 2006 *Prog Energy Combustion Science* **32** 162
[11] Sazhin S S 2005 *J Phys.: Conf. Series* **22** 174
[12] Dent J C 1971 *SAE report 710571*
[13] Hay N and Jones P L 1972 *SAE report 720776*
[14] Sendyka B 1998 *SAE report 980496*
[15] Borman G L and Ragland K W 1998 *Combustion Engineering* (McGraw-Hill)
[16] Sazhin S S, Feng G and Heikal M R 2001 *Fuel* **80** 2171
[17] Ghosh S and Hunt J C R 1994 *Proceedings Royal Society London* **A444** 105
[18] Sazhin S S, Crua C, Kennaird D and Heikal M R 2003 *Fuel* **82** 875
[19] Sazhin S, Crua C, Hwang J-S, No S-Y and Heikal M 2005 *Proc. Estonian Acad. of Sci. Engineering* **11** 154
[20] Pozorski J, Sazhin S S, Wacławczyk M, Crua C, Kennaird D and Heikal M R 2002 *Flow, Turbulence and Combustion* **68** 153
[21] Siebers D L 1999 *SAE report 1999-01-0528*
[22] Cossali G E 2001 *J Fluid Mech.* **439** 353
[23] Iyer V A, Abraham J and Magi V 2002 *Int. J. Heat and Mass Transfer* **45** 519
[24] Sazhin S S, Kaplanski F, Feng G, Heikal M R and Bowen P J 2001 *Fuel* **80** 1871
[25] Lefebvre A H 1989 *Atomization and Sprays* (Taylor & Francis)
[26] Wallis G B 1969 *One-Dimensional Two-Phase Flows* (McGraw-Hill)
[27] Morsi S A and Alexander A J 1972 *J Fluid Mechanics* **55** 193
[28] Yanenko N N, Soloukhin R I, Papyrin A N and Fomin V M 1980 *Supersonic Two-Phase Flows with Non-equilibrium Particle Velocities* (Novosibirsk: Nauka Publishing House, in Russian)
[29] Panton R L 1996 *Incompressible flow* (John Wiley & Sons)
[30] Feng Z-G and Michaelides E E 2001 *Int. J Heat and Mass Transfer* **44** 4445
[31] Douglas J F, Gasiorek J M, Swaffield J A and Jack L B 2005 *Fluid Mechanics* (Longman)

[32] Reitz R D and Diwakar R 1986 *SAE report 860469*
[33] Reitz R D and Diwakar R 1987 *SAE report 870598*
[34] Reitz R D 1987 *Atomisation and Spray Technology* **3** 309
[35] O'Rourke P J and Amsden A A 1987 *SAE report 872089*
[36] Gorokhovski M 2001 *Atomization and Sprays* **11** 505
[37] Gorokhovski M and Saveliev V L 2003 *Physics of Fluids* **15** 184
[38] Tanner F X 2004 *Atomization and Sprays* **14** 211
[39] Martynov S, Mason D and Heikal M 2006 *International Journal of Engine Research* (in press)
[40] Martynov S, Mason D, Heikal M, Sazhin S and Gorokhovski 2006 In *Proceedings of 13th International Heat Transfer Conference (13–18 August 2006)* (Sydney, Australia)
[41] Abramzon B and Sazhin S S 2005 *Int. J. Heat Mass Transfer* **48** 1868
[42] Abramzon B and Sazhin S S 2006 *Fuel* **85** 32
[43] Sazhin S S, Abdelghaffar W A, Krutitskii P A, Sazhina E M and Heikal M R 2005 *Int. J. Heat Mass Transfer* **48** 4215
[44] Sazhin S S, Abdelghaffar W A, Sazhina E M and Heikal M R 2005 *Int. J. Thermal Science* **44** 610
[45] Sazhin S S, Kristyadi T, Abdelghaffar W A and Heikal M R 2006 *Fuel* **85** 1613
[46] Cooper F 1977 *Int. J. Heat and Mass Transfer* **20** 991
[47] Sazhin S S, Abdelghaffar W A, Martynov S B, Sazhina E M, Heikal M R and Krutitskii P A 2005 In *Proceedings of '5rd International Symposium on Multiphase Flow, Heat Mass Transfer and Energy Conversion', Xi'an, China, 3-6 July 2005* (CD-ROM)
[48] Sazhin S S, Krutitskii P A, Martynov S B, Mason D, Heikal M R and Sazhina E M 2005 In *Proceedings of HEFAT2005 (4th International Conference on Heat Transfer, Fluid Mechanics and Thermodynamics), Cairo, Egypt 2005* Paper number: SS1.
[49] Sazhin S S, Feng G, Heikal M R, Goldfarb I, Goldshtein V and Kuzmenko G 2001 *Combustion and Flame* **124** 684
[50] Todes O M 1968 In *Problems of evaporation, combustion and gas dynamics in disperse systems* (Editor V A Fedoseev) Proceedings of the Sixths Conference on Evaporation, Combustion and Gas Dynamics in Disperse Systems (October 1966), Odessa: Odessa University Publishing House; 1968, pp. 151-9 (in Russian).
[51] Feng Z-G and Michaelides E E 1996 *ASME J Fluids Engineering* **118** 96
[52] Sazhin S S, Goldshtein V and Heikal M R 2001 *ASME J. Heat Transfer* **123** 63
[53] Shishkova I N and Sazhin S S 2006 *J Computational Physics* (in press).
[54] Kryukov A P, Levashov V Yu and Sazhin S S 2004 *Int J Heat Mass Transfer* **47** 2541
[55] Muratova T M and Labuntsov D A 1969 *Thermal, Physics of High temperatures* **7** 959 (in Russian)
[56] Bykov V, Goldfarb I, Gol'dshtein V, Sazhin S and Sazhina E 2006 *Computers and Fluids* (in press).
[57] Sazhina E M, Bykov V, Goldfarb I, Goldshtein V, Sazhin S S and Heikal M R 2005 In *Proceedings of HEFAT2005 (4th International Conference on Heat Transfer, Fluid Mechanics and Thermodynamics), Cairo, Egypt 2005* Paper number: SE2.
[58] Bykov V, Goldfarb I, Gol'dshtein V, Sazhin S and Sazhina E 2006 In *Model Reduction and Coarse-Graining Approaches for Multiscale Phenomena* ed A N Gorban, N Kazantzis, Y G Kevrekidis, H C Ottinger and C Theodoropoulos (Berlin–Heidelberg–New York: Springer) (in press).
[59] Halstead M, Kirsh L and Quinn C P 1970 *Combustion and Flame* **30** 45
[60] Sazhina E M, Sazhin S S, Heikal M R and Marooney C 1999 *Fuel* **78** 389
[61] Sazhina E M, Sazhin S S, Heikal M R, Babushok V I and Johns R 2000 *Comb. Sci. Tech.* **160** 317
[62] Hamosfakidis V and Reitz R D 2003 *Combustion and Flame* **132** 433

Institute of Physics Publishing
doi:10.1088/1742-6596/55/1/019
Journal of Physics: Conference Series **55** (2006) 203–213
International Workshop on Multi-Rate Processes and Hysteresis

The Richardson extrapolation technique for quasilinear parabolic singularly perturbed convection-diffusion equations

G I Shishkin and L P Shishkina

Institute of Mathematics and Mechanics, Russian Academy of Science, S.Kovalevskaya str., 620219 Yekaterinburg, Russia

E-mail: shishkin@imm.uran.ru, Lida@convex.ru

Abstract. The Dirichlet problem is considered for a quasilinear singularly perturbed parabolic convection-diffusion equation on a rectangular domain. For this problem classical finite difference (nonlinear) schemes on *piecewise uniform* meshes condensing in the boundary layer converge ε-uniformly at a rate that is at best first-order. Using a Richardson extrapolation technique, we construct an improved (nonlinear) scheme that is *ε-uniformly* convergent at the rate $\mathcal{O}((N^{-1}\ln N)^2 + N_0^{-2})$, where N and N_0 define the number of nodes in the spatial and time meshes, respectively.

This nonlinear scheme is used in the construction of a linearized scheme where at each time level the nonlinear term is evaluated using the computed solution from the previous time level. Furthermore, using the linearized and nonlinear improved schemes, we construct a linearized improved Richardson scheme that converges ε-uniformly at the rate $\mathcal{O}((N^{-1}\ln N)^2 + N_0^{-q})$, where $q \geq 2$.

1. Introduction

At present, for a wide class of singularly perturbed boundary value problems, it is known that special finite difference methods allow us to obtain ε-uniformly convergent (in the maximum norm) computed solutions. In the case of boundary value problems for parabolic convection-diffusion equations, the order of ε-uniform convergence with respect to the spatial and time variables does not exceed 1 (see, e.g., [1, 4, 8, 13, 14, 17] and also their bibliographies). Thus for parabolic convection-diffusion problems it is important to construct special schemes whose ε-uniform convergence rate is greater than 1 with respect to both variables.

To increase the accuracy of computed approximations of the solutions of classical boundary value problems, defect correction and Richardson extrapolation (see, e.g., [2, 11, 12] and their bibliographies) are used. The same methods are used to improve the ε-uniform rates of convergence of computed solutions for linear singularly perturbed problems (see, e.g., [5, 6, 7, 16, 20]). Recently, using Richardson extrapolation, ε-uniformly convergent finite difference schemes with improved accuracy were constructed also for quasilinear singularly perturbed reaction-diffusion parabolic [21] and elliptic [19] problems.

In [9], Richardson extrapolation is considered for a convection-diffusion problem on the rectangle. Note that the technique used in [9] (which differs from the technique of the present paper) cannot be used to justify the improvement of convergence (and in particular ε-uniform convergence) for Richardson extrapolation when the parameter ε is not small compared to the step-size in the x variable. In [9], references can be found to papers where the Richardson method is studied for ordinary differential equations.

In the present paper, we apply Richardson extrapolation to an initial-boundary value problem for a quasilinear singularly perturbed parabolic convection-diffusion equation on a rectangular domain (the problem formulation is given in Section 2). It is known that the use of meshes condensing in the boundary layer proves to be necessary (see, e.g., [3, 13, 18]) for the construction of ε-uniformly convergent methods for nonlinear (and in particular quasilinear) singularly perturbed problems, just as in the case of linear problems when parabolic layers are present. Therefore, it is of interest to develop special schemes of increased ε-uniform accuracy on nonuniform meshes.

In this connection, to construct ε-uniformly convergent schemes for the quasilinear problem under investigation we apply an approach based on meshes condensing in the boundary layer (for a description of this approach, see, e.g., [1, 13, 17] and their bibliographies); when constructing the schemes, monotone classical approximations [15] of the problem are used. This approach leads to a basic special scheme on a piecewise uniform rectangular mesh that is convergent ε-uniformly at the rate $\mathcal{O}(N^{-1}\ln N + N_0^{-1})$, where $N+1$ and N_0+1 are the number of nodes in the spatial and temporal meshes respectively (see Section 3). This basic nonlinear scheme is used for the construction of a nonlinear scheme with improved accuracy by means of Richardson extrapolation (see Section 4) and a linearized basic scheme where at each time level the nonlinear term is evaluated using the computed function from the previous time level (see Section 5). The improved nonlinear Richardson scheme converges ε-uniformly at the rate $\mathcal{O}((N^{-1}\ln N)^2 + N_0^{-2})$. Finally, these improved nonlinear and linearized schemes are used to construct a linearized improved Richardson scheme that converges ε-uniformly at the rate $\mathcal{O}((N^{-1}\ln N)^2 + N_0^{-q})$, $q \geq 2$ (see Section 6), thereby preserving the accuracy of ε-uniformly convergent nonlinear schemes. A preliminary version of these results appeared in [22].

2. Problem formulation and aim of research

Set

$$\overline{G} = G \bigcup S, \quad G = D \times (0, T], \tag{1}$$

where $D = (0, d)$. On $\overline{G}$ we consider the Dirichlet problem for a quasilinear singularly perturbed parabolic convection-diffusion equation:

$$\begin{aligned} L\,(u(x,t)) &\equiv L^2 u(x,t) - f(x,t,\,u(x,t)) = 0, \quad (x,t) \in G, \\ u(x,t) &= \varphi(x,t), \quad (x,t) \in S. \end{aligned} \tag{2}$$

Here

$$L^2 \equiv \varepsilon\, a(x,t)\frac{\partial^2}{\partial x^2} + b(x,t)\frac{\partial}{\partial x} - c(x,t) - p(x,t)\frac{\partial}{\partial t}, \quad (x,t) \in G,$$

and the functions

$$a(x,t), \quad b(x,t), \quad c(x,t), \quad p(x,t), \quad f(x,t,\,u) \;\; and \;\; \varphi(x,t)$$

are assumed to be sufficiently smooth on $\overline{G}$, $\overline{G} \times R$ and S respectively; moreover [1]

$$\begin{aligned} &a_0 \leq a(x,t) \leq a^0, \quad b_0 \leq b(x,t) \leq b^0, \quad |c(x,t)| \leq c^0, \\ &p_0 \leq p(x,t) \leq p^0, \quad (x,t) \in \overline{G}; \end{aligned} \tag{3}$$

[1] Here and below, M, M_i (or m) denote sufficiently large (small) positive constants independent of the parameter ε and of the discretization parameters.

$$|f(x,t,\,u)| \le M, \quad c_1 \le c(x,t) + \frac{\partial}{\partial u} f(x,t,\,u) \le c^1, \quad (x,t,\,u) \in \overline{G} \times R;$$

$$|\varphi(x,t)| \le M, \quad x \in S; \qquad a_0,\, b_0,\, c_1,\, p_0 > 0;$$

and the parameter ε takes arbitrary values in the half-interval $(0,1]$.

Assume that the data of problem (2), (1) on the set of corner points $S^* = S_0 \bigcap \overline{S}^L$ satisfy compatibility conditions that ensure the required smoothness of the solution on $\overline{G}$ (see, e.g., [10]). Here $S = S_0 \bigcup S^L$, where S_0 and S^L are the lower and lateral parts of the boundary; $S_0 = \overline{S}_0$.

For small values of ε, a regular boundary layer appears in a neighbourhood of the set

$$S_1^L = \{(x,t): \; x = 0, \; 0 < t \le T\}.$$

Here S_1^L and S_2^L are the left and right parts of the lateral boundary; $S^L = S_1^L \bigcup S_2^L$.

Our aim is to construct a finite difference scheme that is ε-uniformly convergent, with order of accuracy greater than 1 with respect to both variables x and t, for the boundary value problem (2), (1); this will be done via Richardson extrapolation.

3. Basic finite difference scheme

Let us give an ε-uniformly convergent finite difference scheme based on a classical approximation of problem (2), (1).

On the set $\overline{G}$, we introduce the rectangular mesh

$$\overline{G}_h = \overline{\omega} \times \overline{\omega}_0, \tag{4}$$

where $\overline{\omega}$ and $\overline{\omega}_0$ are, in general, arbitrary nonuniform meshes on the intervals $[0,d]$ and $[0,T]$ respectively. Let

$$h^i = x^{i+1} - x^i \;\; for \;\; x^i,\, x^{i+1} \in \overline{\omega}, \quad h = \max_i h^i,$$

$$h_t^k = t^{k+1} - t^k \;\; for \;\; t^k,\, t^{k+1} \in \overline{\omega}_0, \quad h_t = \max_k h_t^k$$

Assume that the conditions $h \le M\, N^{-1}$ and $h_t \le M\, N_0^{-1}$ are satisfied, where $N+1$ and N_0+1 are the number of nodes in the meshes $\overline{\omega}$ and $\overline{\omega}_0$ respectively.

We approximate problem (2), (1) by the finite difference scheme [15]

$$\begin{aligned} \Lambda\,(z(x,t)) &\equiv \Lambda^2\, z(x,t) - f(x,t,\, z(x,t)) = 0, \quad (x,t) \in G_h, \\ z(x,t) &= \varphi(x,t), \quad (x,t) \in S_h. \end{aligned} \tag{5}$$

Here,

$$G_h = G \bigcap \overline{G}_h, \quad S_h = S \bigcap \overline{G},$$

$$\Lambda^2 \equiv \varepsilon\, a(x,t)\, \delta_{\overline{x}\widehat{x}} + b(x,t)\, \delta_x - c(x,t) - p(x,t)\, \delta_{\overline{t}}, \quad (x,t) \in G_h,$$

$\delta_{\overline{x}\widehat{x}}\, z(x,t)$ is the second-order central difference derivative on the nonuniform mesh, i.e.,

$$\delta_{\overline{x}\widehat{x}}\, z(x,t) = 2(h^i + h^{i-1})^{-1}[\delta_x\, z(x,t) - \delta_{\overline{x}}\, z(x,t)], \quad (x,t) = (x^i,t) \in G_h,$$

while $\delta_x\, z(x,t)$ and $\delta_{\overline{x}}\, z(x,t)$ are the first-order forward and backward difference derivatives.

The scheme (5), (4) is ε-uniformly monotone [15]. The following version of the comparison theorem holds.

Theorem 1 *Let the functions* $z^1(x,t)$ *and* $z^2(x,t)$ *satisfy the conditions*

$$\Lambda\,(z^1(x,t)) < \Lambda\,(z^2(x,t)), \quad (x,t) \in G_h, \quad z^1(x,t) > z^2(x,t), \quad (x,t) \in S_h.$$

Then $z^1(x,t) > z^2(x,t)$ *for all* $(x,t) \in \overline{G}_h$.

In the case where the mesh

$$\overline{G}_h = \overline{\omega} \times \overline{\omega}_0 \tag{6}$$

is uniform in both variables, by using the maximum principle we can show that

$$|u(x,t) - z(x,t)| \le M\left[\left(\varepsilon + N^{-1}\right)^{-1} N^{-1} + N_0^{-1}\right], \quad (x,t) \in \overline{G}_h. \tag{7}$$

Let us construct the basic ε-uniformly convergent scheme (see, e.g., [13, 17]). On the set $\overline{G}$, we introduce the mesh

$$\overline{G}_h = \overline{\omega}^* \times \overline{\omega}_0, \tag{8a}$$

where $\overline{\omega}_0 = \overline{\omega}_{0(6)}$ and $\overline{\omega}^*$ is a special *piecewise uniform* mesh constructed as follows. The interval $[0,d]$ is divided into two parts $[0,\sigma]$ and $[\sigma,d]$, while the step size in each part is constant and equal to

$$h^{(1)} = 2\,\sigma\,N^{-1} \quad \textit{and} \quad h^{(2)} = 2(d-\sigma)N^{-1},$$

respectively. The parameter σ is defined by

$$\sigma = \sigma(\varepsilon,\, N,\, l) = \min\left[2^{-1}\,d,\; l\,m^{-1}\,\varepsilon\,\ln N\right], \tag{8b}$$

where m is an arbitrary number in the interval $(0, m_0)$,

$$m_0 = \min_{\overline{G}}[a^{-1}(x,t)\,b(x,t)].$$

Here,

$$l = 1; \tag{8c}$$

for later meshes, this parameter will be chosen differently. The construction of the meshes $\overline{\omega}^*$ and $\overline{G}_h = \overline{G}_h(l=1)$ is now complete.

For solutions of the difference scheme (5), (8), we obtain the estimate

$$|u(x,t) - z(x,t)| \le M\left[N^{-1}\min\left[\varepsilon^{-1},\, \ln N\right] + N_0^{-1}\right], \quad (x,t) \in \overline{G}_h, \tag{9}$$

and also the ε-uniform estimate

$$|u(x,t) - z(x,t)| \le M\left[N^{-1}\ln N + N_0^{-1}\right], \quad (x,t) \in \overline{G}_h. \tag{10}$$

Theorem 2 *Let the solution* $u(x,t)$ *of problem* (2), (1) *satisfy the estimates in Theorem* 6, *where* $K=4$. *Then the difference scheme* (5), (6) *converges for fixed values of the parameter* ε, *while the scheme* (5), (8) *converges* ε*-uniformly. Their discrete solutions satisfy the respective estimates* (7), (9) *and* (10).

4. Richardson extrapolation for (2), (1)

Let us describe the Richardson extrapolation method used to increase the accuracy of computed solutions of the special basic scheme.

1. On the set $\overline{G}$, we construct the meshes

$$\overline{G}_h^i = \overline{\omega}^{*i} \times \overline{\omega}_0^i, \quad i = 1, 2, \tag{11a}$$

which are *uniform* in the t direction and *piecewise uniform* in the x direction. Here $\overline{G}_h^2$ is $\overline{G}_{h(8a)}$, where

$$\sigma = \sigma_{(8b)}(\varepsilon, N, l) \quad with \quad l = 2; \tag{11b}$$

$\overline{G}_h^1$ is a *coarsened* mesh. For the parameters σ^i that define piecewise uniform meshes $\overline{\omega}^{*i} = \overline{\omega}^{*i}(\sigma^i)$, we impose the condition $\sigma^1 = \sigma^2$, where $\sigma^2 = \sigma_{(11b)}$; that is, the intervals on which the meshes $\overline{\omega}^{*1}$ and $\overline{\omega}^{*2}$ have a constant step-size are identical. The step-sizes in the mesh $\overline{\omega}^{*1}$ on the intervals $[0, \sigma]$ and $[\sigma, d]$ are k times larger than the step-sizes in the mesh $\overline{\omega}^{*2}$, and the step-size in the mesh $\overline{\omega}_0^1$ (on the interval $[0, T]$) is k times larger than the step-size in the mesh $\overline{\omega}_0^2$; here $k^{-1} N + 1$ and $k^{-1} N_0 + 1$ are the number of nodes in the meshes $\overline{\omega}^{*1}$ and $\overline{\omega}_0^1$ respectively. Let

$$\overline{G}_h^0 = \overline{G}_h^1 \bigcap \overline{G}_h^2. \tag{11c}$$

We have $\overline{G}_h^0 = \overline{G}_h^1$ for integer $k \geq 2$, and $\overline{G}_h^0 \neq \overline{G}_h^1$ for noninteger k.

Let $z^i(x,t)$, $(x,t) \in \overline{G}_h^i$, for $i = 1, 2$, be the solutions of the difference schemes[2]

$$\Lambda_{(5)} \left(z^i(x,t) \right) = 0, \qquad (x,t) \in G_h^i, \tag{12a}$$

$$z^i(x,t) = \varphi(x,t), \quad (x,t) \in S_h^i, \quad i = 1, 2. \tag{12b}$$

We set

$$z^0(x,t) = \gamma\, z^1(x,t) + (1-\gamma)\, z^2(x,t), \quad (x,t) \in \overline{G}_h^0, \tag{12c}$$

where

$$\gamma = \gamma(k) = -(k-1)^{-1}.$$

The difference scheme (12), (11) is constructed on the basis of the nonlinear scheme (5), (8) using a Richardson extrapolation technique on two embedded meshes. We call this scheme the Richardson scheme (12), (11). We call the function $z_{(12)}^0(x,t)$, $(x,t) \in \overline{G}_h^0$, the solution of the Richardson scheme (12), (11); we call the functions $z_{(12)}^1(x,t)$, $(x,t) \in \overline{G}_h^1$ and $z_{(12)}^2(x,t)$, $(x,t) \in \overline{G}_h^2$ the components that generate the solution of the Richardson scheme (12), (11).

2. To justify the convergence of the Richardson scheme (12), (11), we apply a technique similar to one used in [7, 19, 21]. It is convenient to consider the expansions of the functions $z^i(x,t)$, $(x,t) \in \overline{G}_h^i$, $i = 1, 2$, with respect to the values N^{-1} and N_0^{-1}:

$$z^i(x,t) = u(x,t) + k^{(i-1)} \left[N^{-1} u_1(x,t) + N_0^{-1} u_0(x,t) \right] + v^i(x,t), \tag{13}$$

$$(x,t) \in \overline{G}_h^i, \quad i = 1, 2,$$

[2] Throughout the paper, the notation $L_{(j.k)}$ ($M_{(j.k)}$, $G_{h(j.k)}$) means that these operators (constants, grids) are introduced in formula $(j.k)$.

where $v^i(x,t)$ is the remainder term. Here the function $u_0(x,t)$ is the solution of the problem

$$\begin{aligned} L_{(14)}\, u_0(x,t) &\equiv \left\{ L^2_{(2)} - \frac{\partial}{\partial u} f(x,t,\,u(x,t)) \right\} u_0(x,t) \\ &= -2^{-1}\, T\, p(x,t)\, \frac{\partial^2}{\partial t^2}\, u(x,t), \quad (x,t) \in G, \\ u_0(x,t) &= 0, \quad (x,t) \in S. \end{aligned} \tag{14}$$

The function $u_1(x,t)$ can be written as the sum of functions

$$u_1(x,t) = u_{11}(x,t) + u_{12}(x,t) + u_{13}(x,t), \quad (x,t) \in \overline{G}. \tag{15}$$

Here the $u_{1j}(x,t)$, $(x,t) \in \overline{G}$, are the solutions of the problems

$$\begin{aligned} L_{(14)}\, u_{11}(x,t) &= -\sigma\, b(x,t)\, \frac{\partial^2}{\partial x^2} V(x,t), \quad (x,t) \in G, \\ u_{11}(x,t) &= 0, \quad (x,t) \in S; \end{aligned} \tag{16}$$

$$\begin{aligned} L_{(14)}\, u_{12}(x,t) &= -\sigma\, b(x,t)\, \frac{\partial^2}{\partial x^2} U(x,t), \quad (x,t) \in G, \\ u_{12}(x,t) &= 0, \quad (x,t) \in S; \end{aligned} \tag{17}$$

$$\begin{aligned} L_{(14)}\, u_{13}(x,t) &= \left\{ \begin{array}{cc} 0, & x < \sigma \\ -(d - 2\sigma)\, b(x,t) \dfrac{\partial^2}{\partial x^2}\, U(x,t), & x > \sigma \end{array} \right\}, \quad (x,t) \in G, \\ u_{13}(x,t) &= 0, \quad (x,t) \in S, \end{aligned} \tag{18}$$

with

$$U(x,t) = U_{(27)}(x,t), \quad V(x,t) = V_{(27)}(x,t), \quad (x,t) \in \overline{G}, \quad \sigma = \sigma_{(11)}.$$

The functions $u_0(x,t)$, $u_{11}(x,t)$, $u_{12}(x,t)$, $u_{13}(x,t)$ satisfy the estimates

$$\begin{aligned} &|u_0(x,t)|, \ |u_{12}(x,t)| \le M, \\ &|u_{11}(x,t)| \le M\, \ln N, \\ &|u_{13}(x,t)| \le M\, \sigma, \quad (x,t) \in \overline{G}. \end{aligned}$$

The components $u_0(x,t)$, $u_{11}(x,t)$, $u_{12}(x,t)$ are sufficiently smooth on $\overline{G}$, and the component $u_{13}(x,t)$ is sufficiently smooth on $\overline{G}$ in the sets $x \le \sigma$ and $x \ge \sigma$. Taking into account estimates for the components $u_1(x,t)$ and $u_0(x,t)$, we obtain the estimate for the function $v^i(x,t)$:

$$|v^i(x,t)| \le M\, \left[(N^{-1}\, \ln N)^2 + N_0^{-2} \right], \quad (x,t) \in \overline{G}^i_h, \quad i = 1, 2. \tag{19}$$

Then the estimate (19) yields

$$|u(x,t) - z^0(x,t)| \le M\, \left[(N^{-1}\, \ln N)^2 + N_0^{-2} \right], \quad (x,t) \in \overline{G}^0_h. \tag{20}$$

Theorem 3 *In the boundary value problem* (2), (1), *assume that* a, b, c, $p \in C^{6+\alpha}(\overline{G})$, $f \in C^{6+\alpha}(\overline{G} \times R)$, $\varphi \in C^{6+\alpha}(S)$, $\alpha > 0$, *that the conditions* (3) *and* (31) *are satisfied for* $l = 6$, *and that for the components in the representation* (27) *of solutions to the problem* (2), (1), *the a priori estimates* (33) *are satisfied for* $K = 6$. *Then the solution* $z^0(x,t)$ *of the Richardson scheme* (12), (11) *converges to the solution* $u(x,t)$ *of the boundary value problem at an* ε*-uniform rate of* $\mathcal{O}((N^{-1}\ln N)^2 + N_0^{-2})$: *the computed solution satisfies the estimate* (20).

5. Linearized basic scheme

On the mesh (4), we consider a difference scheme in which the nonlinear term of the differential equation is evaluated at each time level by using the computed solution at the previous time level. Thus to the boundary value problem (2), (1) we assign the linearized difference scheme (see [15])

$$\begin{aligned} \Lambda_{(21)}(z(x,\, t)) &\equiv \Lambda^2_{(5)} z(x,\, t) - f(x, t,\, \check{z}(x,\, t)) = 0, \quad (x,\, t) \in G_h, \\ z(x,\, t) &= \varphi(x,\, t), \quad (x,\, t) \in S_h. \end{aligned} \tag{21}$$

Here

$$\check{z}(x,\, t) = z(x, t - h_t), \quad (x, t) \in \overline{G}_h, \quad t > 0$$

Under the condition

$$\frac{\partial}{\partial u} f(x, t,\, u) \le c(x, t), \quad (x, t,\, u) \in \overline{G} \times R, \tag{22}$$

the difference scheme (21), (4) is monotone.

For simplicity, we assume that the condition (22) is satisfied.

Taking into account estimates of the solution to problem (2), (1) for the linearized difference scheme (21) on the special mesh (8), we obtain an estimate similar to estimate (10):

$$|u(x, t) - z(x, t)| \le M\left[N^{-1} \ln N + N_0^{-1}\right], \quad (x, t) \in \overline{G}_h. \tag{23}$$

If the condition (22) is not satisfied, then in problem (21), (4), we pass from the function $z(x,\, t)$ to the function $z^*(x,\, t)$ defined by

$$z(x,\, t) = z^*(x,\, t) \exp(\alpha\, t),$$

and we choose the constant α sufficiently large so that

$$\frac{\partial}{\partial u} f(x, t,\, u) \le c(x, t) + \delta_{\bar{t}}\left[\exp(\alpha\, t)\right] p(x, t), \quad (x, t,\, u) \in \overline{G} \times R;$$

this will ensure the monotonicity of the discrete problem obtained. Furthermore, we can then establish the convergence of the function $z^*(x,\, t)$ to the function $u^*(x,\, t)$ defined by

$$u(x,\, t) = u^*(x,\, t) \exp(\alpha\, t).$$

Returning to the function $z(x,\, t)$, we finally obtain the estimate (23).

Theorem 4 *Let the hypotheses of Theorem* 2 *be fulfilled. Then the solution of the linearized difference scheme* (21), (8) *converges to the solution of the boundary value problem* (2), (1) *ε-uniformly: the estimate* (23) *holds for the computed solution.*

6. Linearized scheme of improved accuracy

Let us present a Richardson scheme of improved accuracy whose construction is based on the linearized scheme (21), (8).

Let $z^i(x, t)$, $(x, t) \in \overline{G}^i_h$, $i = 1, 2$, be the solutions of the difference schemes

$$\begin{aligned} \Lambda_{(21)}\left(z^i(x,\, t)\right) &= 0, \qquad (x,\, t) \in G^i_h, \qquad\qquad (24a) \\ z^i(x,\, t) &= \varphi(x,\, t), \quad (x,\, t) \in S^i_h, \quad i = 1, 2, \end{aligned}$$

where $\overline{G}_h^{\,i} = \overline{G}_{h(11)}^{\,i}$. On the set $\overline{G}_h^{\,0} = \overline{G}_{h(11)}^{\,0}$, we introduce the function

$$z^0(x,\,t) = \gamma\, z^1(x,\,t) + (1-\gamma)\, z^2(x,\,t), \quad (x,\,t) \in \overline{G}_h^{\,0}, \tag{24b}$$

where $\gamma = \gamma_{(12)}$; here the functions

$$z^i(x,\,t), \quad (x,\,t) \in \overline{G}_h^{\,i}, \quad i = 1, 2,$$

are the solutions of the problem (24), (11).

We call the linearized difference scheme (24), (11) the linearized Richardson scheme (24), (11). We call the function $z_{(24)}^0(x,\,t)$, $(x,\,t) \in \overline{G}_h^{\,0}$, the solution of the linearized Richardson scheme (24), (11).

The analysis of the scheme (24), (11) is similar to that for the scheme (12), (11). In the representation (13) of the solution of scheme (24), (11), the component $u_0(x,t)$ is the solution of a problem similar to problem (14):

$$\begin{aligned} L_{(14)} u_0(x,t) &= -T\left[2^{-1} p(x,t)\,\frac{\partial^2}{\partial t^2}\, u(x,t) + \frac{\partial}{\partial u}\, f(x,t,\,u(x,t))\,\frac{\partial}{\partial t}\, u(x,t)\right], \\ &\quad\ (x,t) \in G, \\ u_0(x,t) &= 0, \quad (x,t) \in S. \end{aligned}$$

In the same representation (13), the component $u_1(x,t)$ has the form (15). In (15) the functions $u_{1j}(x,t)$, $(x,t) \in \overline{G}$, are the solutions of problems (16)–(18).

For the solution of problem (24), (11) we obtain an estimate similar to (20):

$$|u(x,t) - z^0(x,t)| \le M\,\left[(N^{-1}\ln N)^2 + N_0^{-2}\right], \quad (x,t) \in \overline{G}_h^0. \tag{25}$$

Theorem 5 *Let the hypotheses of Theorem* 3 *be fulfilled. Then the solution of the linearized Richardson scheme* (24), (11) *converges to the solution of the boundary value problem* (2), (1) *ε-uniformly at the rate $\mathcal{O}((N^{-1}\ln N)^2 + N_0^{-2})$: the estimate* (25) *is valid.*

Remark 1 Using Richardson extrapolation on $q > 2$ embedded meshes in the time variable t allows us to obtain schemes that are ε-uniformly convergent at the rate $\mathcal{O}((N^{-1}\ln N)^2 + N_0^{-q})$ (see, e.g., [16]). The consideration of examples—even for linear problems—shows however that increasing the number of embedded meshes in the spatial variable x does not allow us to construct schemes that are ε-uniformly convergent with accuracy in N^{-1} greater than 2.

7. *A priori* estimates of the solution and its derivatives

Let us give *a priori* estimates for the solution of the boundary value problem (2), (1) and its derivatives. The derivation of these estimates is similar to [17].

Using the majorant function technique (see, e.g., [10]), we obtain

$$|u(x,t)| \le M, \quad (x,t) \in \overline{G}. \tag{26}$$

Write the solution of the problem as the sum of functions

$$u(x,t) = U(x,t) + V(x,t), \quad (x,t) \in \overline{G}, \tag{27a}$$

where $U(x,t)$ and $V(x,t)$ are the regular and singular parts of the solution. The function $U(x,t)$, $(x,t) \in \overline{G}$, is the restriction to $\overline{G}$ of the function $U^0(x,t)$, $(x,t) \in \overline{G}^0$. Here the function $U^0(x,t)$, $(x,t) \in \overline{G}^0$, is the solution of the boundary value problem

$$\begin{aligned} L^0\left(U^0(x,t)\right) &\equiv L^{20}\,U^0(x,t) - f^0\left(x,t,\,U^0(x,t)\right) = 0, \quad (x,t) \in \overline{G}^0, \\ U^0(x,t) &= \varphi^0(x,t), \quad (x,t) \in S^0, \end{aligned} \tag{28}$$

and $\overline{G}^0$ is the half-strip $(x \le d)$, which is an extension of $\overline{G}$ beyond the side S_1^L; the data of problem (28) are smooth continuations of the data of problem (2), (1) that preserve properties (3) on $\overline{G}^0$. Assume that the functions

$$f^0(x,t,\,u), \quad (x,t,\,u) \in \overline{G} \times R, \quad \textit{and} \quad \varphi^0(x,t), \quad (x,t) \in S^0,$$

are equal to zero outside the m_1-neighbourhood of the set $\overline{G}$; finally, the operator L^{20} is defined by

$$L^{20} \equiv \varepsilon\,\frac{\partial^2}{\partial x^2} + \frac{\partial}{\partial x} - 1 - \frac{\partial}{\partial t}.$$

The function $V(x,t)$, $(x,t) \in \overline{G}$, is the solution of the problem

$$\begin{aligned} L^2_{(2)}\,V(x,t) &- \left[f(x,t,\,U(x,t)+V(x,t)) - f(x,t,\,U(x,t))\right] = 0, \\ &\qquad (x,t) \in G, \\ V(x,t) &= \varphi(x,t) - U(x,t), \quad (x,t) \in S. \end{aligned} \tag{29}$$

We write the function $U(x,t)$ as a sum of functions:

$$U(x,t) = U_0(x,t) + \varepsilon U_1(x,t) + \ldots + \varepsilon^n U_n(x,t) + v_U(x,t), \quad (x,t) \in \overline{G}. \tag{27b}$$

This is simply the restriction to $\overline{G}$ of the following representation of the function $U^0(x,t)$:

$$U^0(x,t) = U_0^0(x,t) + \varepsilon U_1^0(x,t) + \ldots + \varepsilon^n U_n^0(x,t) + v_U^0(x,t), \quad (x,t) \in \overline{G}^0, \tag{30a}$$

which is the solution of the boundary value problem (28). Here $v_U^0(x,t)$ is the remainder term and

$$U(x,t) = U^0(x,t), \;\ldots,\; v_U(x,t) = v_U^0(x,t), \quad (x,t) \in \overline{G}.$$

In (30a), the functions

$$U_0^0(x,t), \quad U_i^0(x,t) \quad \textit{for} \quad i = 1,\ldots,n,$$

are the solutions of the problems

$$\begin{aligned} L^0_{(30)}U_0^0(x,t) &- f^0(x,t,\,U_0^0(x,t)) = 0, \quad (x,t) \in G^0, \\ U_0^0(x,t) &= \varphi^0(x,t), \quad (x,t) \in S^0; \\ L^0_{(30)}U_i^0(x,t) &- \varepsilon^{-i}\left[f^0\Big(x,t,\,\sum_{j=0}^{i}\varepsilon^j\,U_j^0(x,t)\Big) - f\Big(x,t,\,\sum_{j=0}^{i-1}\varepsilon^j\,U_j^0(x,t)\Big)\right] = \\ &= -a^0(x,t)\,\frac{\partial^2}{\partial x^2}\,U_{i-1}^0(x,t), \quad (x,t) \in G^0, \\ U_i^0(x,t) &= 0, \quad (x,t) \in S^0, \quad i = 1,\ldots,n. \end{aligned} \tag{30b}$$

Here $L^0_{(30)}$ is the operator $L^{20}_{(28)}$ for $\varepsilon = 0$, i.e.,

$$L^0_{(30)} \equiv b^0(x,t)\,\frac{\partial}{\partial x} - c^0(x,t) - p^0(x,t)\,\frac{\partial}{\partial t}, \quad (x,t) \in G^0,$$

and the function $a^0(x,t)$ is the coefficient of the derivative $\partial^2/\partial x^2$ in the operator $L^{20}_{(28)}$; it is an extension of $a(x,t)$ to $\overline{G}^0$ that preserves the properties (3).

Assume that the data of problem (2), (1) satisfy additional conditions on the set $S_2^* = S_0 \bigcap \overline{S}_2^L$ that ensure sufficient smoothness of the functions $U_0^0(x,t)$ and $U_i^0(x,t)$ for $i = 1,\dots,n$. It is not difficult to write down such conditions, for example in the case when the boundary function $\varphi(x,0)$ and its derivatives vanish on the set S^*.

For simplicity, we assume that the following conditions are valid:

$$\frac{\partial^k}{\partial x^k}\,\varphi(x,t),\ \frac{\partial^{k_0}}{\partial t^{k_0}}\,\varphi(x,t) = 0, \quad (x,t) \in S^*, \tag{31}$$

$$\frac{\partial^{k+k_0+k_u}}{\partial x^k\,\partial t^{k_0}\,\partial u^{k_u}}\,f(x,t,\,u) = 0, \quad (x,t) \in S^*, \quad u = 0, \quad k,\,k_0,\,k_u \le l,$$

where $l = 3n+2$, $n \ge 1$.

Assuming sufficient smoothness of the coefficients of the operator $L^2_{(2)}$ on $\overline{G}$, and of the functions $\varphi(x,t)$ on S_0, $\overline{S}^L$ and $f(x,t,\,u)$ on $\overline{G} \times R$, one has

$$u,\,U \in C^{l^1,l^1}(\overline{G}), \quad U_i \in C^{l^2,l^2}(\overline{G}), \quad i = 0,1,\dots,n, \tag{32}$$

where

$$l^1 = n+1+\alpha, \quad l^2 = 3n+2-2i, \quad n \ge 1, \quad \alpha > 0.$$

In this case, $V \in C^{l_1,l_1}(\overline{G})$.

For $U(x,t)$ and $V(x,t)$ we obtain the estimates

$$\left|\frac{\partial^{k+k_0}}{\partial x^k \partial t^{k_0}}\,U(x,t)\right| \le M\,[1+\varepsilon^{n+1-k-k_0}], \tag{33a}$$

$$\left|\frac{\partial^{k+k_0}}{\partial x^k \partial t^{k_0}}\,V(x,t)\right| \le M\,\varepsilon^{-k}\,\exp(-m\,\varepsilon^{-1}\,x), \quad (x,t) \in \overline{G}, \quad k+k_0 \le K, \tag{33b}$$

where m is an arbitrary number in the interval $(0,m_0)$ and $m_0 = \min_{\overline{G}}[a^{-1}(x,t)\,b(x,t)]$; here $K = n+1$ when the data of the boundary value problem (2), (1) are sufficiently smooth.

The function $u(x,t)$ satisfies the estimate

$$\left|\frac{\partial^{k+k_0}}{\partial x^k \partial t^{k_0}}\,u(x,t)\right| \le M\,\varepsilon^{-k-k_0}, \quad (x,t) \in \overline{G}, \quad k+k_0 \le K. \tag{34}$$

Theorem 6 *In the boundary value problem (2), (1) assume that $a, b, c, p \in C^{l,l}(\overline{G})$, $f \in C^{l,l,l}(\overline{G} \times R)$, $\varphi \in C^{l,l}(\overline{G})$, $l = 3n+2$, $n \ge 1$, and also that condition (31) is satisfied. Then the estimates (26), (33) and (34), where $K = n+1$, are satisfied for the solution of the boundary value problem and its components in the representation (27).*

Acknowledgments
The authors would like to express their gratitude to Martin Stynes for interesting discussions of numerical methods with improved accuracy for singularly perturbed problems.

This research was supported by the Netherlands Research Organisation NWO under grant No. 047.016.008, the Russian Foundation for Basic Research under grants No. 04–01–00578, 04–01–89007–NWO_a., the Boole Centre for Research in Informatics (BCRI) at the National University of Ireland, Cork and the Mathematics Applications Consortium for Science and Industry in Ireland (MACSI) under the Science Foundation Ireland (SFI) mathematics initiative.

References

[1] Bakhvalov N S 1969 On the optimization of methods for boundary-value problems with boundary layers *Zh. Vychisl. Mat. Mat. Fiz.* **9** 841–59 (in Russian)

[2] Bochmer K, Stetter H 1984 *Defect correction methods. Theory and applications*(Springer-Verlag, Wien-New: Computing Supplementum 5)

[3] Farrell P A, Miller J J H, O'Riordan E and Shishkin G I 1998 On the non-existence of e-uniform finite difference methods on uniform meshes for semilinear two-point boundary value problems *Math. Comp.* **67** 603–17

[4] Farrell P A, Hegarty A F, Miller J J H, O'Riordan E and Shishkin G I 2000 *Robust Computational Techniques for Boundary Layers* (Boca Raton: Chapman and Hall/CRC)

[5] Hemker P W, Shishkin G I and Shishkina L P 1997 The use of defect correction for the solution of parabolic singular perturbation problems *ZAMM Z. Angew. Math. Mech* **77** 59–74

[6] Hemker P W, Shishkin G I and Shishkina L P 2000 ε-uniform schemes with high-order time-accuracy for parabolic singular perturbation problems *IMA J. Numer. Anal.* **20** 99–121

[7] Hemker P W, Shishkin G I and Shishkina L P 2004 High-order accurate decomposition of Richardson's method for a singularly perturbed elliptic reaction-diffusion equation *Comp. Math. Math. Phys.* **44** 309–16

[8] Il'in A M 1969 Differencing Scheme for a Differential Equation with a Small Parameter Affecting the Highest Derivative *Math. Notes* **6** 596–602

[9] Kopteva N V 2003 Error expansion for an upwind scheme applied to a two-dimensional convection-diffusion problem *SIAM J. Numer. Anal.* **41** 1851–69

[10] Ladyzhenskaya O A, Solonnikov V A and Ural'tseva N N 1968 *Linear and Quasilinear Equations of Parabolic Type* (Translations of Mathematical Monographs vol 23) (American Mathematical Society USA)

[11] Marchuk G I 1982 *Methods of numerical mathematics* (New York: Springer)

[12] Marchuk G I and Shaidurov V V 1983 *Difference methods and their extrapolations* (New York: Springer)

[13] Miller J J H, O'Riordan E and Shishkin G I 1996 *Fitted numerical methods for singular perturbation problems* (Singapore: World Scientific)

[14] Roos H-G, Stynes M and Tobiska L Numerical 1996 *Methods for Singularly Perturbed Differential Equations. Convection-Diffusion and Flow Problems* (Berlin: Springer-Verlag)

[15] Samarskii A A 2001 *The Theory of Difference Schemes* (New York: Marcel Dekker)

[16] Shishkin G I 1984 The method of increasing the accuracy of solutions of difference schemes for parabolic equations with a small parameter at the highest derivative *USSR Comput. Maths. Math. Phys.* **24** 864–75

[17] Shishkin G I 1992 *Grid Approximations of Singularly Perturbed Elliptic and Parabolic Equations* (Ekaterinburg: Ural Branch of Russian Academy of Sciences) (in Russian) 150–7

[18] Shishkin G I 2002 Grid approximations with an improved rate of convergence for singularly perturbed elliptic equations in domains with characteristic boundaries *Siberian J. Numer. Math.* **5** 71–92

[19] Shishkin G I and Shishkina L P 2005 A High-Order Richardson Method for a Quasilinear Singularly Perturbed Elliptic Reaction-Diffusion Equation *Differential Equations* **41** 1030–9

[20] Shishkin G I 2006 The Richardson method of increasing the accuracy of discrete solutions for singularly perturbed elliptic equations with convective terms *Izv. Vyssh. Uchebn. Zaved. Mat.* **2** 57–71 (in Russian)

[21] Shishkina L P 2004 The Richardson method of high-order accuracy in *t* for a semilinear singularly perturbed parabolic reaction-diffusion equation on a strip *Proc. Int. Conf. on ICCM'2004 (Novosibirsk, Russia)* vol 2 (Novosibirsk: ICM&MG Publisher) 927–31

[22] Shishkina L Shishkin G 2005 The discrete Richardson method for semilinear parabolic singularly perturbed convection-diffusion equations *Proc. Int. Conf. on Mathematical Modelling and Analysis 2005 (Trakai)* (Technika, Vilnius) 259–64

Institute of Physics Publishing
doi:10.1088/1742-6596/55/1/020

Journal of Physics: Conference Series **55** (2006) 214–231
International Workshop on Multi-Rate Processes and Hysteresis

Sub-Hopf/fold-cycle bursting and its relation to (quasi-)periodic oscillations

R Straube[1], D Flockerzi[2] and M J B Hauser[3]

[1] Abteilung Theoretische Physik, Hahn-Meitner-Institut Berlin, Glienicker Str. 100, D-14109 Berlin, Germany

[2] Max-Planck-Institut für Dynamik komplexer technischer Systeme, Sandtorstr. 1, D-39106 Magdeburg, Germany

[3] Abteilung Biophysik, Institut für Experimentelle Physik, Otto-von-Guericke Universität, Universitätsplatz 2, D-39106 Magdeburg, Germany

E-mail: ronny.straube@hmi.de

Abstract. We investigate the emergence of bursting oscillations and their relation to (quasi-) periodic behaviour in two different model systems, a pH oscillator and a calcium oscillator. Both systems are described by 3-dimensional ODE systems and follow different 'routes' to bursting oscillations as parameters are varied. In the first part, we exploit the slow-fast structure of the 3-dimensional ODE systems to show that the bursting oscillations are of sub-Hopf/fold-cycle type over a wide range of parameters. For this purpose, a slow variable is used as a quasi-static bifurcation parameter for the remaining 2-dimensional fast subsystem. For the pH oscillator, a subsequent two-parameter continuation reveals a transition in the bursting behaviour from sub-Hopf/fold-cycle to fold/sub-Hopf type. In the second part, we argue that the particular constellation of a subcritical Hopf bifurcation together with a saddle-node bifurcation of a periodic orbit (fold-cycle) may not only account for the bursting oscillations, but also allows for (quasi-)periodic oscillations on a 2-torus suggesting a common origin for both types of dynamics.

1. Introduction

Bursting or mixed-mode oscillations are frequently observed as asymptotic states in spatially homogeneous chemical [1, 2, 3, 4] and biological [5, 6, 7, 8, 9, 10, 11, 12] reaction systems where they are of potential relevance for signal transduction processes [13, 14]. Biological signal transduction networks frequently involve so-called second messengers, i.e. small and unspecific molecules, for instance calcium ions, which are able to transport different information contents. In such cases information is coded in the geometric outline of the concentration oscillations of the second messenger [15], while the intensity of a stimulus is stored in the frequency of the oscillations [16]. Bursting oscillations are periodic patterns consisting of L large amplitude (spiking) oscillations and S small amplitude oscillations per oscillation period for which the nomenclature L^S has prevailed. Such oscillations are highly suitable for the encoding of different information contents, and, in fact, bursting oscillations are often observed to occur in signal transduction processes.

Alternatively, bursting behaviour can be classified by a slow-fast analysis introduced by Rinzel and Ermentrout [17]. Systems exhibiting bursting oscillations typically involve a fast oscillatory subsystem which is coupled to at least one slow variable which then acts as a quasi-static

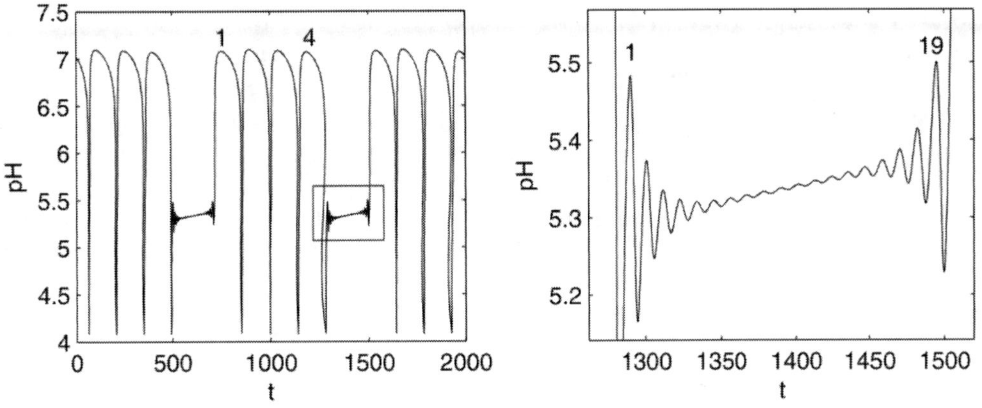

Figure 1. Time series obtained from a numerical simulation of the bursting pH oscillations in the hemin system (left image): 4 large amplitude oscillations alternate with 19 small ones. The rectangular region is magnified in the right image showing the small amplitude oscillations.

bifurcation parameter for the fast subsystem. Subsequently, Izhikevich [18] has extended this approach to classify the bursting behaviour according to the type of bifurcations occurring in the fast subsystem that lead to the emergence and disappearance of the small amplitude oscillations.

Recently, the hemin – hydrogen peroxide – sulfite reaction system (henceforth called 'hemin system') has been introduced as a biocompatible enzyme model system [19, 20] that exhibits bursting oscillations in the pH value of the reaction medium (Figure 1). It belongs to a larger family of so-called pH oscillators [24, 25, 26, 27, 28] that induce periodic changes in the pH value of the corresponding reaction medium which may affect physiological parameters in the environment of the system, e.g. in the permeability of membranes or the activity of enzymes.

The dynamics of the hemin system were originally modelled by a 6-dimensional ODE system [29]. Recently, we have derived a 3-dimensional approximation to the 6-dimensional system using the method of quasi integrals [30] which preserved the local codimension two bifurcation structure of the original ODE system such that both systems are identical at this level. The method of quasi integrals comprises a rather general algorithmic procedure to search for slow invariant manifolds, particularly in higher dimensional reaction networks where analytical methods usually fail. Thus, it is not required that the different time scales of the individual reaction steps of a reaction mechanism are known in advance. On the contrary, the method identifies them. Once the different time scales (or the fast reactions) are known, it is more or less straightforward to apply singular perturbation techniques [32, 33, 34, 35] to obtain a lower dimensional approximation of the original dynamics on a slow manifold of the system which, in its simplest form, yields just a quasi-steady-state approximation [36, 37]. On the other hand, since the method of quasi integrals leads to the identification of the fast reactions of a given system, it can be used as a supplement to existing methods that rely on an a priori grouping of the individual reaction steps into fast and slow ones before singular perturbation techniques may successfully be applied.

In the present article we investigate the emergence of bursting oscillations and their relation to oscillations on a 2-torus in two different model systems, the hemin system and a calcium oscillator model [38]. Since oscillations on a 2-torus can be either periodic or quasiperiodic, we shall refer to this kind of behaviour collectively as (quasi-)periodic. The dynamics of both models are well described by 3-dimensional ODE systems on which a slow-fast analysis is performed to characterize the bursting behaviour of the systems.

The basic ideas underlying a slow-fast analysis are introduced in section 2. The following section is devoted to a detailed study of the bursting oscillations in the hemin system. It will be shown that they are of sub-Hopf/fold-cycle type over a wide range of parameters. A subsequent two-parameter continuation reveals a transition in the bursting behaviour from sub-Hopf/fold-cycle to fold/sub-Hopf type which can be attributed to a codimension two bifurcation (saddle-node separatrix-loop) in the fast subsystem. The relation between (quasi-)periodic and bursting oscillations is analyzed in section 4 where we will demonstrate that the bifurcations in the fast subsystem leading to sub-Hopf/fold-cycle bursting may equally lead to (quasi-) periodic oscillations on a 2-torus. For the calcium oscillator model we will see that the bursting oscillations arise as a 2-torus continuously deforms. In contrast, the bursting oscillations in the hemin system appear as periodic orbits in narrow parameter windows in a chaotic domain far away from the parameter region where (quasi-)periodic behaviour is observed. The final section summarizes the results and discusses some of their consequences.

2. Slow-fast analysis: The method

The slow-fast analysis has been introduced by Rinzel and Ermentrout to describe the bursting behaviour of neural systems [17]. Slow-fast bursters typically involve a fast oscillatory subsystem (generating spiking oscillations) and a slow subsystem which triggers the emergence of small amplitude oscillations as the trajectory of the full system passes certain codimension one bifurcation points in the fast subsystem. Therefore, the fast subsystem must be at least 2-dimensional whereas the slow subsystem may comprise only a single degree of freedom. In the following, we will consider only slow-fast bursters with a 2-dimensional fast and a 1-dimensional slow subsystem of the form

$$\begin{aligned}\dot{x} &= f(x,y,p)\\ \dot{y} &= g(x,y,p)\\ \dot{p} &= \varepsilon h(x,y,p)\end{aligned}\tag{1}$$

where ε is a small parameter indicating that p evolves on a longer time scale than the fast x-y subsystem.

Consider Figure 2 where the slow-fast structure of the 3-dimensional hemin system is illustrated in a cartoon. For the time being, let us consider the limit $\varepsilon \to 0$ in (1) which corresponds to the assumption that p is not a dynamical variable, but a parameter for the 2-dimensional fast x-y subsystem. Then, for each fixed parameter value p_0, there exist certain invariant sets in the 2-dimensional fast subsystem. In Figure 2, for example, three invariant sets coexist in the plane F_{p_0}: A stable limit cycle (black circle), an unstable limit cycle (red circle) and a stable fixed point (black dot).

At the next step, the dynamical nature of p is taken into account. For this purpose, we again consider the dynamics of the ODE system (1), but this time for small nonzero ε. As p slowly evolves in time according to the third equation in (1), the type and stability of the invariant sets of the fast x-y subsystem will also change. In this sense, p now acts as a quasi-static bifurcation parameter for the fast subsystem. For example, as p moves to the left, the stable and the unstable limit cycle merge in a saddle-node bifurcation (SNP) and disappear leaving the stable stationary state (thin black line) as the only invariant set. On the other hand, as p moves to the right, the unstable limit cycle shrinks and vanishes in a subcritical Hopf bifurcation (SH). Henceforth, a large amplitude limit cycle coexists with a saddle point. Eventually, a saddle-node bifurcation (SN_1) occurs on the large amplitude limit cycle to form a saddle-node homoclinic orbit by which the periodic solution of the fast subsystem ceases to exist.

The basic assumption underlying the slow-fast analysis is that the time scale separation between the fast and the slow subsystems is such that the stationary and oscillatory states,

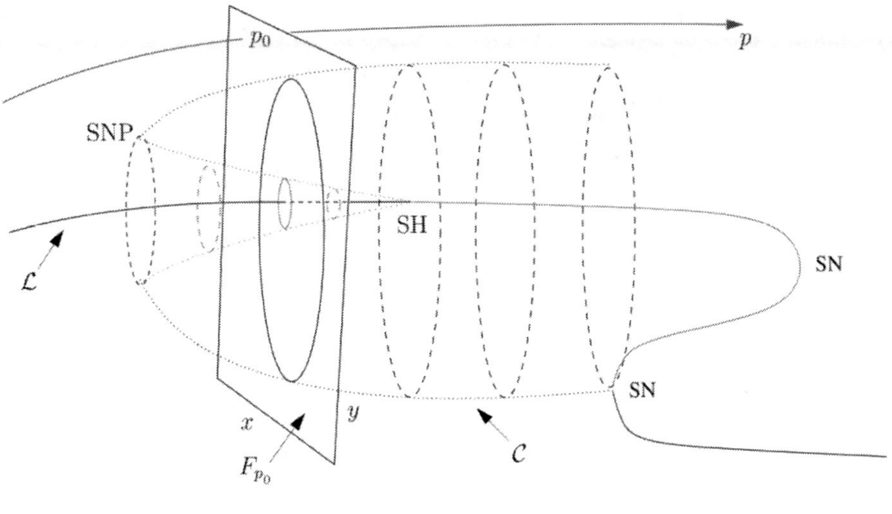

Figure 2. Cartoon of the slow-fast structure of the hemin system (2). SNP – saddle-node bifurcation of periodic orbits, SH – subcritical Hopf bifurcation, SN – saddle-node bifurcation of stationary points.

which exist in the fast subsystem at a particular value of p, extend along the p direction to quasi-stationary manifolds in the 3-dimensional phase space (cf. Fig. 2). Thus, the stationary states become line-like quasi-stationary manifolds ($\mathcal{L}$) while the oscillatory states form cylinder-like manifolds ($\mathcal{C}$). This means that the flow of the full 3-dimensional ODE system always remains in the neighbourhood of the stable invariant sets of the fast subsystem. For example, the large amplitude (spiking) oscillations in Figure 1 occur close to the cylinder-like manifold $\mathcal{C}$ of Figure 2 while the small amplitude (bursting) oscillations occur along the line-like manifold ($\mathcal{L}$). Note that the cylinder-like manifold is terminated at the left side by a saddle-node bifurcation (SNP) and at the right side by a saddle-node homoclinic orbit such that large amplitude oscillations are only possible in a finite region of the phase space.

3. Bursting oscillations in the hemin system

3.1. *The hemin system*

The bursting oscillations observed experimentally [19, 20] are qualitatively well reproduced by a 6-dimensional ODE system [29]. Recently, we derived a 3-dimensional approximation to the 6-dimensional ODE system using quasi integrals [30]. It was shown that the local codimension two bifurcation diagrams of both systems are virtually identical. However, to analyze the slow-fast structure of the 3-dimensional ODE system, it was convenient to perform a suitable change of coordinates [31]. The resulting 3-dimensional ODE system will be used as the starting point for the slow-fast analysis of the bursting oscillations in subsection 3.2. It reads

$$\begin{aligned} \dot{x} &= k_0 x_2^0 - x\Big\{k_0 + k_1 s(x,y) + (k_2 + k_3(a - x - y + s(x,y)))(x_1^0 - x_2^0 + x - s(x,y))\Big\} \\ \dot{y} &= -(k_6 + k_0)y + k_7(p - y)(a - x - y + s(x,y)) \\ \dot{p} &= k_0(p^0 - p) - k_8(p - y). \end{aligned} \tag{2}$$

Table 1. Parameter set for the hemin system (2).

$k_1 = 0.2\ \mathrm{M^{-1}s^{-1}}$	$k_2 = 1.5\ \mathrm{M^{-1}s^{-1}}$	$k_3 = 8.5 \cdot 10^6\ \mathrm{M^{-2}s^{-1}}$	$k_4 = 1000\ \mathrm{s^{-1}}$
$k_5 = 10^{10}\ \mathrm{M^{-1}s^{-1}}$	$k_6 = 0.011\ \mathrm{s^{-1}}$	$k_7 = 2.5 \cdot 10^4\ \mathrm{M^{-1}s^{-1}}$	$k_8 = 2.5 \cdot 10^{-4}\ \mathrm{s^{-1}}$
$x_1^0 = 0.025\ \mathrm{M}$	$x_2^0 = 0.045\ \mathrm{M}$	$x_4^0 = 2.2 \cdot 10^{-4}\ \mathrm{M}$	$p^0 = 3 \cdot 10^{-4}\ \mathrm{M}$
$\mathrm{a} \equiv x_4^0 - x_1^0 + x_2^0$	$k_0 \in [1,5] \cdot 10^{-4}\ \mathrm{s^{-1}}$		

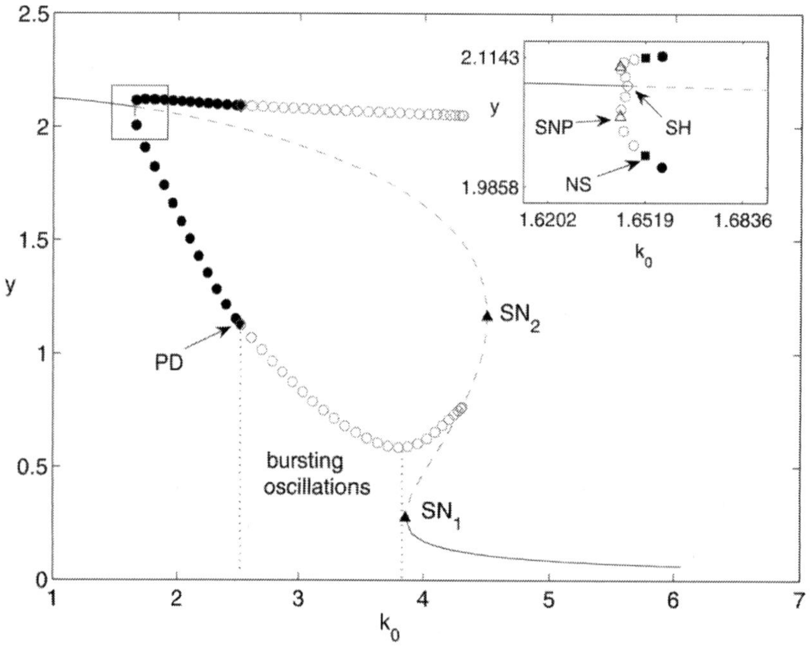

Figure 3. One-parameter bifurcation diagram for the ODE system (2). Black and red circles denote minima and maxima of stable and unstable oscillations, respectively.

where $s(x,y)$ stands for the slow manifold of the 6-dimensional ODE system given by

$$s \;=\; \frac{1}{2}(x+y-a-\frac{k_4}{k_5}) + \frac{1}{2}\sqrt{(x+y-a+\frac{k_4}{k_5})^2 + 4\frac{k_4}{k_5}(x_4^0-y)}. \qquad (3)$$

In (2) all parameters but k_0 are fixed. Their dimensional numerical values are compiled in Table 1. Similar to the experimental situation, the flow rate k_0 is the principal control parameter for the system. The higher the flow rate, the higher is the matter flow through the system. Thus, the flow rate is a suitable parameter to drive the system away from equilibrium.

As the flow rate is increased, one observes successively the following asymptotic states (cf. Fig. 3)[1]: A stable stationary state (black line) becomes unstable at $k_0 = 1.6461$ via a subcritical Hopf bifurcation (Fig. 3, inset) giving rise to an unstable limit cycle (red circles).

[1] Variables in the Figures and in the text are rescaled according to $x \to 10^2\mathrm{M}^{-1}x$, $y \to 10^4\mathrm{M}^{-1}y$, $p \to 10^4\mathrm{M}^{-1}p$ and $k_0 \to 10^4 s\ k_0$

This unstable cycle merges with a second one in a saddle-node bifurcation SNP (inset, open triangle) at $k_0 = 1.6438$. However, stable oscillations (black circles) arise by an inverse Neimark-Sacker bifurcation NS (inset, filled square) at $k_0 = 1.6519$ where a 2-torus bifurcates towards lower values of k_0 where it coexists with a saddle point (corresponding to the red dashed line in the inset of Fig. 3) in a very narrow parameter interval before it becomes homoclinic to the saddle and vanishes.

The simple periodic oscillations are stable up to $k_0 = 2.5169$ where the first of a whole cascade of period doubling bifurcations PD occurs and the primary periodic orbit (red circles) becomes unstable (cf. Fig. 3). For increasing k_0 a chaotic attractor is formed where the bursting oscillations appear as embedded periodic orbits. As the flow rate k_0 is increased beyond the period doubling bifurcation point PD, one observes periodic-chaotic progressions of bursting oscillations and chaotic behaviour, i.e. the large parameter interval $k_0 \in (2.517, 3.858)$ (bounded by the dotted lines in Fig. 3) is interspersed with smaller intervals where bursting oscillations arise as stable periodic orbits which differ in the number of large and small amplitude oscillations (for details, see [31]).

3.2. SubHopf/fold-cycle bursting

In the following, we shall investigate the type of bursting of the hemin system at a fixed value of the bifurcation parameter $k_0 = 2.8$ which is chosen such that it lies within one of the periodic windows described in the last subsection (cf. Fig. 3) where a 4^{19} state is stable (Fig. 4). Here, the trajectory performs 4 large amplitude oscillations which are followed by 19 small amplitude oscillations. We perform a slow-fast analysis by treating the variable p as a (quasi-static) bifurcation parameter for the 2-dimensional fast subsystem

$$\begin{aligned} \dot{x} &= k_0 x_2^0 - x\Big\{k_0 + k_1 s(x,y) + \Big(k_2 + k_3(a - x - y + s(x,y))\Big)(x_1^0 - x_2^0 + x - s(x,y))\Big\} \\ \dot{y} &= -(k_6 + k_0)y + k_7(p - y)(a - x - y + s(x,y)). \end{aligned} \quad (4)$$

Note that this *fast* subsystem is simply obtained by omitting the third equation

$$\dot{p} = -(k_0 + k_8)p + k_8 y + k_0 p^0 \quad (5)$$

of the 3-dimensional ODE system (2). The justification of this procedure is provided by inspection of dimensionless variables. After scaling the variables $\tilde{x} = 10^2 M^{-1} x$, $\tilde{y} = 10^4 M^{-1} y$ and $\tilde{p} = 10^4 M^{-1} p$ according to their inflow concentrations (x_2^0 and p^0, cf. Table 1), these new variables are $\mathcal{O}(1)$. With the additional scalings $\tilde{s} = 10^2 M^{-1} s$ and $\tilde{a} = 10^2 M^{-1} x$ we also get $\mathcal{O}(\tilde{s}) = \mathcal{O}(1)$ and $\mathcal{O}(\tilde{a}) = \mathcal{O}(1)$. Therefore the right hand sides of $\dot{\tilde{x}}$ and $\dot{\tilde{y}}$ will be at least of order $\mathcal{O}(10^2 s^{-1})$ whereas the one of $\dot{\tilde{p}}$ will turn out to be of order $\mathcal{O}(10^{-4} s^{-1})$.

The one-parameter bifurcation diagram of the fast subsystem (4) is shown in Figure 5 which, clearly, is topologically equivalent to the cartoon in Figure 2 in the region close to the subcritical Hopf bifurcation SH. In particular, there is a stable quasi-stationary line-like manifold (black line) that is composed of stable fixed points of the fast subsystem. This manifold becomes unstable (red dashed line) via a subcritical Hopf bifurcation SH where an unstable limit cycle (red circles) bifurcates towards lower values of p. The cylinder-like manifold is created by a saddle-node bifurcation of periodic orbits SNP from which a stable (black circles) and an unstable limit cycle bifurcate towards higher values of p.

In order to determine the nature of the bursting behaviour of the hemin ODE system at $k_0 = 2.8$, we overlay the bifurcation diagram of the fast subsystem (Fig. 5) with a projection of the 4^{19} orbit (Fig. 4) onto the p-x plane as shown in Figure 6. Note that the (blue) trajectory always remains in the neighbourhood of the quasi-stationary manifolds of the fast subsystem. More precisely, the large amplitude oscillations occur close to the cylinder-like manifold while

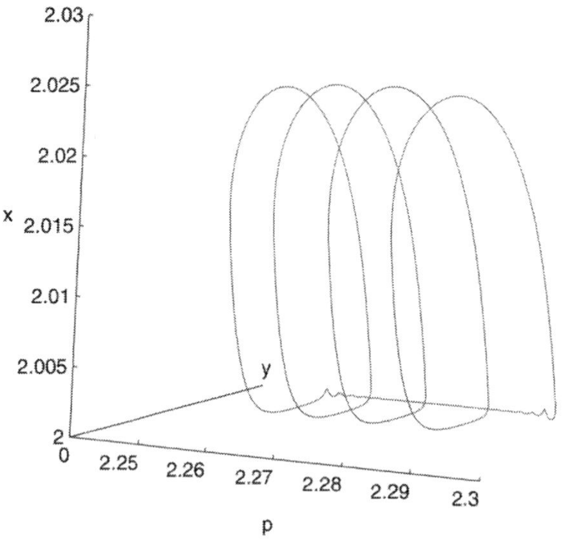

Figure 4. 3-dimensional phase portrait of a 4^{19} bursting state at $k_0 = 2.8$.

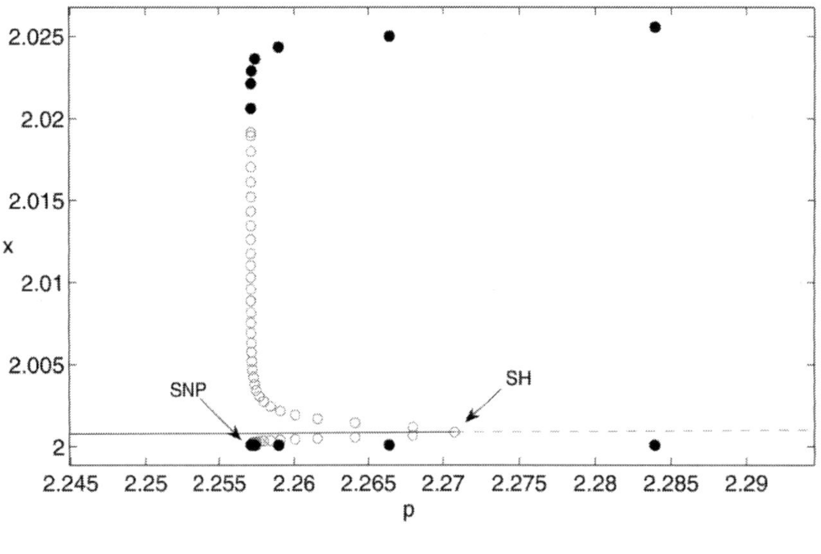

Figure 5. One-parameter bifurcation diagram of the fast subsystem (4).

the small amplitude oscillations occur along the line-like manifold. This can be seen in Figure 7 which shows a magnification of the dashed rectangular region of Figure 6.

In Figure 6 we will describe one revolution of the trajectory in detail: The blue arrow indicates

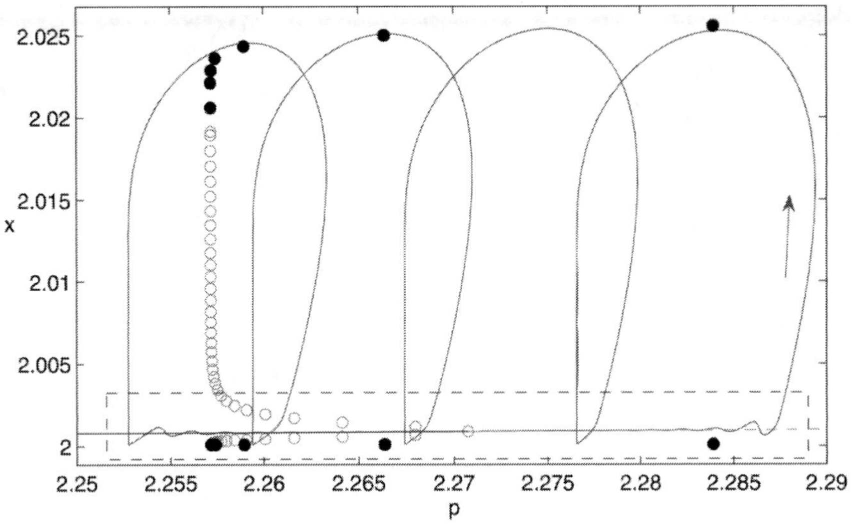

Figure 6. Overlay of the one-parameter bifurcation diagram of the fast subsystem (Fig. 5) with a projection of the 4^{19} orbit (Fig. 4) onto the p-x plane.

the direction of the flow. To the left of the SNP point, the line-like quasi-stationary manifold is the only attractor. Since the eigenvalues along this manifold are complex, the trajectory performs damped oscillations as it approaches the subcritical Hopf bifurcation point SH. Beyond that point, the quasi-stationary manifold becomes unstable causing the subsequent oscillations to increase until the orbit is finally attracted by the cylinder-like manifold (near the stable limit cycle solutions (black circles) of the fast subsystem). Then 4 large amplitude oscillations are performed in the vicinity of the cylinder-like manifold (Fig. 6). When the trajectory passes the saddle-node bifurcation point SNP the orbit 'jumps back' to the line-like manifold to begin the next revolution.

According to the classification scheme of bursting mechanisms given in [18], which is based on the nature of the bifurcations in the fast subsystem delimiting the domain of bursting oscillations, the hemin system is a sub-Hopf/fold-cycle burster at $k_0 = 2.8$ since the large amplitude (spiking) oscillations terminate at a fold-cycle bifurcation (which is also called as saddle-node bifurcation of periodic orbits, SNP) while the small amplitude (bursting) oscillations disappear via a subcritial Hopf bifurcation (SH).

In general, a L^S state found in a periodic window of the chaotic domain described in subsection 3.1 wraps around the cylinder-like manifold L times and oscillates S times along the line-like manifold before the next period begins.

3.3. A transition between sub-Hopf/fold-cycle and fold/sub-Hopf bursting

In the last subsection, we studied the bursting type of the ODE system (2) at one particular value of the control parameter k_0. However, it can be presumed that the quasi-stationary manifolds of the fast subsystem (4) somehow deform as the bifurcation parameter k_0 is varied which, in turn, may cause changes in the bursting behaviour. In this sense, one can consider the flow rate k_0 as an external bifurcation parameter for the fast subsystem while the slow variable p acts as an internal parameter since it cannot be tuned externally, but instead evolves dynamically according to (5). Hence, we are led to consider the two-parameter bifurcation diagram shown

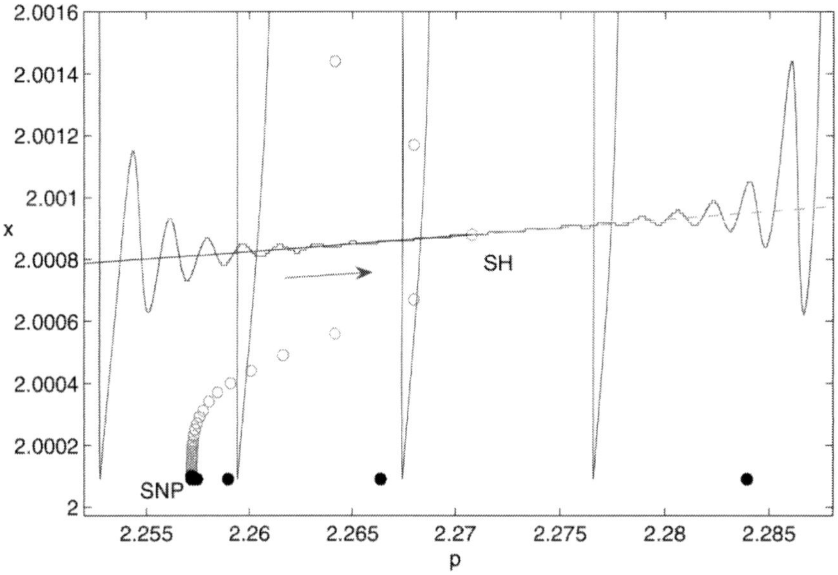

Figure 7. Magnification of the dashed rectangular region in Figure 6.

in Figure 8 where the external parameter k_0 and the slow variable p are used as the relevant parameters in order to detect changes in the bursting behaviour of the hemin system.

The blue dashed line marks the section $k_0 = 2.8$ along which the one-parameter bifurcation diagram of Figure 5 has been calculated. The subcritical Hopf bifurcation SH and the saddle-node bifurcation of periodic orbits SNP, shown there, have now become curves in the two-parameter plane (SH – red, dashed; SNP – black, dash-dot) which emanate from a codimension two bifurcation point, the generalized Hopf (GH, open triangle) or Bautin bifurcation [39]. Note that these two curves remain close together over a wide range of values of the flow rate parameter k_0 which is, in fact, a prerequisite for sub-Hopf/fold-cycle bursting to occur. In addition, there is a second curve of subcritical Hopf bifurcations SH_2 and a second curve of saddle-node bifurcations (not shown) at higher values of the slow variable which emanate from GH_2. However, they occur in a different region of phase space and thus have no direct impact on the actual dynamics of the 3-dimensional system. Whether the actual dynamics are (quasi-)periodic (for $k_0 \in (1.6461, 16519)$), simple periodic (for $k_0 \in (1.6519, 2.5169)$), or of bursting type (for $k_0 \in (2.5169, 3, 8580)$) is eventually determined by the interplay between the fast and slow subsystem and therefore cannot be decided exclusively from the bifurcation structure of the fast subsystem.

However, the two-parameter bifurcation diagram shown in Figure 8 can be used to identify transitions in the bursting behaviour of the ODE system (2). For this purpose, consider the intersection point $(p, k_0) = (2.211, 3.773)$, marked with the red arrow, where the curve of subcritical Hopf bifurcations SH and the curve of saddle-node bifurcations of stationary points SN_1 cross each other. This opens the possibility that in a neighbourhood of the intersection point the (unstable) oscillatory states emanating from the subcritical Hopf bifurcation interact with the branch of (unstable) stationary states that originate in the saddle-node bifurcation. In order to show that this does indeed lead to a transition in the bursting behaviour, we compare the bifurcation diagrams of the fast subsystem for two neighbouring values of the flow rate k_0 corresponding to the two blue dotted lines in Figure 8.

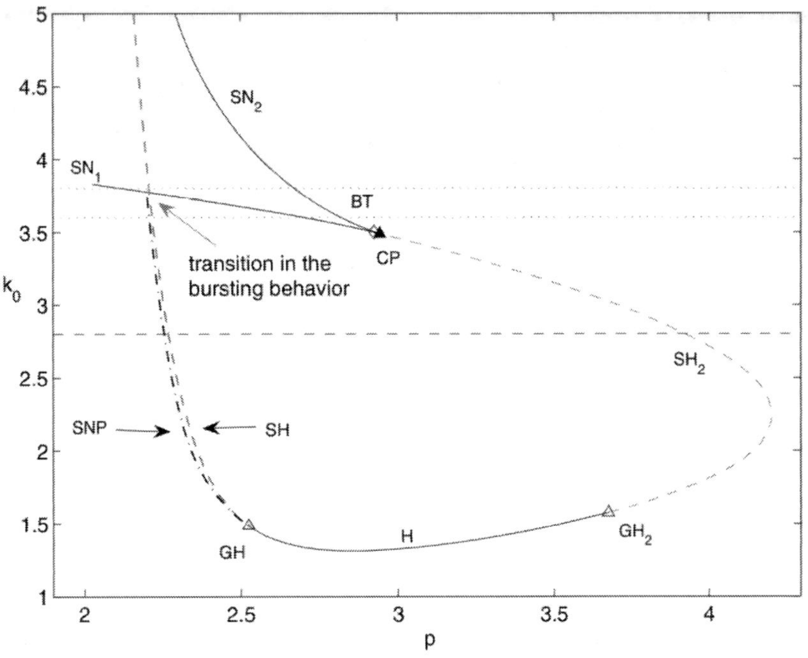

Figure 8. Two-parameter bifurcation diagram of the fast subsystem (4). SH – subcritical Hopf bifurcation, H – supercritical Hopf bifurcation, SNP – saddle-node bifurcation of periodic orbits, BT – Bogdanov-Takens point, CP – cusp singularity, GH – generalized Hopf bifurcation, SN – saddle-node bifurcation of fixed points.

Figure 9 shows an overlay of the codimension one bifurcation diagrams of the fast subsystem with the trajectories (blue lines) for $k_0 = 3.6$ (Fig. 9a) and $k_0 = 3.8$ (Fig. 9b), respectively. These diagrams are similar to the one presented in Figure 5, but now for the variable y instead of x. Note, however, that the range of the axis of the slow variable has been increased to visualize the global aspects of the bifurcation structure of the fast subsystem before and after the transition in the bursting behaviour. At $k_0 = 3.6$, the cylinder-like manifold (Fig. 9a, black circles) is bounded by the saddle-node bifurcation SNP at the left side and the saddle-node homoclinic orbit SNHC at the right side (at $p \sim 2.7$) where the saddle-node bifurcation SN_1 occurs on the large amplitude limit cycle. The two saddle-node bifurcations SN_1 and SN_2 are the same as those found at the intersections of the blue dotted lines with the black solid lines SN_1 and SN_2 in Figure 8.

As the flow rate k_0 increases from 3.6 to 3.8, the saddle-node homoclinic orbit as well as the two saddle-node bifurcation points SN_1 and SN_2 move towards lower values of p until the first of them (SN_1) collides with the unstable limit cycle (red circles) at approximately $k_0 \sim 3.778$ (not shown). This leads to a codimension two bifurcation in the fast subsystem, a so-called saddle-node separatrix-loop bifurcation [40], where the saddle-node homoclinic orbit turns into a saddle homoclinic orbit (SHC) and, in addition, the saddle-node bifurcation SNP disappears (Fig. 9b). The SNHC is an orbit homoclinic to a non-hyperbolic equilibrium whereas the SHC is an orbit homoclinic to a hyperbolic equilibrium. After the codimension two bifurcation has occured ($k_0 > 3.778$), the cylinder-like manifold disappears and the fast subsystem becomes

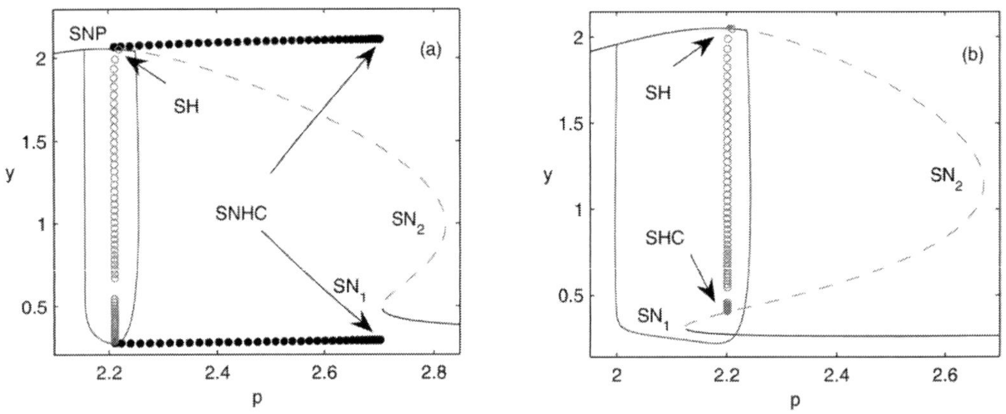

Figure 9. Transition in the bursting behaviour from sub-Hopf/fold-cycle type at $k_0 = 3.6$ (a) to fold/sub-Hopf type at $k_0 = 3.8$ (b) due to a saddle-node separatrix-loop codimension two bifurcation, SHC – saddle homoclinic orbit, SNHC – saddle-node homoclinic orbit.

bistable. Henceforth, the bursting behaviour of the hemin system is of fold/sub-Hopf type since the upper stationary state disappears via the subcritical Hopf bifurcation SH while the lower stationary state undergoes a fold bifurcation at SN_1. A typical trajectory (blue line, Fig. 9b) jumps back and forth between the two quasi-stationary states causing a strong relaxational character to the oscillations.

4. Sub-Hopf/fold-cycle bursting and (quasi-)periodic oscillations on a 2-torus

In this section we wish to investigate the relation between bursting oscillations of sub-Hopf/fold-cycle type and (quasi-)periodic oscillations on a 2-torus. Therefore, we consider two model systems: The hemin system and, in addition, a recently introduced model for a Ca^{2+} oscillator system [38]. The latter also exhibits bursting oscillations of sub-Hopf/fold-cycle type. Again using the slow-fast method described in section 2, it will be shown that in both systems the (quasi-)periodic oscillations are associated with the same bifurcations of the fast subsystem that also account for the emergence of bursting oscillations.

4.1. The hemin system

As described in subsection 3.1 the hemin system has a Neimark-Sacker point NS at $k_0 = 1.6519$ (cf. inset Fig. 3) where a smooth 2-torus bifurcates towards lower values of k_0 (Figs. 10a,c). As the flow rate k_0 is slightly decreased, the torus rapidly deforms (Figs. 10b,d) and approaches the saddle point S (open triangle) which emerges after the subcritical Hopf bifurcation SH corresponding to the red dashed line in the inset of Figure 3. The saddle point acts as an organizing center for the torus (Fig. 10b). The flow approaches the 'outer part' of the torus along the 2-dimensional unstable manifold of the saddle point. Then it moves to the left (i.e. towards lower p values) until it changes direction and returns along the 1-dimensional stable manifold of the saddle. Note that in Figures 10(a,b) the numerical integration was stopped before the trajectory made a full revolution on the torus in order to reveal a portion of the flow along the 'inner part' of the torus.

The reason for the trajectory to change its direction in Figures 10(a,b) can be grasped from the slow-fast analysis of (4). For this purpose, we superimpose the orbits shown in Figures 10(a,b) on the bifurcation diagram of the fast subsystem (4). Note that the bifurcation structure is the

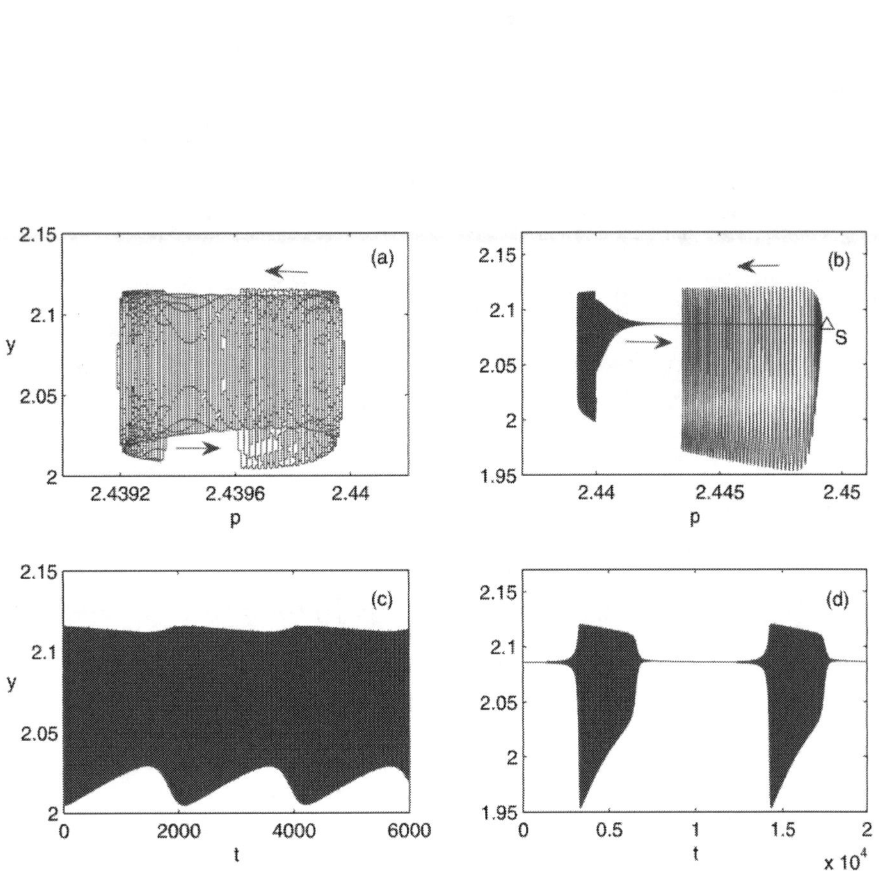

Figure 10. Transition from a smooth torus at $k_0 = 1.65189$ (a,c) to a distorted one at $k_0 = 1.65180$ (b,d). Blue arrows denote the direction of the flow. 'S' marks a saddle point.

same as the one generating bursting oscillations. From Figure 11a it becomes apparent that the trajectory (blue line) on the smooth torus sweeps back and forth the saddle-node bifurcation point SNP of the fast subsystem. This suggests that the SNP point acts as a trap for the flow as long as the torus remains sufficiently small, i.e. as long as k_0 is close to the Neimark-Sacker bifurcation point $k_0 = 1.6519$. However, as the flow rate becomes smaller, the torus increases in size until the flow gets near the stable manifold of the saddle point S. When this happens the smooth torus almost instantaneously deforms into the torus shown in Figure 11b.

The flow on the deformed torus also closely follows the invariant sets of the fast subsystem. The large amplitude oscillations occur in the vicinity of the cylinder-like manifold (bold black lines) until the saddle-node bifurcation point SNP is passed to the left and the trajectory gets attracted by the line-like manifold (black solid line) along which the orbit returns to the neighbourhood of the saddle point S to begin the next revolution.

4.2. A calcium oscillator model

Recently, a calcium oscillator model was proposed which describes the coupling between Ca^{2+} transport from the cytoplasm into the endoplasmic reticulum due to Ca^{2+}-ATPases (SERCA pumps) and intracellular energy-generation processes delivering ATP for the active transport [38]. The dynamics of the system were modeled using a 3-dimensional ODE system of the form

$$\begin{aligned} \frac{\mathrm{d}}{\mathrm{d}t}[Ca^{2+}] &= \nu_0 - \Phi([Ca^{2+}],[ATP]) \\ \frac{\mathrm{d}}{\mathrm{d}t}[ATP] &= q\Psi([GLC],\Phi) - \Phi([Ca^{2+}],[ATP]) - k[ATP] \end{aligned} \qquad (6)$$

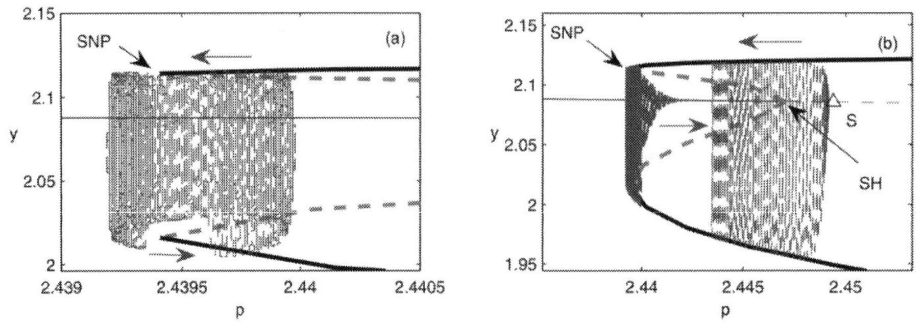

Figure 11. Overlay of the smooth torus (a) at $k_0 = 1.65189$ and the distorted torus (b) at $k_0 = 1.65180$ with the bifurcation diagram of the fast subsystem (4). Bold black/red lines denote stable/unstable limit cycles.

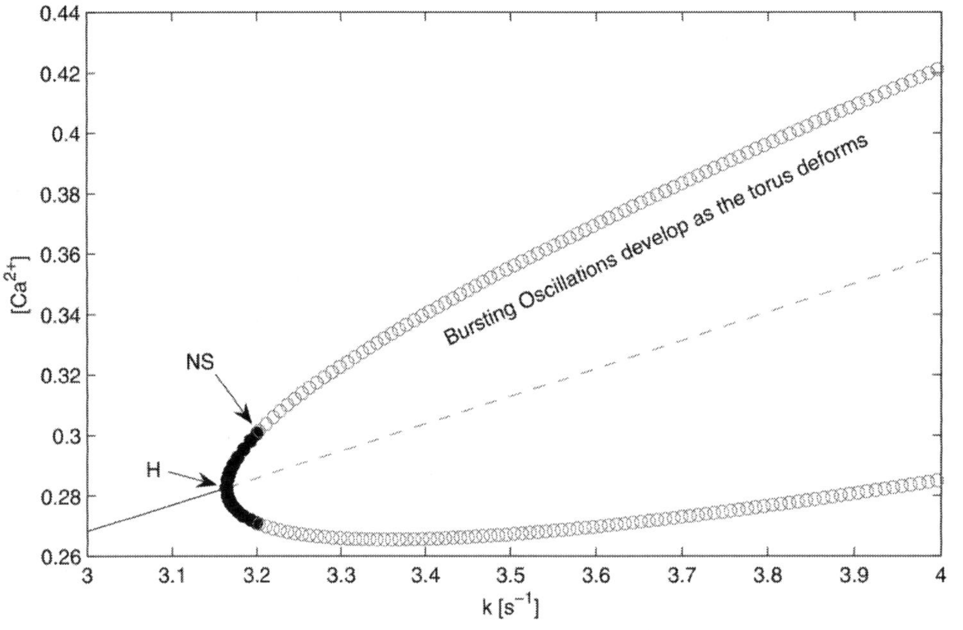

Figure 12. One-parameter bifurcation diagram of the calcium oscillator model (6).

$$\frac{\mathrm{d}}{\mathrm{d}t}[GLC] \quad = \quad \nu_1 - \Psi([GLC], \Phi)$$

where Φ and Ψ are nonlinear functions describing the rate of Ca^{2+} transport and the ATP-generation rate, respectively. [GLC] denotes the glucose concentration in the system. The two main control parameters are the Ca^{2+} influx rate ν_0 and the ATP-degradation rate k. (For details, see [38]).

A one-parameter bifurcation diagram of the 3-dimensional ODE system (6) is displayed in Figure 12. As the ATP-degradation rate k is increased from $3s^{-1}$ to $4s^{-1}$, the system undergoes a supercritical Hopf bifurcation H at $k = 3.163s^{-1}$ which is followed by a supercritical Neimark-Sacker bifurcation NS at $k = 3.199s^{-1}$ where the primary periodic orbit (black circles) becomes unstable (red circles) and a stable 2-torus is created (Fig. 13). As k is increased from the NS point, the torus continuously deforms (a,b) and eventually bursting oscillations arise (c,d). Figures 13(e,f,g,h) show the corresponding time series.

The flow on the torus in Figure 13a and the bursting oscillations in Figure 13d are strongly reminescent of their corresponding counterparts in the hemin system (Fig. 10a and Fig. 6) which suggests that the ODE system (6) exhibits a slow-fast structure similar to that of the hemin system (2) where the glucose concentration [GLC] now plays the role of the slow variable p of the hemin system. In order to test this hypothesis, we calculated the bifurcation diagram of the fast subsystem

$$\begin{aligned}\frac{\mathrm{d}}{\mathrm{d}t}[Ca^{2+}] &= \nu_0 - \Phi([Ca^{2+}],[ATP]) \\ \frac{\mathrm{d}}{\mathrm{d}t}[ATP] &= q\Psi([GLC],\Phi) - \Phi([Ca^{2+}],[ATP]) - k[ATP] \qquad (7)\end{aligned}$$

by treating the glucose concentration [GLC] as a quasi-static bifurcation parameter. The result is shown in Figure 14 where the bifurcation diagrams are superimposed on the phase portraits displayed in Figures 13(a-d).

Unlike the torus in the hemin system, the torus close to the Neimark-Sacker point in Figure 14a is now located near the subcritical Hopf bifurcation SH. As long as the torus is small, i.e. for k close to $k_{NS} = 3.199s^{-1}$, it remains in the neighbourhood of the SH point and the flow can be described as follows (Fig. 14b, blue arrow): The trajectory (blue line) oscillates back and forth around the SH point. It approaches the SH point with decreasing amplitude following the stable part of the line-like quasi-stationary manifold (black solid line) in an oscillatory manner. Beyond the subcritical Hopf point the line-like manifold becomes unstable and the amplitude of the oscillations increases while the trajectory re-enters the basin of attraction of the stable part of the line-like manifold. Consequently, it changes direction and completes one revolution along the torus.

As the ATP-degradation rate k is increased, the torus continuously grows in size until the trajectory reaches the basin of attraction of the cylinder-like manifold near the large amplitude limit cycles (black circles) of the fast subsystem. When this happens, bursting oscillations become possible (Figs. 14c,d) which are of sub-Hopf/fold-cycle type since the bifurcation structure of the fast subsystem is identical to that of the hemin system shown in Figure 5. In contrast to the oscillations on the 2-torus shown in Figures 14(a,b), the large and small amplitude oscillations of the bursting oscillations have different frequencies.

5. Summary and Discussion

We have investigated the bursting oscillations in two model systems, the hemin system and a Ca^{2+} oscillator system, both of which exhibit sub-Hopf/fold-cycle bursting over a wide range of parameters. The dynamics of these two systems are described by 3-dimensional ODE systems (eqs. 2 and 6) with a slow-fast substructure, i.e. each of them comprises a 2-dimensional fast and a 1-dimensional slow subsystem where the slow variable acts as a quasi-static bifurcation parameter for the fast subsystem. The merit of such a decomposition is that it can be used to classify the bursting behaviour of the full 3-dimensional system according to the bifurcations occurring in the fast subsystem that lead to the emergence and disappearance of the bursting oscillations.

A prerequisite for sub-Hopf/fold-cycle bursting is that a subcritical Hopf bifurcation and a saddle-node bifurcation of a periodic orbit take place at nearby values of the slow variable in the fast subsystem, as is the case if the fast subsystem is close to a generalized Hopf bifurcation

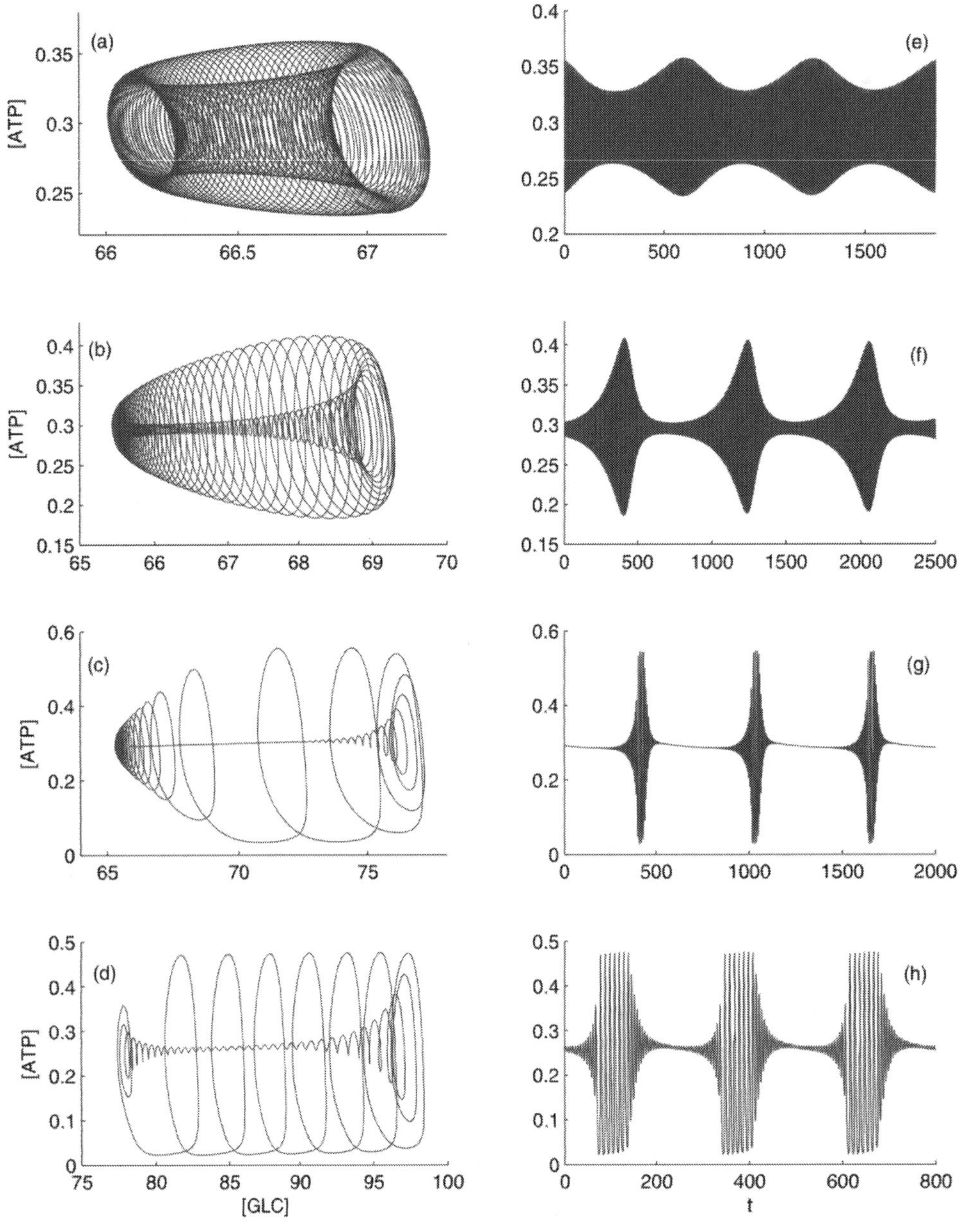

Figure 13. Deformation of the 2-torus: (Quasi-)periodic oscillations close to the Neimark-Sacker point at $k = 3.1991s^{-1}$ (a,e) and $k = 3.2s^{-1}$ (b,f) and the transition to bursting oscillations at $k = 3.23s^{-1}$ (c,g) and $k = 4s^{-1}$ (d,h).

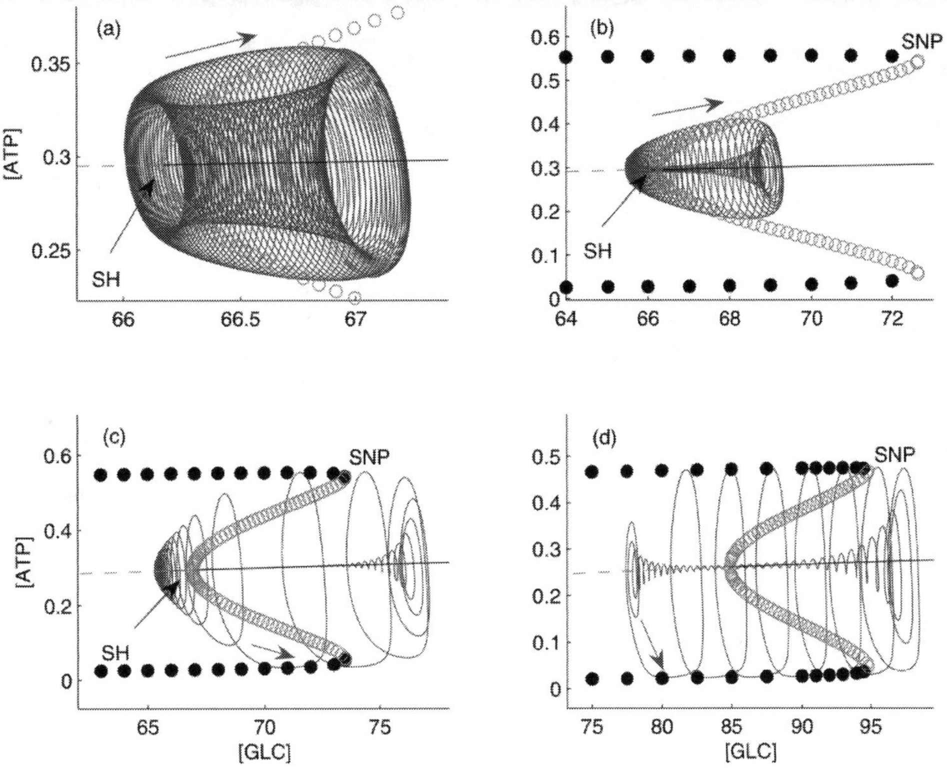

Figure 14. Overlay of the bifurcation diagram of the fast subsystem (7) with the tori (a,b) and the bursting oscillations (c,d) shown in Figures 13(a-d). Black and red circles denote stable and unstable limit cycles, respectively.

(GH, Fig. 8). For the hemin system, we explicitly verified that this condition is fulfilled over a wide range of the external control parameter k_0. For this purpose, a two-parameter continuation was performed using the flow rate k_0 and the slow variable p (Fig. 8).

We further argued that the two-parameter bifurcation diagram can be useful to detect qualitative changes in the bursting behaviour of the system. For the hemin system, we found a transition in the bursting behaviour from sub-Hopf/fold-cycle to fold/sub-Hopf type which is caused by a codimension two bifurcation in the fast subsystem. As a result of the bifurcation, the non-hyperbolic homoclinic orbit SNHC becomes hyperbolic (SHC) and the invariant cylinder-like manifold is destroyed (Fig. 9). Henceforth, the fast subsystem is bistable and leads to relaxational oscillations.

A further issue that has been addressed is the relation between bursting oscillations of sub-Hopf/fold-cycle type and (quasi-)periodic oscillations. In both of the model systems investigated we observed tori with special flows (cf. Fig. 11 and Figures 14a,b) suggesting an explanation in terms of the slow-fast structure of the corresponding ODE systems. Clearly, the flow on these tori is such that a typical trajectory oscillates faster in the direction perpendicular to the slow variable. If we consider a 2-torus as a direct product of two circles with different radii (Fig. 15), then the angular velocity ω_L along the circle with the larger radius is much larger than that of the circle with the smaller radius (Fig. 15a). In other words, the 'unusual' phase flow in

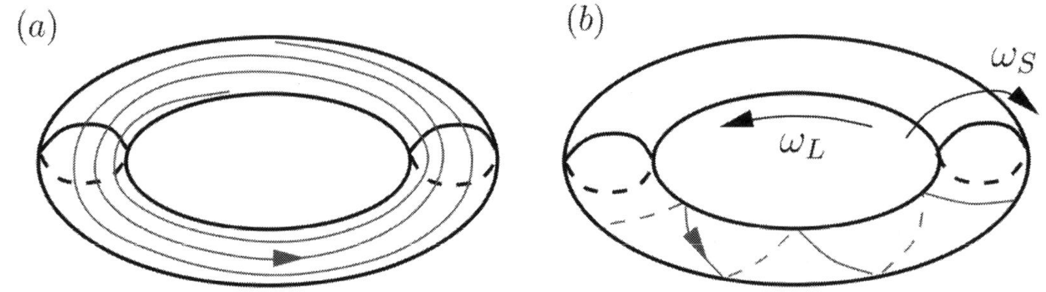

Figure 15. Sketch of the flow on 2-tori with different ratios of angular velocities: $\omega_L > \omega_S$ ('unusual') (a) and $\omega_L < \omega_S$ ('usual') (b).

Figure 11 and Figure 14 is a result of the time scale separation in the ODE systems (2) and (6).

A subsequent slow-fast analysis of the flow on the tori supports this picture. In particular, we observed the same bifurcation structure in the fast subsystem as for the bursting oscillations. Thus, the particular constellation of a subcritical Hopf bifurcation and a saddle-node bifurcation of a periodic orbit also allows for (quasi-)periodic oscillations on a 2-torus. While this holds true for both systems, there is, however, an important difference: The (quasi-)periodic oscillations in the hemin system emerge near the saddle-node bifurcation point (SNP, Fig. 11a) which leads to the blow-up effect, i.e. the torus rapidly increases in size in a very narrow parameter interval as the flow rate k_0 is decreased below the Neimark-Sacker point (cf. Fig. 11b). In contrast, the torus in the calcium oscillator model is generated near the subcritical Hopf bifurcation (Fig. 14a) which leads to a continuous deformation into bursting oscillations as the torus increases in size (Fig. 13). Thus, the two model systems follow different 'routes' to bursting oscillations. While the bursting oscillations in the hemin system arise as periodic orbits in narrow parameter windows of a chaotic domain far away from the region in parameter space where (quasi-)periodic behaviour is observed, the bursting oscillations in the calcium oscillator model emerge as a 2-torus continuously deforms. Both of these routes are well-known. However, our analysis suggests that the quasi-periodic route to bursting oscillations is followed if the corresponding torus is created near the subcritical Hopf bifurcation of the fast subsystem.

References

[1] Schmitz R A, Graziani K R and Hudson J L 1977 Experimental evidence of chaotic states in the Belousov–Zhabotinskii reaction *J. Chem. Phys.* **67** 3040

[2] Maselko J and Swinney H L 1986 Complex periodic oscillations and Farey arithmetic in the Belousov–Zhabotinskii reaction *J. Chem. Phys.* **85** 6430

[3] Hudson J L, Hart M and Marinko D 1979 An experimental study of multiple peak periodic and nonperiodic oscillations in the Belousov–Zhabotinskii reaction *J. Chem. Phys.* **71** 1601

[4] Koper M T M 1995 Bifurcations of mixed-mode oscillations in a three-variable autonomous Van der Pol–Duffing model with a cross-shaped phase diagram *Physica* D **80** 72

[5] Hauser M J B and Olsen L F 1996 Mixed-mode oscillations and homoclinic chaos in an enzyme reaction *J. Chem. Soc., Faraday Trans.* **92** 2857

[6] Steinmetz C G and Larter R 1991 The quasiperiodic route to chaos in a model of the peroxidase-oxidase reaction *J. Chem. Phys.* **94** 1388

[7] Hauck T and Schneider F W 1993 Mixed-mode and quasiperiodic oscillations in the peroxidase-oxidase reaction *J. Phys. Chem.* **97** 391

[8] Brusch L, Lorenz W, Or-Guil M, Bär M and Kummer U 2002 Fold-Hopf bursting in a model for calcium signal transduction *Z. Phys. Chem.* **216** 487

[9] Falcke M and Malchow D 2003 *Understanding Calcium Dynamics* (Berlin: Springer)

[10] Dean P M and Mathews E K 1970 Glucose-induced electrical activity in pancreatic islet cells *J. Physiol.* (London) **210** 255
[11] Wierschem K and Bertram R 2004 Complex bursting in pancreatic islets: a potential glycolytic mechanism *J. Theor. Biol.* **228** 513
[12] Goldbeter A 1996 *Biochemical Oscillations and Cellular Rhythms* (Cambridge, UK: Cambridge University Press)
[13] Wang X J and Rinzel J 1995 *The Handbook of Brain Theory and Neural Networks* ed M Arbib (Cambridge, MA: The MIT Press)
[14] Glass L and Mackey M C 1988 *From Clocks to Chaos: The Rhythms of Life* (Princeton, NJ: Princeton University Press)
[15] Dixon C J, Woods N M, Cuthbertson K S R and Cobbold P H 1990 Evidence for two Ca^{2+}-mobilizing purinoceptors on rat hepatocytes *Biochem. J.* **269** 499
[16] Woods N M, Cuthbertson K S R and Cobbold P H 1986 Repetitive transient rises in cytoplasmic free calcium in hormone-stimulated hepatocytes *Nature* **319** 600
[17] Rinzel J and Ermentrout G B 1989 *Methods in Neuronal Modeling* ed Koch C and Segev I (Cambridge, MA: The MIT Press) chapter: Analysis of neural excitability and oscillations
[18] Izhikevich E M 2000 Neural excitablitiy, spiking and bursting *Int. J. Bifurcat. Chaos* **10** 1171
[19] Hauser M J B, Strich A, Bakos R, Nagy-Ungvárai Z and Müller S C 2001 pH oscillations in the hemin – hydrogen peroxide – sulfite reaction *Faraday Discuss.* **120** 229
[20] Hauser M J B, Fricke N, Storb U and Müller S C 2002 Periodic and bursting pH oscillations in an enzyme model reaction *Z. Phys. Chem.* **216** 375
[21] Traylor T G and Xu F, 1987 The mechanisms of reaction of hydrogen peroxide and hydroperoxides with iron(III) porphyrins *J. Am. Chem. Soc.* **109** 6202
[22] Zhang G and Dasgupta P K 1992 Hematin as a peroxidase substitute in hydrogen peroxide determinations *Anal. Chem.* **64** 517
[23] Zhu Q-Z, Liu F-H, Xu J-G and Huang J-W 1998 Fluorescence immunossay of $\alpha - 1$-fetoprotein with hemin as a mimetic enzyme labelling reagent *Fresenius J. Anal. Chem.* **362** 537
[24] Rabai G, Kustin K and Epstein I R 1989 A systematically designed pH oscillator: the hydrogen peroxide – sulfite – ferrocyanide reaction in a continuous-flow stirred tank reactor *J. Am. Chem. Soc.* **111** 3870
[25] Rabai G 1997 Period-doubling route to chaos in the hydrogen peroxide – sulfur(IV) – hydrogen carbonate flow system *J. Phys. Chem.* A **101** 7085
[26] Frerichs G A and Thompson R C 1998 A pH-regulated chemical oscillator: the homogeneous system of hydrogen peroxide – sulfite – carbonate – sulfuric acid in a CSTR *J. Phys. Chem.* A **102** 8142
[27] Hanazaki I, Ishibashi N, Mori H and Tanimoto Y 2000 Bifurcation structure in the hydrogen peroxide – sulfite system *J. Phys. Chem.* A **104** 7695
[28] Rabai G, Okazaki N and Hanazaki I 1999 Kinetic role of CO_2 escape in the oscillatory H_2O_2 – HSO_3 – HCO_3 – flow system *J. Phys. Chem.* A **103** (1999) 7224
[29] Straube R, Müller S C and Hauser M J B 2003 Bursting oscillations in the revised mechanism of the hemin – hydrogen peroxid – sulfite oscillator *J. Phys. Chem.* **217** 1427
[30] Straube R, Flockerzi D, Müller S C and Hauser M J B 2005 Reduction of chemical reaction networks using quasi integrals *J. Phys. Chem.* A **109** 441
[31] Straube R, Flockerzi D, Müller S C and Hauser M J B 2005 Origin of bursting pH oscillations in an enzyme model system *Phys. Rev.* E **72** 066205
[32] Fenichel N 1979 Geometric singular perturbation theory for ordinary differential equations *J. Diff. Eq.* **31** 53
[33] O'Malley R E 1991 *Singular Perturbation Methods for Ordinary Differential Equations* (New York: Springer) p 83
[34] Lin C C and Segel L A 1994 *Mathematics Applied to Deterministic Problems in the Natural Sciences* (Philadelphia: SIAM) chapter 10
[35] Roussel M R and Fraser S J 2001 Invariant manifold methods for metabolic model reduction *Chaos* **11** 196
[36] Schauer M and Heinrich R 1983 Quasi-steady-state approximation in the mathematical modeling of biochemical reaction networks *Math. Biosci.* **65** 155
[37] Stiefenhofer M 1998 Quasi-steady-state approximation for chemical reaction networks *J. Math. Biol.* **36** 593
[38] Dellen B K, Barber M J, Ristig M L, Hescheler J, Sauer H and Wartenberg M 2005 [Ca^{2+}] oscillations in a model of energy-dependent Ca^{2+} uptake by the endoplasmic reticulum *J. Theor. Biol.* **237** 279
[39] Kuznetsov Y 1998 *Elements of Applied Bifurcation Theory* (Heidelberg: Springer)
[40] Hoppensteadt F C and Izhikevich E M 1997 *Weakly Connected Neural Networks* (New York: Springer)

Institute of Physics Publishing
doi:10.1088/1742-6596/55/1/021

Journal of Physics: Conference Series **55** (2006) 232–242
International Workshop on Multi-Rate Processes and Hysteresis

ODEs with Preisach operator under the derivative and with discontinuous in time right-hand side

A Zhezherun[1] and D Flynn[2]

[1,2]Dept of Civil & Environmental Engineering, University College Cork, Ireland

E-mail: [1]a.zhezherun@mars.ucc.ie, [2]d.flynn@mars.ucc.ie

Abstract. We consider ordinary differential equations with a Preisach operator under the derivative. A special case when the right-hand side has discontinuities in time is studied. We present theorems about the existence and uniqueness of solutions. We also prove a theorem which describes the behavior of a solution at the points of discontinuity of the right-hand side.

1. Introduction

This paper is devoted to a special kind of ordinary differential equation with a Preisach operator. This type of equation has been used in various areas of science to describe systems with hysteresis. The Preisach operator was originally used to describe magnetization phenomena in physics [6], but was understood to have a more generic nature, so that it can be applied in various areas [14]. Recent applications can be found, for example, in hydrology [9, 10, 11] and economics [12, 13].

Differential equations with a derivative of a Preisach operator and a smooth right-hand side have been studied in [1]. The operator introduces singularities in the equation, which can lead to non-uniqueness of solutions and a necessity to consider singular points in a special way. These questions have been addressed in [1]. In this paper we extend the class of right-hand sides of such equations to those containing discontinuities in time. Such equations naturally arise, for example, in hydrological models where discontinuous terms describe the rainfall [11]. These discontinuities lead to another set of singular points, compared to those of [1]. Such points are analyzed below.

The results of this paper provide a mathematical background for an algorithm to solve equations with a Preisach operator under the derivative, see companion paper [2]. This algorithm takes into account points of discontinuity of the right-hand side, so it can be viewed as an extended version of an algorithm published earlier [3]. The companion paper [2] also contains several numerical experiments that demonstrate the use of the algorithm for a particular hydrological problem.

The paper is organised as follows. In section 2 we describe the equation that we are interested in and give a definition of a solution of such an equation. Section 3 contains a brief definition of a Preisach operator and describes its properties that are necessary to formulate the results of the paper. Section 4 contains the main results about existence, uniqueness and extendability of solutions, and about the behavior of solutions at singular points. Then in section 5 we provide a simple explanation of how to come up with the main formula (4) which describes the behavior of a solution at the points of discontinuity of right-hand side. Finally, in section 6 we give the proofs of the theorems.

2. Problem

Consider the equation

$$y'(t) = f(t, x(t)) + g(t) = F(t, x(t)), \qquad (1)$$
$$y(t) = \mathcal{P}[\eta(t)]x(t),$$

where $x(t)$ and $y(t)$ are the input and the output respectively of the Preisach operator $\mathcal{P}$ with a variable state $\eta(t)$, and consider an initial value problem for (1) with

$$x(t_0) = x_0, \quad \eta(t_0) = \eta_0. \qquad (2)$$

Suppose that the function $f(t, x)$ is continuously differentiable with respect to t and x, and $g(t)$ is continuously differentiable except for points $\mathbb{T} = \{\tau_i\}$ at which all of the values $g(\tau_i - 0)$, $g(\tau_i + 0)$, $g'(\tau_i - 0)$, $g'(\tau_i + 0)$ are defined and finite, but $g(t)$ or $g'(t)$ may have finite jumps at τ_i. We also assume that any finite interval contains only a finite number of points τ_i. The measure density μ of the Preisach operator is assumed to satisfy properties (p1)–(p3) (see section 3).

Definition 1. A continuous function $x(t)$ is called a *solution* of (1)–(2) on the interval $[t_0, T)$, if $x(t_0) = x_0$ and the following conditions hold:

- (i) function $y(t) = \mathcal{P}[\eta(t)]x(t)$ with $\eta(t_0) = \eta_0$ is continuously differentiable on $(t_0, T) \setminus \mathbb{T}$, has finite left and right derivatives for all $t \in (t_0, T) \cap \mathbb{T}$, and has a finite right derivative at t_0 if $t_0 \notin \mathbb{T}$,
- (ii) equation (1) holds for $t \in (t_0, T) \setminus \mathbb{T}$,

There is a connection between this problem and a problem with a smooth right-hand side. Let (t_1, t_2) be any interval that does not contain points from $\mathbb{T}$, so that $F(t, x)$ is continuously differentiable for $t \in (t_1, t_2)$, $x \in \mathbb{R}$. Consider a modified initial value problem with a continuously differentiable right-hand side

$$y'(t) = f(t, x(t)) + \tilde{g}(t) = \tilde{F}(t, x(t)), \qquad (1')$$

where $\tilde{g} : \mathbb{R} \to \mathbb{R}$ is a continuously differentiable extension of g from (t_1, t_2) to $\mathbb{R}$. Then any solution $\tilde{x}$ of the modified problem on the segment $[t_1, t_2]$ will be a solution of the original problem (1)–(2) on the same interval, and vice versa.

3. Preisach operator

The Preisach nonlinearity is a parallel connection of a continuous bundle of nonideal relays. In this section we give a compact definition of this nonlinearity and state all the properties that are required for subsequent sections. For a more detailed definition and discussion of a Preisach operator, see [7, 8].

The nonideal relay $R_{\alpha,\beta}$ (with threshold values $\alpha \leq \beta$) is an elementary building block from which the Preisach nonlinearity is constructed. Consider a family $\mathcal{R}$ of relays $R_{\alpha,\beta}$ with threshold values in the half-plane $\{\alpha \leq \beta\}$. The plane of all thresholds α and β is referred to as the Preisach plane, and each point above the line $\{\alpha = \beta\}$ in this plane corresponds to a single relay from this family.

Introduce a probability measure on the half-plane $\{\alpha \leq \beta\}$ with a density $\mu(\alpha, \beta)$, and a measurable function $\omega_0(\alpha, \beta)$ which we call the initial state of the family $\mathcal{R}$. Then for any initial state $\omega_0(\alpha, \beta)$ and any continuous input $x(t)$ for $t \geq t_0$ we define the function

$$y(t) = \mathcal{P}x(t) = \iint_{\alpha \leq \beta} \mu(\alpha, \beta) R_{\alpha,\beta}[t_0, \omega_0(\alpha, \beta)]x(t)\, d\alpha\, d\beta. \qquad (3)$$

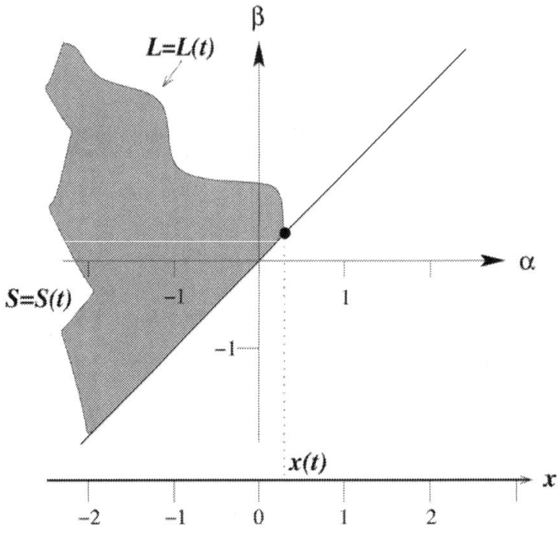

Figure 1. The state of Preisach operator

This model is called the Preisach nonlinearity (or, when convenient, Preisach model or Preisach operator). Here $y(t)$ is the output of the model with initial state ω_0 and input $x(t)$.

Throughout this paper, we assume the following properties of the measure density μ:

(p1) μ is zero outside some strip $0 \leq \beta - \alpha \leq d$,

(p2) The function $\mu = \mu(\alpha, \beta)$ is continuously differentiable in the strip $0 \leq \beta - \alpha \leq d$,

(p3) There is a strip $0 \leq \beta - \alpha \leq d_1$ with $d_1 \leq d$ where μ is positive.

The same assumptions are used in [1], where differential equations with a smooth right-hand side are considered.

For $t > t_0$ the function

$$\omega(\alpha, \beta) = R_{\alpha,\beta}[t_0, \omega_0(\alpha, \beta)]x(t).$$

is interpreted as the state of the Preisach nonlinearity at the moment t.

It is natural to limit the set of possible initial states to those defined by

$$\omega_0(\alpha, \beta) = \begin{cases} 0, & \text{if } \alpha + \beta > \eta_0(\beta - \alpha) + 2x_0, \\ 1, & \text{if } \alpha + \beta \leq \eta_0(\beta - \alpha) + 2x_0. \end{cases}$$

where a function $\eta_0 : [0, d] \to \mathbb{R}$ satisfies

$$\eta_0(0) = 0, \quad |\eta_0(\xi_1) - \eta_0(\xi_2)| \leq |\xi_1 - \xi_2|.$$

We will identify this function η_0 with the initial state ω_0, Then for any initial state $\eta_0 = \eta_0(\xi)$ and any continuous input $x : [t_0, \infty) \to \mathbb{R}$ the variable state $\eta(t) = \eta(t; \xi)$ at any monent $t \geq t_0$ is defined. We will refer to this function $\eta(t)$ as to the state of the Preisach nonlinearity.

Geometrically this definition of a state means that the relays with parameter values (α, β) below the curve $L(t) = \{\alpha + \beta = \eta(t; \beta - \alpha) + 2x(t)\}$ are switched on, and the relays that correspond to points above this curve are switched off, see figure 1. Then the output $y(t)$ of the operator equals the measure of the shaded set $S(t) = \{\alpha + \beta \leq \eta(t; \beta - \alpha) + 2x(t)\}$.

The evolution of the Preisach state admits a simple geometrical interpretation [8], see figure 2. Here the input $x(t)$ moves along the horizontal line and controls the point on the diagonal

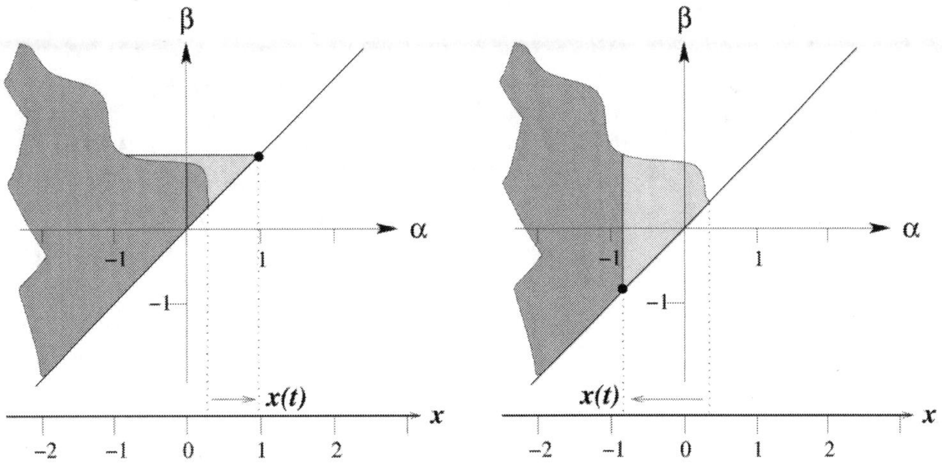

(a) Increasing input

(b) Decreasing input

Figure 2. Dynamics of the Preisach state

$\alpha = \beta$ immediately above itself. When the input increases, this point on the diagonal drags the horizontal line and shades the domain below this line and above the diagonal, see figure 2(a). When the input decreases, the diagonal point drags the vertical line, and 'clears' everything to the right of this line and above the diagonal, see figure 2(b).

The situations depicted in figure 2 are important for the subsequent text. Namely, we will say that the Preisach state has a *horizontal last segment*, if it looks like in the figure 2(a), and a *vertical last segment*, if it looks like in the figure 2(b).

The Preisach operator with a measure density satisfying (p1)–(p3) has some important properties, which were proved in [1] and which are required for the proofs in section 6. For convenience, they are reproduced below.

Proposition 1. *Suppose that the output $y(t)$ of Preisach operator $\mathcal{P}[\eta(t)]x(t)$ is increasing (decreasing) on the interval $[t_1, t_2]$. Then the input $x(t)$ is also increasing (decreasing) on the same interval.*

Proposition 2. *Suppose that the output $y(t)$ of Preisach operator $\mathcal{P}[\eta(t)]x(t)$ is increasing (decreasing) and absolutely continuous on the interval $[t_1, t_2]$. Then the input $x(t)$ is also absolutely continuous on the same interval.*

4. Main Results

Let

$$\Omega = \{(t,x) \colon t \notin \mathbb{T} \wedge F(t,x) = 0 \wedge F_t(t,x) = 0\},$$
$$\Omega_1 = \{(t,x) \colon t \in \mathbb{T} \wedge F(t-0,x)F(t+0,x) = 0\},$$

where F_t denotes $\frac{\partial F}{\partial t}$.

Definition 2. A triple (t_0, x_0, η_0) is called *admissible*, if $(t_0, x_0) \notin \tilde{\Omega} = \Omega \cup \Omega_1$, and the initial state η_0 has the following property: if $F(t_0, x_0) = 0$, $F_t(t_0, x_0) > 0$, then η_0 has a vertical last segment, and if $F(t_0, x_0) = 0$, $F_t(t_0, x_0) < 0$, then η_0 has a horizontal last segment. If $t_0 \in \mathbb{T}$, then η_0 has a nonzero last segment (horizontal or vertical).

Theorem 1. *For any admissible initial data, the initial value problem (1)–(2) has a unique solution on some interval* $t_0 \le t < t_1$.

Theorem 2. *For any admissible initial data the solution of initial value problem (1)–(2) can be uniquely extended to the maximal interval* $t_0 \le t < T$ *(finite or infinite) such that* $(t, x(t)) \in \mathbb{R}^2 \setminus \tilde{\Omega}$ *for all* t *from this interval. If this interval is finite* ($T < \infty$), *then either* $x(t) \to +\infty$, $x(t) \to -\infty$ *or* $x(t) \to x_*$ *as* $t \to T - 0$, *and* $(T, x_*) \in \tilde{\Omega}$ *in the latter case.*

The proofs of these two theorems are obtained by considering equivalent modified problems with smooth right hand sides $\tilde{F}$ on intervals (τ_i, τ_{i+1}), and then applying the corresponding existence, uniqueness and extension theorems from [1].

If $(T, x_*) \in \Omega_1$, it does not necessarily mean that the solution cannot be uniquely continued past that point. Here we omit the analysis of such points for simplicity.

We should also note that the triple $(t, x(t), \eta(t))$ remains admissible along the trajectory of a solution $x(t)$, $t \in [t_0, T)$.

Let (t_0, x_0, η_0) be an admissible initial data, and $x\colon [0, T) \to \mathbb{R}$ be a solution of the initial value problem (1)–(2). Then for any $t \in (t_0, T), t \notin \mathbb{T}$: $x(t)$ has finite left and right derivatives. For $t_* \in \mathbb{T}$ such that $F(t_* - 0, x_*)F(t_* + 0, x_*) > 0$, $x(t)$ also has finite one-sided derivatives. All these facts are proved by considering modified problems with smooth right-hand sides and applying Theorem 4.3 from [1] which gives values for $x'(t)$. The same is true for the left derivative of x at the points $t_* \in \mathbb{T}$ such that $F(t_* - 0, x_*)F(t_* + 0, x_*) < 0$. The right derivative at such points, however, is infinite, and the behavior of the solution to the right of such a time point is described by the following theorem.

Theorem 3. *Let* $t_* \in \mathbb{T}$ *be such that* $F(t_* - 0, x_*)F(t_* + 0, x_*) < 0$. *Then the solution* $x(t)$ *has an infinite right derivative at* t_*, *and there exist* $\delta > 0$ *and* $C > 0$ *such that for any* $t_* \le t < t_* + \delta$ *the following inequality holds:*

$$\left| (x(t) - x_*) - \operatorname{sgn}(F_*)\sqrt{2(t - t_*)\frac{|F_*|}{\mu_*}} \right| \le C(t - t_*), \tag{4}$$

where

$$F_* = F(t_* + 0, x_*), \quad \mu_* = \mu(x_*, x_*), \quad x_* = x(t_*).$$

The same inequality holds at the moment t_0, if $t_0 \in \mathbb{T}$, $F(t_0 + 0, x_0) > 0$ and η_0 has a vertical last segment (or $F(t_0 + 0, x_0) < 0$ and η_0 has a horizontal last segment).

The last theorem gives explicit formulas for the constants δ and C.

Theorem 4. *Let* $t_* \in \mathbb{T}$ *be such that* $F(t_* - 0, x_*)F(t_* + 0, x_*) < 0$. *Denote by* Δx *the nonzero length of the last segment of the Preisach state* $\eta(t_*)$. *Denote by* (t_*, T) *an interval that does not contain any points from* $\mathbb{T}$ *such that there exists a constant* $\frac{C_1}{2}$ *which bounds the first derivatives of* F *and* μ:

$$\left|\frac{\partial F}{\partial t}\right|, \left|\frac{\partial F}{\partial x}\right|, \left|\frac{\partial \mu}{\partial \alpha}\right|, \left|\frac{\partial \mu}{\partial \beta}\right| \le \frac{C_1}{2}. \tag{5}$$

for all $(t, x) \in [t_*, T) \times [x_* - \Delta x, x_* + \Delta x]$. *Introduce constants*

$$\tilde{C} = \frac{4\sqrt{3}C_1}{\mu_*^2}(\mu_* + F_*), \tag{6}$$

$$\delta_* = \min\left\{T - t_*, \Delta x, \frac{F_*}{2C_1}, \frac{\mu_*}{2C_1}, \frac{6F_*}{\mu_*}\right\}, \tag{7}$$

$$\tilde{\delta} = \delta_*^2 \frac{\mu_*}{6F_*} \le \delta_*. \tag{8}$$

Then Theorem 3 holds with $\tilde{\delta}$ *and* $\tilde{C}$.

It is possible to use Theorem 4 to obtain guaranteed error estimates for numerical methods for solving differential equations with a Preisach operator under the derivative and a discontinuous in t right-hand side. However, such estimates lie beyond the scope of this paper, and obtaining them will require additional work.

The above theorems were formulated under the assumption that the Preisach measure density μ satisfies properties (p1)–(p3). It should be noted that in important applications this may not be the case. For example, in the wedge model for hysteresis in hydrology (see [4, 5]) the measure density $\mu(\alpha, \beta)$ is equal to zero for any $\alpha > 0$. However, the results of this paper can still be applied to the wedge model after a simple modification. Namely, if there exists a set $S \subset \mathbb{R}$ such that properties (p1)–(p3) hold for any $(\alpha, \beta) \in \{\alpha, \beta \in S \wedge \alpha \leq \beta\}$, then the definition of an admissible triple (t_0, x_0, η_0) can be extended to include a requirement that $x_0 \in S$, and all the theorems will hold on an interval $[t_0, T)$ such that $x(t) \in S$ for all t from this interval.

5. Simple derivation of the main formula (4)

Consider the problem (1)–(2) with admissible triple (t_0, x_0, η_0) such that $t_0 \in \mathbb{T}$, $F(t_0{+}0, x_0) > 0$ and the initial state η_0 has a vertical last segment. We will construct a local approximation of the solution $x(t)$ in a vicinity of point t_0.

First of all, we rewrite the equation (1) in the following form:

$$y'(t) = Q(x(t))x'(t) = F(t, x(t)), \tag{9}$$

where $Q(x) = \frac{\partial Px}{\partial x}$. Note that the output $y(t)$ of the Preisach operator is increasing in some vicinity of t_0 due to $F(t_0 + 0, x_0) > 0$, and, according to Proposition 1, the input $x(t)$ is also increasing in this vicinity. Thus the Preisach state $\eta(t)$ has a horizontal last segment for $t > t_0$ in a vicinity of t_0. Then $Q(x)$ is the measured length of this segment, which has endpoints (x_0, x) and (x, x) in Preisach plane. At the moment t_0 this horizontal segment has zero length, thus creating a singularity.

In order to find an approximation we write the Taylor series for $Q(x)$ and $F(t, x)$ at the point (t_0, x_0) for $t > t_0$ (i.e. after the jump of $F(t, x)$ had occurred). Namely,

$$F(t, x) = F(t_0 + 0, x_0) + o(t - t_0) + o(x - x_0) \approx F_0,$$

and

$$Q(x) = 0 + \mu(x_0, x_0)(x - x_0) + o(x - x_0) \approx \mu_0(x - x_0).$$

where $F_0 = F(t_0 + 0, x_0)$ and $\mu_0 = \mu(x_0, x_0)$. This allows us to approximate (9) by

$$\mu_0 z(t) z'(t) \approx F_0,$$

where $z(t) = x(t) - x_0$. Integrating both sides and solving for z gives

$$z(t) \approx \sqrt{2(t - t_0)\frac{F_0}{\mu_0}}. \tag{10}$$

This is exactly the approximation from (4). Theorems 3 and 4 give a means by which this formula can be interpreted.

The computer algorithm in the companion paper [2] uses approximation (10) to make a special step when crossing a discontinuity point. It uses a standard numerical ODE solver prior to the jump, then a single time-step is made in accordance with (10), after which the standard ODE solver can continue once more with new initial conditions.

6. Proof of theorems

6.1. Proof of Theorem 1

If $t_0 \notin \mathbb{T}$, then the function $F(t, x)$ is smooth in the vicinity of the point (t_0, x_0), and local existence and uniqueness follow from the corresponding theorem for a smooth right-hand side (see Theorem 4.1 from [1]).

If $t_0 \in \mathbb{T}$, then consider the modified problem (1')–(2) on a sufficiently small interval $[t_0, t_0+\varepsilon]$ with a smooth right-hand side function $\tilde{F}$. According to definition 2, $\tilde{F}(t_0, x_0) \neq 0$, which implies that the initial data will be admissible for this modified problem, so it also has a unique solution. It remains to note that any solution of problem (1')–(2) on the interval $[t_0, t_0 + \varepsilon]$ is a solution of the original system (1)–(2) on the same interval and vice versa, and therefore the original problem also has a unique local solution. □

6.2. Proof of Theorem 2

Consider the interval $[t_0, \tau_1]$, where $\tau_1 > t_0$, $\tau_1 \in \mathbb{T}$ is a "jump point" of $g(t)$ closest to t_0. On this interval the problem (1)–(2) is equivalent to the modified problem (1')–(2). According to the theorem on the extension of solutions for equations with a smooth right-hand side (see Theorem 4.2 from [1]), the problem (1')–(2) has a unique solution that can be extended to a maximal interval $t_0 \leq t < \tilde{T}$, and there are the following options for $\tilde{T}$:

(i) $\tilde{T} \leq \tau_1$, and $x(t) \to \pm\infty$ or $x(t) \to x_*$ as $t \to \tilde{T} - 0$, where $(\tilde{T}, x_*) \in \Omega \subset \tilde{\Omega}$. In this case we can set $T = \tilde{T}$, and $[t_0, T)$ will be the maximal interval to which the solution of (1)–(2) can be extended.

(ii) $\tilde{T} > \tau_1$. Then either $(\tau_1, x(\tau_1)) \in \tilde{\Omega}$, in which case we can set $T = \tau_1$, or $(\tau_1, x(\tau_1)) \notin \tilde{\Omega}$. In the latter case we consider the initial value problem problem with initial data τ_1, $x(\tau_1)$ and $\eta(\tau_1)$ on the interval $[\tau_1, \tau_2]$, where τ_2 is the next jump point, and repeat the same reasoning as above.

If this process stops on some interval $[\tau_i, \tau_{i+1}]$, then the statement of the theorem will hold with corresponding $T = \tilde{T}$. If it doesn't stop, or if the last interval $[\tau_k, \infty)$ is infinite and for that last interval we get $\tilde{T} = \infty$, then the statement of the theorem will hold with $T = \infty$. □

6.3. Proof of Theorems 3 and 4

Consider the case when $t_* \in \mathbb{T}$, $F(t_* - 0, x_*) < 0$, $F(t_* + 0, x_*) > 0$. This implies that the state $\eta_* = \eta(t_*)$ of the Preisach nonlinearity has a vertical last segment of length Δx.

If $F(t_* + 0, x_*) > 0$ then $y'(t) > 0$ in some interval $t_* \leq t < t_1$. Then $y(t)$ is increasing on this interval, and according to Proposition 1 this implies that $x(t)$ is also increasing on the same interval.

Introduce constants C_1, $\tilde{C}$, δ_* and $\tilde{\delta}$ given by equations (5)–(8) as specified in Theorem 4. The interval $(t_*, t_* + \tilde{\delta})$ does not contain points from $\mathbb{T}$, which implies that if we replace $g(t)$ with its continuous extension from this interval to $\mathbb{R}$, then the modified problem will have the same solution. So, in the following text, the function F will be assumed to be continuously differentiable on $\mathbb{R}^2$.

Denote

$$\bar{x}(t, \varepsilon) = x_* + \sqrt{2(t - t_*)\frac{F_* + C_1\varepsilon}{\mu_* - C_1\varepsilon}}.$$

Select an arbitrary $\delta \in (0, \tilde{\delta})$ and let $\varepsilon = \sqrt{\delta \frac{6F_*}{\mu_*}}$, $\delta < \varepsilon < \delta_*$.

Lemma. *Under these conditions the solution $x(t)$ exists and is increasing on the interval $[t_*, t_* + \delta]$, and for any $t \in [t_*, t_* + \delta]$*

$$x_* \leq x(t) \leq \bar{x}(t, \varepsilon) \leq x_* + \varepsilon. \tag{11}$$

Proof of the lemma will be given after the proof of the theorem.

The function $\bar{x}(t,\varepsilon)$ is differentiable with respect to ε. This implies

$$|\bar{x}(t,\varepsilon)-\bar{x}(t,0)| \le \varepsilon \max_{0\le\theta\le\varepsilon}\left|\frac{\partial \bar{x}}{\partial \varepsilon}(t,\theta)\right|, \quad \text{where } \bar{x}(t,0)=x_*+\sqrt{2(t-t_*)\frac{F_*}{\mu_*}}.$$

Using the inequalities

$$\frac{\mu_*}{2} < \mu_* - C_1\varepsilon \le \mu_*, \quad F_* \le F_* + C_1\varepsilon < \frac{3F_*}{2},$$

we write the following series of relations

$$\begin{aligned}
\left|\frac{\partial \bar{x}}{\partial \varepsilon}\right| &= \left|\sqrt{\frac{(t-t_*)(\mu_*-C_1\varepsilon)}{2(F_*+C_1\varepsilon)}}\cdot\frac{C_1(\mu_*+F_*)}{(\mu_*-C_1\varepsilon)^2}\right| \\
&\le \sqrt{\frac{(t-t_*)\mu_*}{2F_*}}\cdot\frac{4}{\mu_*^2}\cdot C_1(\mu_*+F_*) \\
&= \frac{2\sqrt{2}C_1}{\mu_*\sqrt{\mu_* F_*}}(\mu_*+F_*)\sqrt{t-t_*} = C_2\sqrt{t-t_*}.
\end{aligned}$$

Finally, taking into account that $\varepsilon=\sqrt{\delta\frac{6F_*}{\mu_*}}$,

$$\begin{aligned}
x(t) &\le \bar{x}(t,\varepsilon) \le \bar{x}(t,0)+C_2\sqrt{t-t_*}\varepsilon = \bar{x}(t,0)+C_2\sqrt{\delta\frac{6F_*}{\mu_*}}\sqrt{t-t_*} \\
&\le \bar{x}(t,0)+C_2\sqrt{\frac{6F_*}{\mu_*}}\cdot\delta = x_*+\sqrt{2(t-t_*)\frac{F_*}{\mu_*}}+C\delta,
\end{aligned}$$

where

$$C=C_2\sqrt{\frac{6F_*}{\mu_*}}=\frac{4\sqrt{3}C_1}{\mu_*^2}(\mu_*+F_*)=\tilde{C}.$$

Note that this inequality holds for each $t_*\le t\le t_*+\delta$, so we can write it for $t=t_*+\delta$:

$$x(t_*+\delta)\le x_*+\sqrt{2\delta\frac{F_*}{\mu_*}}+\tilde{C}\delta,$$

which holds for all $0\le\delta<\tilde{\delta}$. Replacing δ with $t-t_*$, $t_*\le t<t_*+\tilde{\delta}$, we get

$$x(t)\le x_*+\sqrt{2(t-t_*)\frac{F_*}{\mu_*}}+\tilde{C}(t-t_*) \quad \text{for all } t_*\le t<t_*+\tilde{\delta}, \tag{12}$$

which provides us with the upper bound for $x(t)$. To get the lower bound, we use a series of similar calculations, so for the rest of the proof we give only the most important equations. First, we introduce a function $\underline{x}(t,\varepsilon)$:

$$\underline{x}(t,\varepsilon)=x_*+\sqrt{2(t-t_*)\frac{F_*-C_1\varepsilon}{\mu_*+C_1\varepsilon}},$$

and prove that $x(t) \geq \underline{x}(t, \varepsilon)$. Then we obtain an estimate for the difference $\underline{x}(t, \varepsilon) - \underline{x}(t, 0)$:

$$|\underline{x}(t,\varepsilon) - \underline{x}(t,0)| \leq \varepsilon \max_{0 \leq \theta \leq \varepsilon} \left| \frac{\partial \underline{x}}{\partial \varepsilon}(t, \theta) \right| \leq C_3 \delta,$$

$$\text{where } C_3 = \frac{3\sqrt{2}C_1}{\mu_*^2}(\mu_* + F_*) < \tilde{C}.$$

Note that $\underline{x}(t, 0) = \bar{x}(t, 0)$, so we obtain the inequality

$$x(t) \geq x_* + \sqrt{2(t - t_*)\frac{F_*}{\mu_*}} - \tilde{C}(t - t_*) \quad \text{for all } t_* \leq t < t_* + \tilde{\delta},$$

which together with (12) proves (4) in the case when $F(t_* + 0, x_*) > 0$. The case when $F(t_* + 0, x_*) < 0$ can be considered in the same way with the appropriate modification of functions $\bar{x}$ and $\underline{x}$. Thus, the statement of Theorem 3 holds with constants $\tilde{\delta}$ and $\tilde{C}$, so both Theorem 3 and Theorem 4 are proved. □

Proof of Lemma. The proof consists of two parts. Firstly, we suppose that the solution $x(t)$ is defined and is increasing on the interval $[t_*, t_* + \delta]$, and that $x(t) \leq x_* + \varepsilon$ for all t from this interval, and prove that (11) holds under these assumptions. Secondly, we prove that these initial assumptions were correct.

Introduce a continuous increasing function φ defined on $[x_*, x_* + \varepsilon]$:

$$\varphi(x) = y_* + \text{mes}_\mu\{(\alpha, \beta) \colon x_* \leq \alpha \leq x, \alpha \leq \beta \leq x\},$$

where $y_* = y(t_*)$. If $x(t)$ is increasing on $[t_*, t_* + \delta]$ and $x(t) \leq x_* + \varepsilon < x_* + \Delta x$, and η_* has a vertical last segment of size Δx, then the output y of the Preisach nonlinearity equals $y = \varphi(x)$. The function $\varphi(x)$ is increasing on $[x_*, x_* + \varepsilon]$, therefore it has an inverse function $\psi(y) = \varphi^{-1}(y)$ defined on $[y_*, \varphi(x + \varepsilon)]$, and the equation

$$y'(t) = F(t, \psi(y))$$

holds for all $t \in [t_*, t_* + \delta]$.

Let $\tilde{F} = F_* + C_1\varepsilon$. Then we get the following inequality for the function $F(t, x)$:

$$\begin{aligned} F(t,x) &\leq F_* + (t - t_*)\max|F_t| + (x - x_*)\max|F_x| \leq F_* + \frac{C_1}{2}\delta + \frac{C_1}{2}\varepsilon \\ &< F_* + C_1\varepsilon = \tilde{F} \end{aligned}$$

for all $(t, x) \in [t_*, t_* + \delta] \times [x_*, x_* + \varepsilon]$. In the same way we obtain an inequality for $\mu(\alpha, \beta)$:

$$\mu(\alpha, \beta) \geq \mu_* - (\alpha - x_*)\max\left|\frac{\partial \mu}{\partial \alpha}\right| + (\beta - x_*)\max\left|\frac{\partial \mu}{\partial \beta}\right| \geq \mu_* - C_1\varepsilon, \tag{13}$$

for $\alpha, \beta \in [x_*, x_* + \varepsilon]$.

Let $\tilde{y}(t)$ be a solution of the equation

$$\tilde{y}'(t) = \tilde{F} = F_* + C_1\varepsilon,$$

with initial data $\tilde{y}(t_*) = y(t_*) = y_*$. Then, according to the theorem on differential inequalities (see, for example, Theorem 4.1 in [15]), $y(t_*) = \tilde{y}(t_*)$ and $F \leq \tilde{F}$ imply that $y(t) \leq \tilde{y}(t)$ for $t \in [t_*, t_* + \delta]$. The function $\tilde{y}$ has the following explicit form:

$$\tilde{y} = y_* + (F_* + C_1\varepsilon)(t - t_*).$$

Thus, we obtain the inequality

$$y(t) \le y_* + (F_* + C_1\varepsilon)(t - t_*) \quad \text{for all } t_* \le t \le t_* + \delta. \tag{14}$$

Now we return to the x variable, and to do so we need a lower estimate for the function $\varphi(x)$. Let $\tilde{\varphi}(x)$ be a continuous increasing function defined for all $x \ge x_*$ as

$$\tilde{\varphi}(x) = y_* + \frac{1}{2}(\mu_* - C_1\varepsilon)(x - x_*)^2.$$

Using the estimate (13) we get

$$\begin{aligned}\varphi(x) &= y_* + \int_{x_*}^{x} d\alpha \int_{\alpha}^{x} \mu(\alpha, \beta) d\beta \\ &\ge y_* + (\mu_* - C_1\varepsilon) \int_{x_*}^{x} d\alpha \int_{\alpha}^{x} d\beta = y_* + \frac{1}{2}(\mu_* - C_1\varepsilon)(x - x_*)^2 = \tilde{\varphi}(x)\end{aligned}$$

for all $x \in [x_*, x_* + \varepsilon]$. Consider a continuous increasing extension of function φ such that the inequality $\varphi(x) \ge \tilde{\varphi}(x)$ holds for all $x \ge x_*$. Then both φ and $\tilde{\varphi}$ will have inverse functions ψ and $\tilde{\psi}$ defined for $y \ge y_*$, and $\psi \le \tilde{\psi}$. The function $\tilde{\psi} = \tilde{\varphi}^{-1}$ can be written explicitly as

$$\tilde{\psi}(y) = x_* + \sqrt{2\frac{y - y_*}{\mu_* - C_1\varepsilon}}.$$

Then, applying ψ to both sides of (14), we obtain

$$\begin{aligned}x(t) &= \psi(y(t)) \le \psi(y_* + (F_* + C_1\varepsilon)(t - t_*)) \\ &\le \tilde{\psi}(y_* + (F_* + C_1\varepsilon)(t - t_*)) = x_* + \sqrt{2(t - t_*)\frac{F_* + C_1\varepsilon}{\mu_* - C_1\varepsilon}} = \bar{x}(t, \varepsilon)\end{aligned}$$

for all $t \in [t_*, t_* + \delta]$. Thus, under our initial assumptions the statement of the lemma holds.

Note that, due to the choice of C_1 and ε, $F_* + C_1\varepsilon < \frac{3}{2}F_*$, $\mu_* - C_1\varepsilon > \frac{1}{2}\mu_*$, and thus $\bar{x}(t, \varepsilon) < x_* + \sqrt{\delta\frac{6F_*}{\mu_*}} = x_* + \varepsilon$.

Now we have to show that our assumptions were correct. We assumed that $x(t) \le x_* + \varepsilon$. If this is not true, then there exists a time moment t such that $t_* < t \le t_* + \delta$ and $x(t) > x_* + \varepsilon$. In this case the continuity of x implies that there exists a moment t_1 such that for $t_* \le t \le t_1$: $x(t) \le x_* + \varepsilon$ and $x(t_1) = x_* + \varepsilon$. Then if we repeat all the above steps for the interval $[t_*, t_1]$, we will get the same inequality $x(t) \le \bar{x}(t, \varepsilon)$ for all $t \in [t_*, t_1]$. But $\bar{x}(t_1, \varepsilon) < x_* + \varepsilon$, and $x(t_1) = x_* + \varepsilon$. This contradiction proves that $x(t) < x_* + \varepsilon$ for all $0 \le t \le \delta$.

We also assumed that $x(t)$ increases on $[t_*, t_* + \delta]$. If this is not true, then there exists a local maximum $t_1 \in (t_*, t_* + \delta]$, such that x(t) increases on $[t_*, t_1]$ and $x'(t_1) = 0$, and thus $F(t_1, x(t_1)) = 0$. But we already proved that $x(t_1) < x_* + \varepsilon$, which means that function $F(t, x)$ is bounded by $F(t, x) > F_* - C_1\varepsilon > \frac{1}{2}F_* > 0$. This contradiction shows that $x(t)$ increases on $[t_*, t_* + \delta]$.

The last assumption was that the local solution $x(t)$ at the point (t_*, x_*) can be continued to the interval $[t_*, t_* + \delta]$. If this is not true, then according to Theorem 2 there exists a moment $t_1 \in (t_*, t_* + \delta]$ such that either $x(t) \to +\infty$, or $x(t) \to \hat{x}$ as $t \to t_1 - 0$ where $(t_1, \hat{x}) \in \tilde{\Omega}$. But neither of these can be true, because $x(t)$ is bounded by $x(t) \le x_* + \varepsilon$, it does not have a local maximum on the interval $(t_1, t_* + \delta]$, and this interval does not contain points from $\mathbb{T}$.

This completes the proof of the lemma. □

Acknowledgments

The authors are grateful to Alexei Pokrovskii, Philip O'Kane, Dmitrii Rachinskii and Oleg Rasskazov for numerous discussions. This research was supported by a private bequest to Cork University Foundation, by SFI grant 05/RFP/ENG062, and Boole Centre for Research in Informatics (BCRI).

References

[1] Krejčí P, O'Kane J P, Pokrovskiĭ A and Rachinskiĭ D 2006 Mathematical models of hydrological systems with Preisach hysteresis *Preprint* http://www.bcri.ucc.ie/BCRI_58.pdf

[2] Flynn D, O'Kane J P and Zhezherun A 2006 Numerical solution of ODEs involving the derivative of a Preisach operator and with discontinuous RHS *J. Phys.: Conf. Series* (this issue)

[3] Flynn D and Rasskazov O 2005 On the integration of an ODE involving the derivative of a Preisach nonlinearity *J. Phys.: Conf. Series* ed M P Mortell, R E O'Malley, A Pokrovskiĭ and V Sobolev **22** 43–55

[4] Flynn D, McNamara H, O'Kane J P and Pokrovskiĭ A 2006 Application of the Preisach model to soil-moisture hysteresis *The Science of Hysteresis* vol 3, ed G Bertotti and I Mayergoyz (Elsevier, Academic Press) pp 689–744

[5] Flynn D 2005 Fitting soil-moisture hysteretic curves with two one-parameter Preisach models *Preisach Memorial Book* (Budapest: Akadémiai Kiadó) pp 197–208

[6] Preisach P 1935 Über die magnetische Nachwirkung *Zeitschrift für Physik* **94** 277–302

[7] Brokate M, Pokrovskiĭ A, Rachinskiĭ D and Rasskazov O 2006 Differential equations with hysteresis via a canonical example *The Science of Hysteresis* vol 1, ed G Bertotti and I Mayergoyz (Elsevier, Academic Press) pp 127–291

[8] Pokrovskiĭ A and Sobolev V 2005 A naive view of time relaxation and hysteresis *Singular Perturbations and Hysteresis* ed M P Mortell, R E O'Malley, A Pokrovskiĭ and V Sobolev (Philadelphia: Soc. for Industrial and Appl. Math.) pp 1–59

[9] O'Kane J P, Pokrovskiĭ A, Krejčí P and Haverkamp R 2003 Hysteresis and terrestrial hydrology *Geophys. Research Abstracts* **5** 11-2-2003

[10] O'Kane J P 2004 Hysteresis, the missing non-linearity *Hydrol. Earth Syst. Sci. Discuss* **1** 41-73

[11] O'Kane J P 2005 The FEST model—a test bed for hysteresis in hydrology and soil physics *J. Phys.: Conf. Series* ed M P Mortell, R E O'Malley, A Pokrovskiĭ and V Sobolev **22** 148–63

[12] Cross R and Allan A 1988 On the history of hysteresis *Unemployment, Hysteresis & Natural Rate Hypothesis* (Blackwell) pp 26-38

[13] Cross R 1995 *Hysteresis, The Handbook of Economic Methodology* (Edward Edgar)

[14] Krasnosel'skiĭ M and Pokrovskiĭ A 1989 *Systems with Hysteresis* (New York: Springer-Verlag)

[15] Hartman P 1964 *Ordinary Differential Equations* (New York: Wiley)

AUTHOR INDEX

Institute of Physics
Dirac House, Temple Back
Bristol BS1 6BE UK